21. Colloquium
der Gesellschaft für Biologische Chemie
9.—11. April 1970 in Mosbach/Baden

Mammalian Reproduction

Edited by H. Gibian and E. J. Plotz

With 255 Figures

Springer-Verlag Berlin · Heidelberg · New York 1970

Professor Dr. HEINZ GIBIAN, Pharmaforschung Schering AG, D-1000 Berlin 65, Müllerstraße 170—172
Professor Dr. ERNST JÜRGEN PLOTZ, Universitätsfrauenklinik, D—5300 Bonn - Venusberg

ISBN-13: 978-3-642-64995-0 e-ISBN-13: 978-3-642-64993-6
DOI: 10.1007/978-3-642-64993-6

Softcover reprint of the hardcover 1st edition 1970

Brühlsche Universitätsdruckerei, Gießen

Preface

It is two years since a general meeting of the Gesellschaft für Biologische Chemie first requested us to organize the 21st Mosbach Colloquium on mammalian reproduction, and one year since we received final authorization to do so.

The present volume contains the papers read at the Colloquium, but the discussions have been omitted because writing and proofreading them would have delayed the appearance of this volume for an unjustifiable long time. Besides, in most cases the discussion was of a relatively specific nature and we did not consider it essential, bearing in mind that the purpose of the Mosbach Colloquia is to provide advanced further education for the non-specialist. One of us has referred to this and to the topical structure of the 21st Colloquium in the introductory and final remarks.

Helpful suggestions for organizing the program were made by some of the invited speakers, but the first important impulses came from Dr. von Berswordt-Wallrabe, Dr. Elger, Dr. Gerhards, Dr. Neumann, and Dr. Ufer to whom we here wish express our thanks.

Thanks are also due to those whose donations, some of which were very generous, made it financially possible to organize the Colloquium.

July 1970

Heinz Gibian
Ernst Jürgen Plotz

Contents

Introduction

HEINZ GIBIAN
Forschungsleitung Pharma, Schering AG., Berlin

Meine Damen und Herren!

Zunächst darf ich eine Anmerkung zur Konferenzsprache machen. Während wir auf den Mosbacher Kolloquien schon seit langem Vorträge von englischsprachigen Gästen kennen, begannen erst vor kurzem auch einige deutsche Vortragende sich des Englischen hier zu bedienen (Abb. 1).

Mosbach	Contributions	from non-German-speaking institutes	English
1965	12	7	7
1966	26	7	7
1967	9	5	4
1968	22	13	17
1969	33	18	32

Fig. 1

Grund hierfür ist zweifellos keine Mißachtung der deutschen Muttersprache, sondern der Wunsch, sich schon hier den Referenten aus dem Ausland für die gemeinsame Diskussion sicher verständlich zu machen und ferner für die gedruckte Vortragsfassung später einen möglichst breiten Leserkreis zu haben. Wir hatten überlegt, ob wir diesmal von vornherein Englisch als Konferenzsprache vorschreiben sollten; wir entschlossen uns dann jedoch, die Wahl freizustellen. Zögern Sie also bitte nicht, nach Belieben englisch oder deutsch zu sprechen. Daß einer unserer Gäste aus den Vereinigten

Staaten, Dr. LIPSETT, seinen Vortragstitel und zumindest das Kurzreferat auf deutsch eingereicht hat, möchte ich als dankenswerte und liebenswürdige Geste gegenüber dem Gastland ausdrücklich vermerken.

Sie gestatten nun sicher, daß ich aus Höflichkeit gegenüber unseren ausländischen Gästen auf englisch fortfahre.*

I have just asked for permission to continue in Englisch, the Latin of the modern scientific world.

When a few years ago I was asked to prepare a Mosbach Colloquium on "Fertility and Sterility", I intended to do so mostly from a biochemical point of view regarding the name and intention of the Gesellschaft für Biologische Chemie. But very soon it turned out that probably too much biochemical information would be missing for a sufficiently broad treatment of the proposed items.

So Dr. PLOTZ, our colleagues, and myself decided to allow for the inclusion of additional physiological and endocrinological viewpoints. At the same time we rounded the theme of the symposium by changing its title to the present one of "Mammalian Reproduction".

This seemed to us permissible for three reasons:

1. The intention of these Mosbach Colloquia generally is *not* to be workshop conferences with detailed technical information but to give surveys on fascinating and topical fields of scientific progress to an audience not necessarily specialized in this field, and as a

---

* Ladies and Gentlemen!

First of all I should like to make a few remarks regarding the language of this conference. Whereas we have long been accustomed to hearing lectures in English at the Mosbach Colloquia from English-speaking guests, it is only recently that some German speakers have also begun to deliver their lectures in English (Fig. 1). The reason for this is certainly not disrespect for the German mother tongue but the desire to make their lectures clearly understandable to the speakers from abroad for the joint discussion, and to have for the printed form of their papers as large a reading public as possible. We did consider making English the official conference language this time, but then we decided to leave it to your option. Therefore, do not hesitate to speak English and German, as you like.

One of our American guests, Dr. LIPSETT, submitted the title of his lecture and at least the abstract in German, and I should not like to let this commendable and kind gesture towards the host country go unmentioned.

I am sure that as a mark of courtesy towards our guests from abroad you will allow me to continue in English.

contribution to the now well-accepted necessity of lifelong learning and education. That human reproduction is an item of very high topicality, considering the threat of global overpopulation and in view of changing attitudes to sexuality, scarcely needs to be stressed.

2. In a scientific world of continuing specialization, with its narrowing of personal horizons, biochemistry seems to be predestined to unite the neighbouring sciences by imparting to them its most highly specialized techniques. It seems to me of importance that at least in Europe most of the biochemists are recruited from students of chemistry, biology, and medicine.

3. It could be hoped that the very lack of biochemical information available would be conspicuous, thus stimulating one or other from the audience to become interested — and actively interested — in entering a field which is rather neglected in this country.

When you look at the program, I hope its rationale will immediately become apparent to you.

The introductory lecture should give the general physiological background; from central to peripheral hormones, from sperm and ovum to the fetus, including various regulatory mechanisms and correlations, furthermore including immunological, genetic, and metabolic aspects. Sexual behaviour and the effects of hormones on it will close the circle and bring us back to the higher centers. I am very hopeful that the sum of contributions of our distinguished speakers will give you a fairly complete survey of mammalian reproduction.

Before ending, I feel obliged to mention the name of David Kirby, who should have been among our speakers. Immediately after accepting our invitation to give a paper last July, he received a disastrous injury as a passenger in a car accident in Seattle; it was not before November that he died in Oxford at the age of thirtyseven. He was already recognized as an outstanding research worker, leader, and teacher, and his death is a serious loss to experimental embryology.

Ladies and Gentlemen, may I ask you to stand up to honor the deceased.

Thank you.

General Outline about Reproductive Physiology and its Developmental Background

ALFRED JOST

Laboratoire de Physiologie comparée, Faculté des Sciences, Paris 5e, France

With 23 Figures

The organizers of this meeting invited me to present a brief and general outline about reproductive physiology and its developmental background, so as to introduce the expert papers which will be presented on selected aspects of the field. This responsability is in the same time a honour and a risky task because reproductive physiology implies an extraordinarily large number of physiological and developmental facets.

The aim of the reproductive function is the production of new individuals and the perpetuation of the lineage. Among all physiological functions it is characterized by a unique feature, it demands two different individuals of the same species, males and females, to fulfill it. Two complementary but discrete anatomical and physiological systems must cooperate to ensure its success. Sex drive and sex behaviour thus are indispensable components of reproductive physiology.

If we provisionally overlook the question of how two different kinds of individuals develop in the same species, and if we take a broad look at reproductive physiology three series of mechanisms emerge:

1) elaborate mechanisms permit the production of specialized male and female reproductive or germ cells, spermatozoa and ovocytes;

2) a series of mechanisms, including sex behaviour, permit one male and one female germ cell to fuse into an egg cell at fertilization, thus adding paternal and maternal genetical information;

3) the egg cell is then fostered during its development until a young individual capable of independant life is produced.

Before surveying these aspects, it seems fit to first insist again on the necessity of two individuals in reproduction and to consider the process of mating which is the first obvious act in reproduction.

Mating Behaviour

Mating necessitates an active participation of both sexes; it can be analyzed in details and quantified in animals. Male rats, for instance, recognize females and display a characteristic behaviour which includes varying numbers of simple mounts, mounts with intromission of the penis and finally intromission with ejaculation.

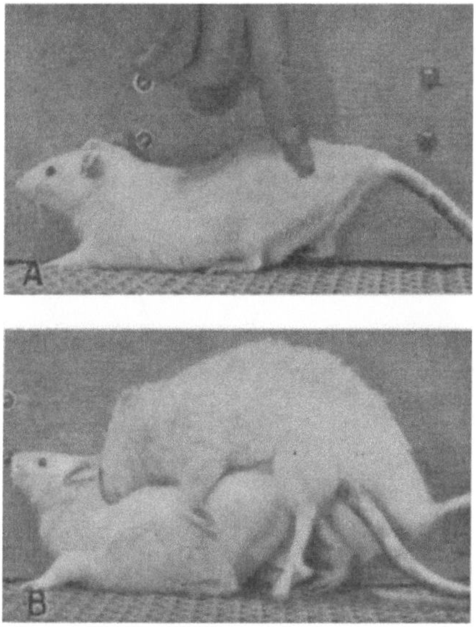

Fig. 1. The lordosis reflex in an oestrous female rat elicited by stroking gently the back (A) or by mounting of a male rat (B). From E. J. Farris and J. Q. Griffith, Jr. (eds.). In: The rat in laboratory investigation. J. B. Lippincott Co., 2nd edition, 1949

The female mating posture or lordosis (Fig. 1) is displayed when the female is in oestrous and ready to accept mating. This postural reflex can be elicited by fingering the back of the animal; the same reflex is well known to owners of cats. In rats the mating behaviour depends upon a precise hormonal control during development and in adulthood. This problem will be considered again a little later in this paper and Dr. F. A. BEACH will give us insights in similar aspects during this Symposium.

Production of Germ Cells

The reproductive cells are produced in the sex glands or gonads according to a pattern which is quite different in females and in males. The genital tract also is very different in both sexes.

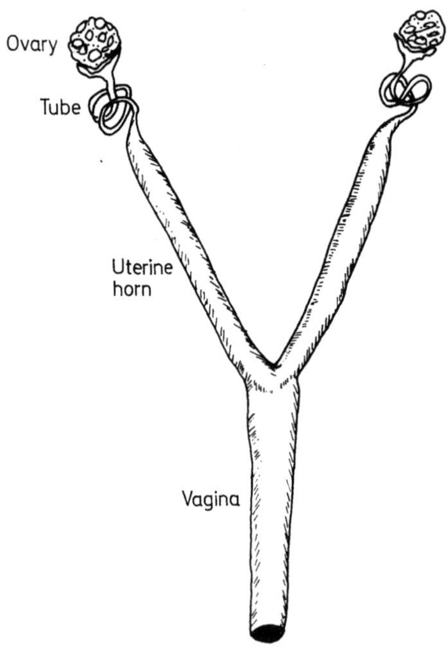

Fig. 2. Schematical drawing of the genital tract of a female rat

The genital tract of the female is simple from the anatomical point of view (Fig. 2). Below the right and the left ovary it comprises the tubes and the uterine horns (these structures are fused in a simple uterus in humans) and a vagina. The ova move down from the ovary through the tube to the uterus where they implant if fertilized;

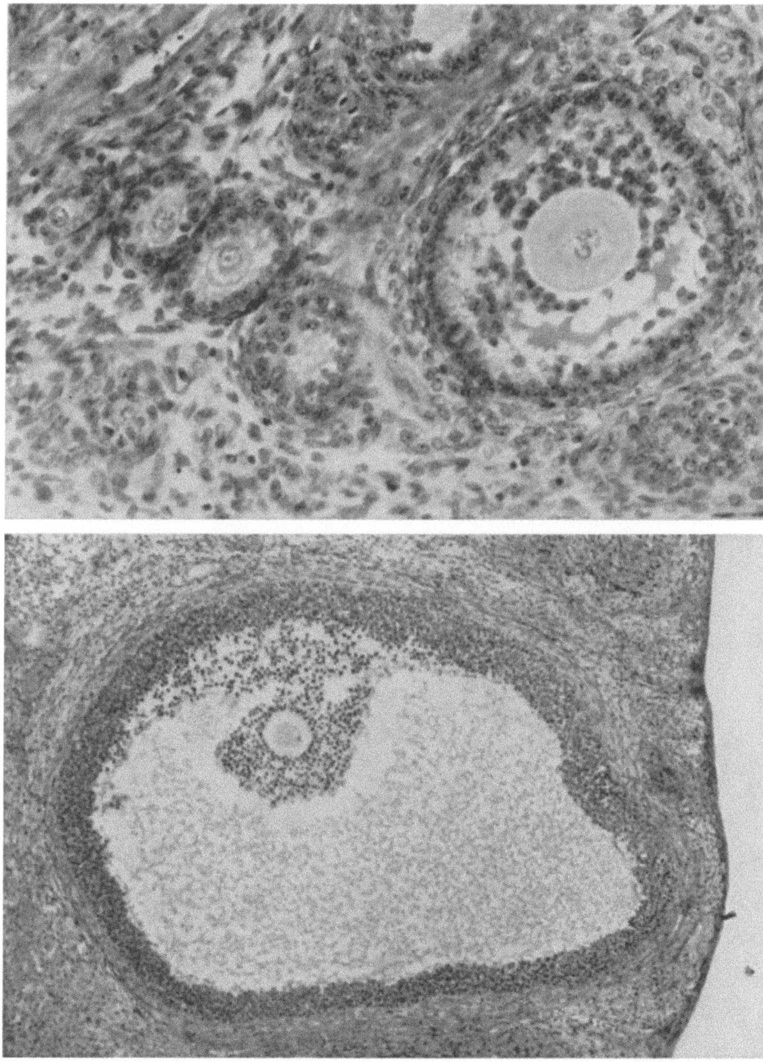

Fig. 3. Histological sections through the ovary of a rat, showing primordial follicles and growing follicles in the upper part (× 250), and a mature Graafian follicle in the lower part (× 80)

the spermatozoa introduced into the genital tract during copulation and ejaculation move up through the uterus toward the tube and the ovary.

The functional unit of the ovary is the ovarian follicle (Fig. 3); in a primordial follicle one female germ cell is surrounded by a layer of follicular cells. Many of these follicles are stored in the neonatal ovary, but most of them never will serve and degenerate. During the process of follicular maturation, which precedes ovulation, one (in humans) or several follicles grow by multiplication of the follicular cells and become hollow Graafian follicles ready to extrude their ovum (Fig. 3 and 8). The ovocyte is discharged superficially before being collected by the tube. Ovulation from a follicle of a rabbit ovary is illustrated in Fig. 4. The female germ cell rapidly degenerates unless it is fertilized.

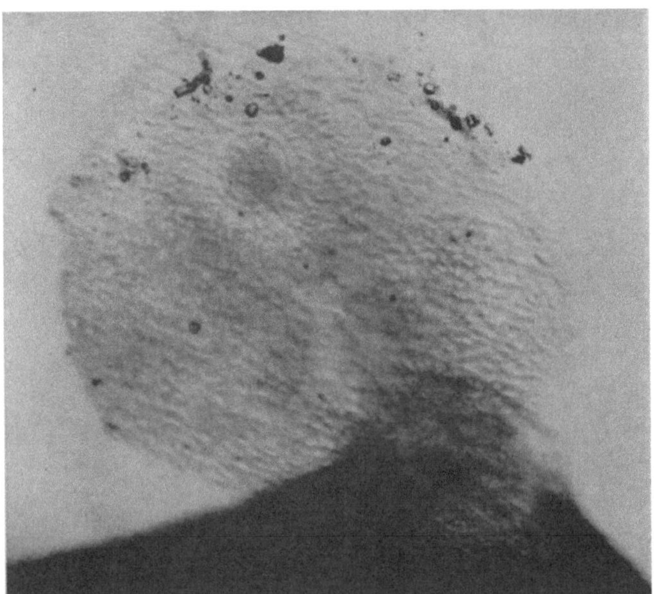

Fig. 4. Newly ovulated rabbit egg surrounded by follicular cells protruding from a ruptured follicle. Photograph taken by M. J. K. HARPER (1963) on fresh material. Ovulation was induced by injecting LH. From M. J. K. HARPER: J. Endocrinology **26**, 307—316 (1963)

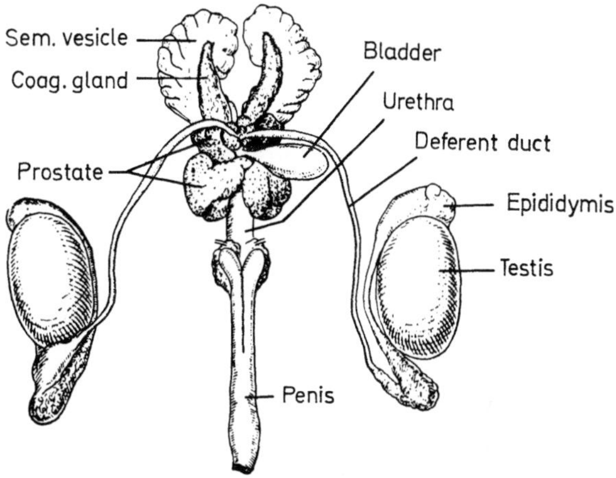

Fig. 5. Dissection of the genital tract of a male rat. Simplified, after C. D. Turner, General Endocrinology, 2nd edition. W. B. Saunders Co., 1955

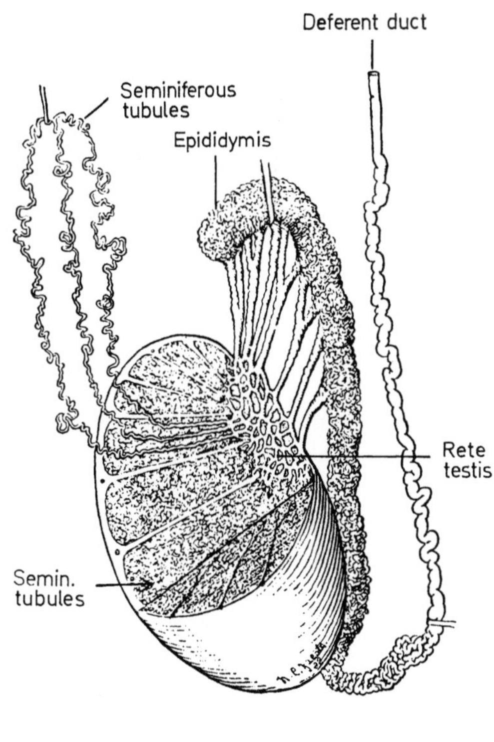

Fig. 6. Schematical drawing of a partially dissected human testis. From G. W. Corner: Les hormones dans la reproduction sexuelle, translated by S. Danysz and A. Bussard, 1949. Corréa, Paris

The remaining intra-ovarian follicle is transformed into an important endocrine gland, the corpus luteum (see Fig. 8).

The male genital tract is a complicated system (Fig. 5). It can be schematically described as follows. Spermatozoa differentiate in the testes, more precisely in a system of tubules contained in the testes, the seminiferous tubules (Fig. 6). The spermatozoa pass through an

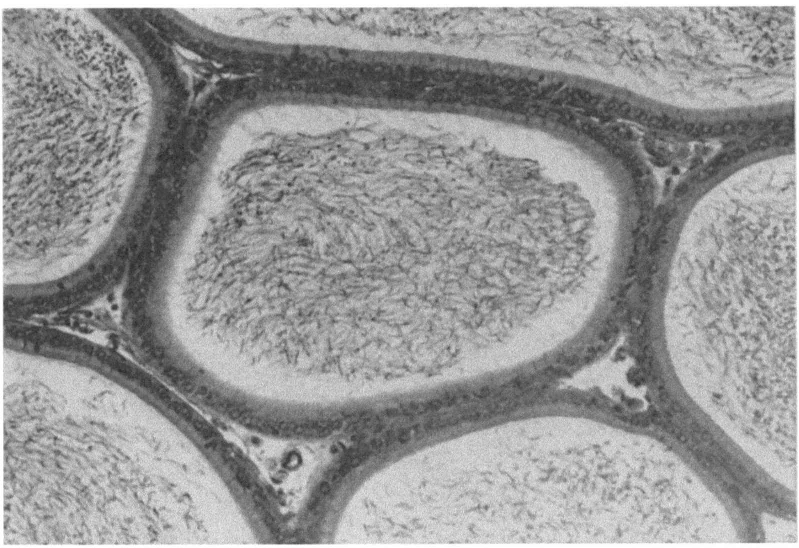

Fig. 7. Histological section through the epididymis of an adult rat showing numerous spermatozoa in the convoluted epididymal duct (× 180)

uninterrupted system of ducts from the seminiferous tubules to the penis, when ejaculation occurs. Since ejaculation is sporadic, whereas spermatogenesis is continuous the spermatozoa are stored in the epididymis (Fig. 7). At ejaculation the sperm contains a suspension of spermatozoa which comes from the epididymis and is diluted in large amounts of secretions produced by the accessory glands of the genital tract (seminal vesicles, prostate, etc.). The seminal plasma plays an important role in the physiology of the spermatozoa as will be discussed by Dr. G. Ruhenstroth.

Fertilization and Early Development of the Egg

Fertilization of the ovocyte by one spermatozoon takes place in the upper part of the tube (see Fig. 8). During fertilization the

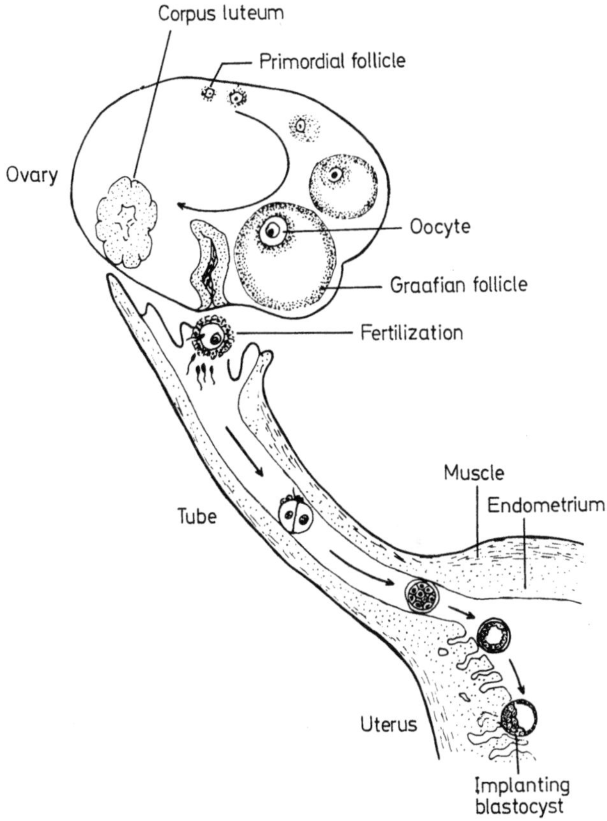

Fig. 8. Composite scheme showing the ovary, the tube and a part of the uterus. The scheme of the ovary is unrealistic; it is intended to show the successive stages (along the arrow) leading from a primordial follicle to a postovulatory follicle and to a corpus luteum. The arrow in the tube indicates the migration of the developing egg (its diameter has been enlarged)

nuclei of the male and of the female germ cell (Fig. 9) meet and add their chromosomes. The fertilized egg then moves down toward the uterus and begins its development (Fig. 8). The tubal migration takes three days approximately in rats and three to four more days are spent in the uterus before the embryonic membranes contact or penetrate the uterine lining, the endometrium, and the placenta is formed. Several problems of great import can easily be imagined by

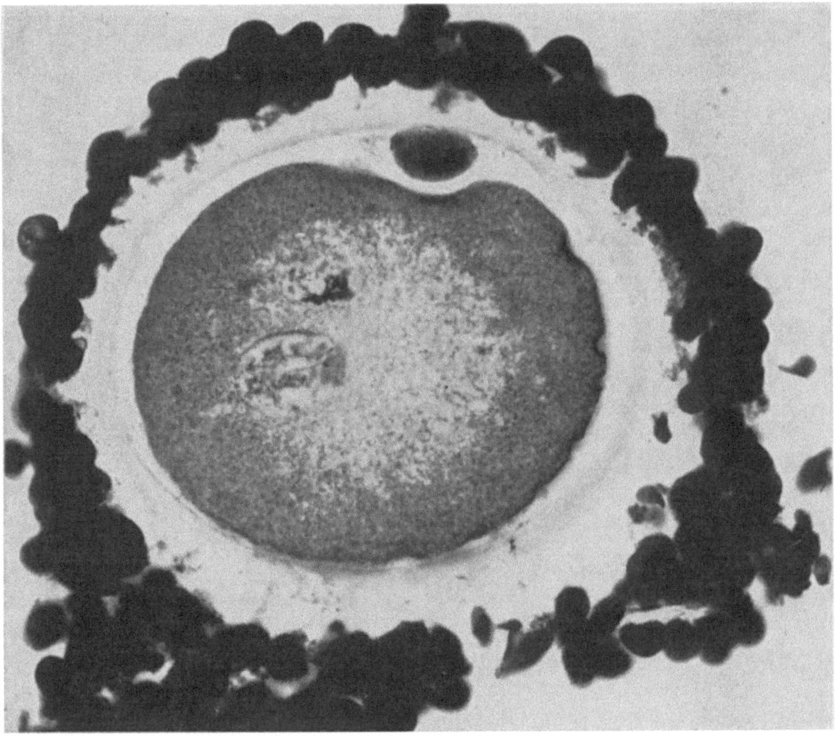

Fig. 9. Histological section through a rabbit egg briefly after fertilization, *in vitro*, showing the female nucleus and the male nucleus close together. In the picture the male nucleus is located beneath the female nucleus. Courtesy of Professor CH. THIBAULT

looking at figure 8. Several aspects will be discussed during this meeting: the "saga" of the sperm in the female genital tract (Dr. J. M. BEDFORD), the transport of the developing egg (Dr. H. KOESTER) and its metabolism (Dr. R. L. BRINSTER), the role of the uterine secretions (Dr. H. M. BEIER), the endocrine activity of the placenta during pregnancy (Dr. R. M. MOOR).

Another point must be emphasized because of its fundamental importance in reproductive physiology. At fertilization the addition of the male and of the female pronuclei determines the genetic assortment of the egg cell, and in particular the sex of the offspring. The germ cells contain half the number of chromosomes characteristic of the species. All the ovocytes carry one so-called X chromosome, whereas the spermatozoa carry either an X or a Y chromosome. The resulting egg may thus contain an XX set of sex chromosomes and be labeled for further feminine development or a XY set of chromosomes and differentiate as a male.

Hormonal Control of Reproductive Function

Historically the discovery of the role of hormones in reproductive physiology proceeded during three steps, when successively the gonadal, the pituitary and finally the hypothalamic hormones were recognized.

The effects of the hormones produced by the testis and the ovaries on the male or on the female genital tract and sexual functions have long been recognized. Testosterone, as a representative of male hormones, and oestradiol and progesterone as the two main ovarian hormones are well known entities. Some aspects concerning the testicular and the ovarian hormones will be considered by Dr. E. STEINBERGER and by Dr. M. B. LIPSETT.

In a second historical period it was discovered that the gonads obey a hypophyseal control (Fig. 10). The same gonado-stimulating or gonadotrophic hormones, the folliculo-stimulating (F.S.H.) and the luteinizing hormone (L.H.), are produced and active in both the male and the female. F.S.H. is generally assumed to control follicular development in females and spermatogenesis in males. L.H. is the hormone which induces ovulation and luteinization (development of the corpus luteum in the ovary) in females and which stimulates the interstitial cells to produce male hormone in the male.

During the early fifties and thereafter it became progressively clear that the gonadostimulating activity of the pituitary gland itself is controlled by the overlying part of the nervous system, the hypothalamus (Fig. 10). In many species testicular activity is continuous whereas the ovarian function is cyclical, ovulation occurring at definite intervals of time. This results from a difference in the pituitary functioning, namely the pattern of F.S.H. and L.H.

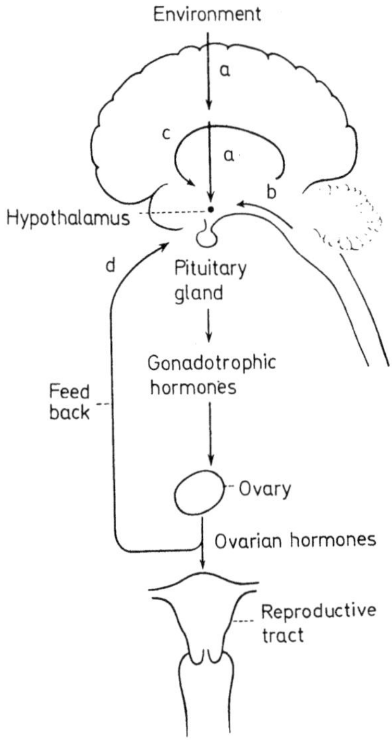

Fig. 10. Composite scheme summarizing the hypothalamic, pituitary and ovarian hormonal influences on the reproductive tract (here schematized as a human uterus). The contour of the brain helps in locating several factors which may modulate the hypothalamic activity: *a* environmental stimuli; *b* the reticular formation of the brain stem; *c* structures located in the limbic lobe; *d* feed-back by ovarian hormones. From G. Harris: Arch. Anat. micr. Morphol. expér. **56**, 385—389 (1967)

release, which is governed by the hypothalamus. Hormones produced by special hypothalamic neurones are transferred by portal blood vessels to the anterior pituitary and induce it to synthetize and to release gonadostimulating hormones. These so-called hypothalamic releasing factors (R.F.) or releasing hormones, have been extensively studied by Dr. A. V. SCHALLY who will report on them. The hypothalamic centers which govern the pituitary have a different pattern of functioning in males and in females; in that sense they may be considered sexualized, as will be seen later again.

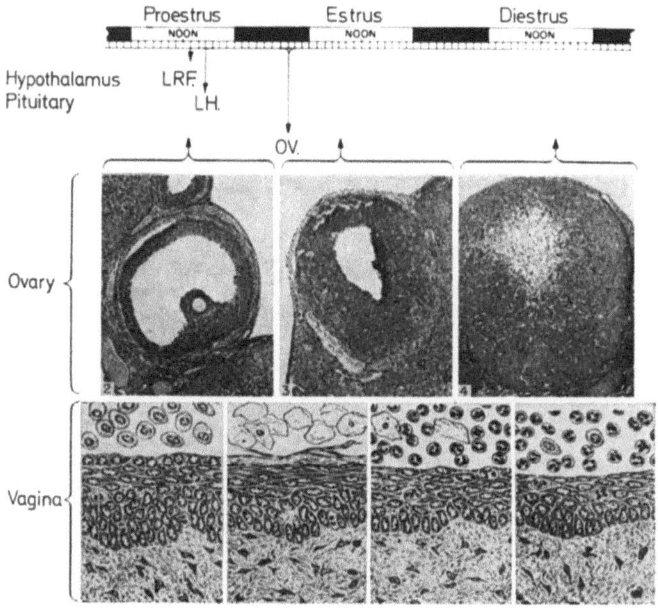

Fig. 11. Composite figure showing some components of a 4 day ovarian cycle in the rat: *On top*, time in hours (in black: light off). The times of the hypothalamic influence on the pituitary release of LH and of ovulation (OV) are indicated. — *In the middle row* are shown a preovulatory follicle and the formation of a corpus luteum; ovaries preserved at times indicated by arrows. — *In the lower row* changes in the vaginal epithelium as induced by the ovarian hormones. Upper part of the figure modified after J. W. EVERETT and CH. H. SAWYER: Endocrinology **47**, 198—218 (1950); lower row after C. D. TURNER, General Endocrinology, 2nd edition. W. B. Saunders Co., 1955

Many structures and many hormones control the sexual cycle of a female animal. Neural centers, several levels of hormonal secretions and feedback effects must be considered. Some of the factors involved in the ovarian cycle of a female rat are summarized in Fig. 11.

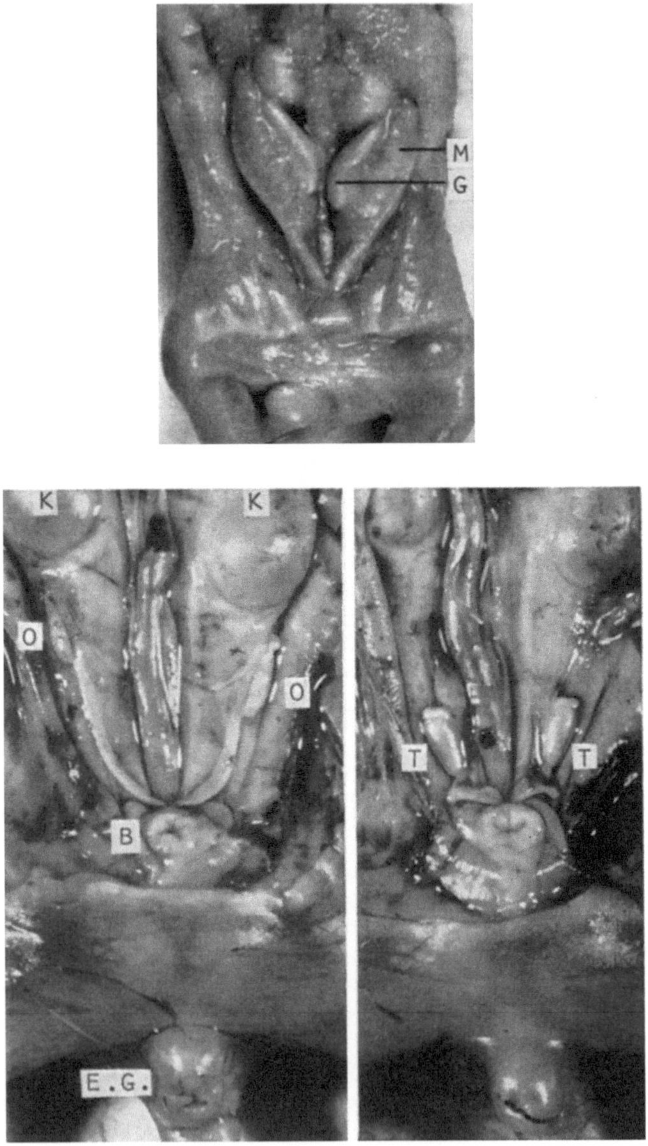

Fig. 12

Some Developmental Homologies between Males and Females

It has long been known that during foetal development the anlagen of the sex organs are first identical in both sexes and that the differentiated male or female organs arise from common undifferentiated primordia. This is summarized in the following very simple scheme:

Undifferentiated primordium

↙ ↘

♀ ♂

This is true for the gonads, for the internal genital organs and for the external genitalia.

The sex glands and ducts develop in close connection with the transitory foetal kidneys (mesonephros), which are later replaced by the definitive kidneys present at birth (Fig. 12). The gonads arise on the inner aspect of the mesonephros as a superficial bulging which increases in size.

Let us look at a composite scheme summarizing some aspects of the development of the gonads and of the internal sex organs (Fig. 13). When the undifferentiated gonadal anlage becomes a testis, seminiferous tubules differentiate in it in a centripetal direction. They connect a net of small ducts (rete testis, see also Fig. 6) which themselves contact urinary tubules. These tubules open in the primitive ureter, the Wolffian ducts. Urogenital connections are thus established and the primitive urinary structures are used in building up the upper male genital system. Ovarian development is superficial, primary follicles being accumulated toward the surface of the ovary; in the adult female the ovocytes are discharged

Fig. 12. Photographs of dissected rabbit foetuses showing the differentiation of the genital system: in the upper part, a 19 day-old foetus showing the mesonephros (M) and the developing gonads (G). In the lower part, at a lower enlargement, two 28 day-old foetuses are seen: a female on the left, and a male on the right. The ovaries (O) are located beneath the definitive kidneys (K) and the uterine horns join behind the bladder (B). In the male the testes (T), the epididymes and the deferent ducts are seen above the bladder (E.G. = external genitalia) superficially. The urogenital connections never completely develop in females and only some remnants of them persist in adult females.

The female sex ducts derive from the Müllerian ducts, a second set of ducts first present in both sexes before disappearing in males (Fig. 13, 15 and 17).

At the level of the external genitalia, the female clitoris is to a large extent homologous to the male penis.

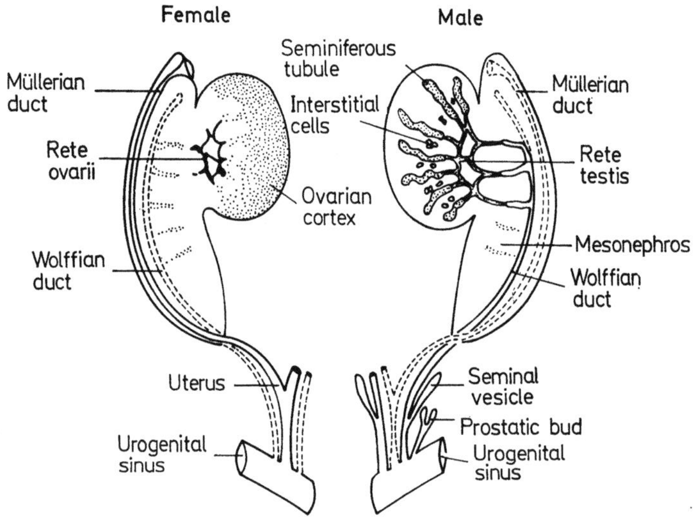

Fig. 13. Composite scheme illustrating some homologies in the development of the male and of the female genital tract (see the text)

Mechanisms Controlling the Sexual Differentiation of the Foetus

The male and female adult individuals who participate in reproduction are theoretically equal even if their role is not identical. Each sex produces its own sex hormones which govern its own genital tract. The differentiation of the genital structures results from a divergent developmental process in common undifferentiated primordia. Many of these foetal structures (not the gonads) can be either masculinized or feminized when androgens or oestrogens are given to the pregnant female.

For all these reasons, it is not astonishing that several early hormonal theories of sex differentiation assumed that similar but discrete mechanisms control the development of the genital system in males and in females. Thus, the sex differentiation of the gonads is often explained on the basis of Witschi's theory of a dual system of intragonadal inductors. A feminizing inductor located peripherically (cortex) and a central masculinizing inductor (medulla)

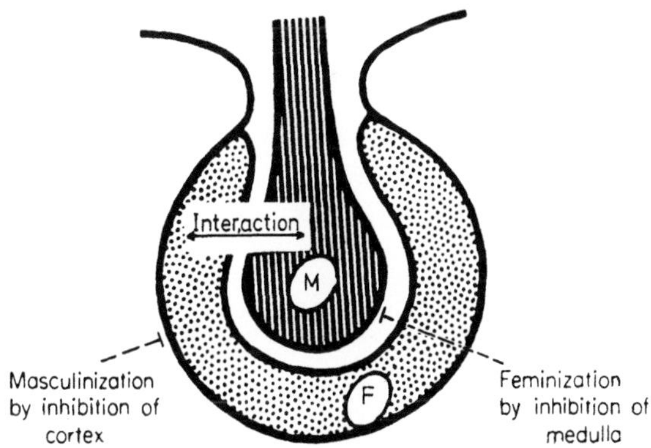

Fig. 14. Witschi's interpretation of the sexual differentiation of the gonads, based mainly on Amphibian material and showing the antagonistic cortex and medulla. From E. Witschi. In: The Biochemistry of Animal Development (R. Weber, ed.), vol. II, chap. 4, 1967. Academic Press, N. Y., originally published in E. Witschi, J. Fac. Sci. Hokkaido Univ. Ser. VI., Zool. 13, pp. 428—439 (1957)

compete until one of them gains prevalence and imposes the development of either an ovary or a testis (Fig. 14).

The differentiation of the genital tract from the undifferentiated condition, takes place in the foetus only after the differentiation of the gonads. It has often been supposed to be controlled by hormones produced by the foetal ovary or by the foetal testis (e.g. Greene, 1944) (Fig. 15). A scheme summarizing the hypothesis of symmetrical male or female factors in the development of the gonads and of the body sex (sex ducts, external genitalia, etc.) is presented in Fig. 16.

The part of this scheme referring to development of the body sex had to be changed when experimental studies were made as will be reported now.

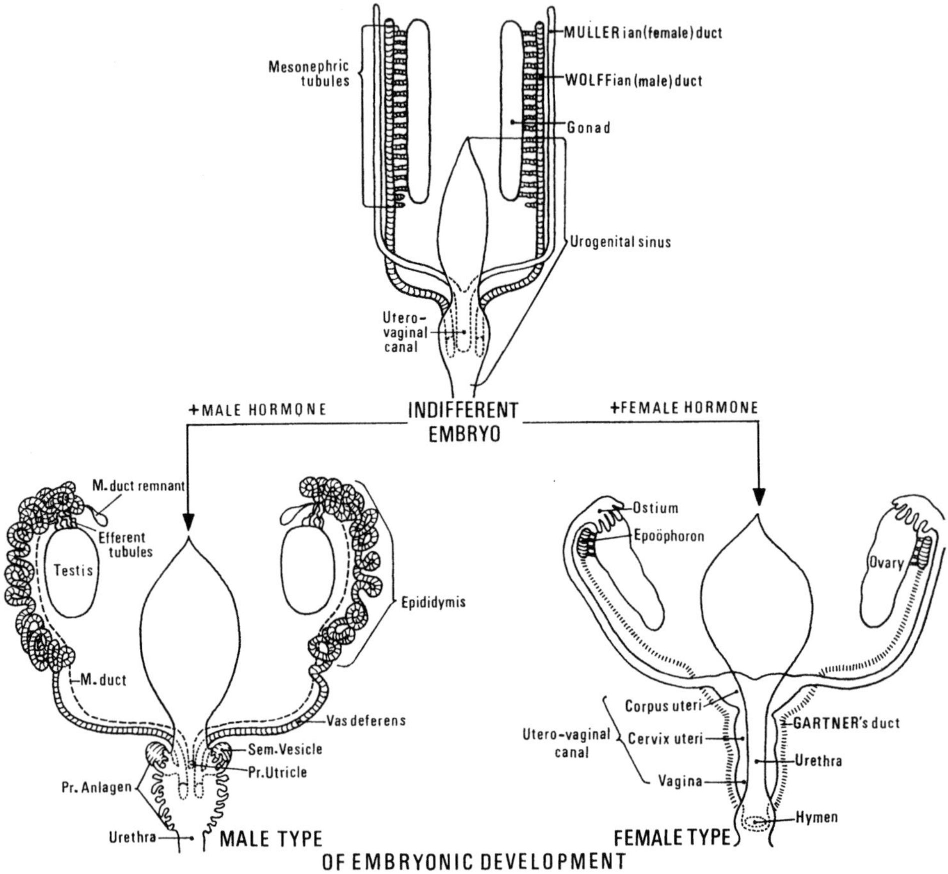

Fig. 15. R. R. Greene's concept of a dual hormonal control of sex differentiation of the genital tract. Redrawn from R. R. Greene, J. clin. Endocrinol. **4**, 335—348 (1944)

Differentiation of Body Sex Characters

A long time ago, I studied the role of the foetal gonads in the control of the differentiation of the genital tract (Jost, 1947, 1953). Rabbit foetuses were castrated *in utero* at a period when the gonads are already recognizable as to sex in histological sections and when the other parts of the genital tract still are identical in both sexes (day 19 of pregnancy). Nine days later, shortly before expected birth,

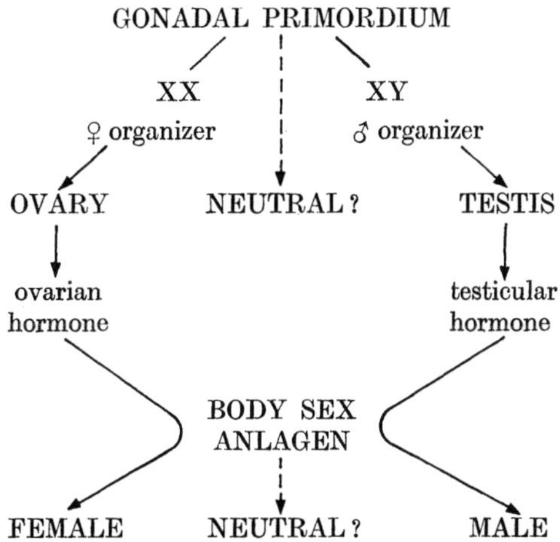

Fig. 16. Theoretical scheme summarizing the concept of symmetrical mechanisms in the sex differentiation of the gonads and of the body sex characters (genital tract, secondary sex characters, etc.)

the foetuses were recovered during a caesarean section and their genital tract was studied. The result was clear: the genital tract of gonadless foetuses developed according to the female type whatever the genetic sex of the young (Fig. 17). This and other additional experiments showed that the foetal testis is the body sex differentiator. It impresses masculinity on an organism which otherwise becomes feminine. The presence of ovaries in females is of no significance in the establishment of feminine body structures.

This result was confirmed in several other experiments *in vivo* or *in vitro*. It permitted to predict that human beings congenitally deprived of gonads should display feminine features whatever their genetic sex (Jost, 1950). This was largely confirmed.

It thus appears that in Fig. 16 the part referring to body sex should be altered so as to show that maleness is actively imposed by the testis on a system which otherwise becomes feminine (Fig. 18).

Some more remarks should be devoted to the mode of action of the foetal testis during sexual differentiation of the genital tract. In order to achieve masculinization, the foetal testis inhibits the Müllerian ducts, stimulates the Wolffian ducts and masculinizes the

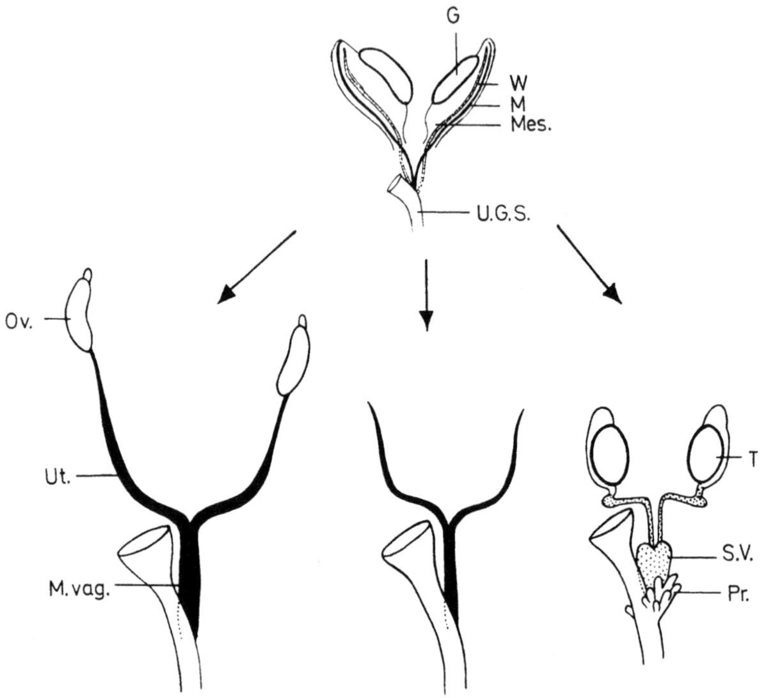

Fig. 17. Schematic presentation of sexual differentiation of the sex ducts in the rabbit embryo. From the undifferentiated condition (top) may arise either the female structure (bottom left), or the male structure (bottom right), or the gonadless feminine structure in castrated embryos of either sex (bottom middle). G., gonad; M., Müllerian duct; Mes., mesonephros; M. vag., Müllerian vagina; Ov., ovary; Pr., prostate; S.V., seminal vesicle; T, testis; U.G.S., urogenital sinus; Ut, uterine horn; W, Wolffian duct (stippled). [After Jost: Mem. Soc. Endocrinol. No. 7, 49—61 (1960)]

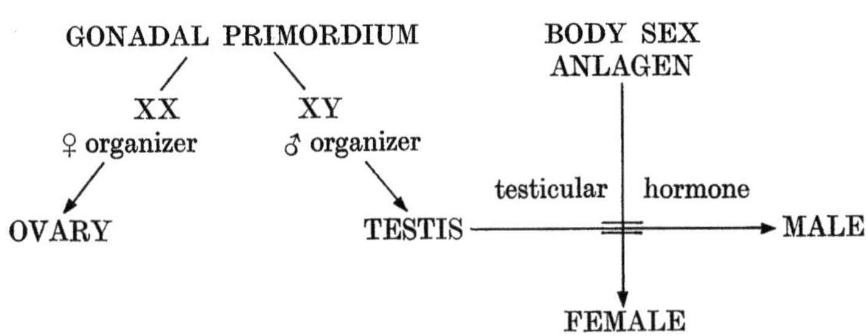

Fig. 18. Theoretical scheme showing how the part of fig. 16 concerning the differentiation of the body sex characters should be modified after experimental studies

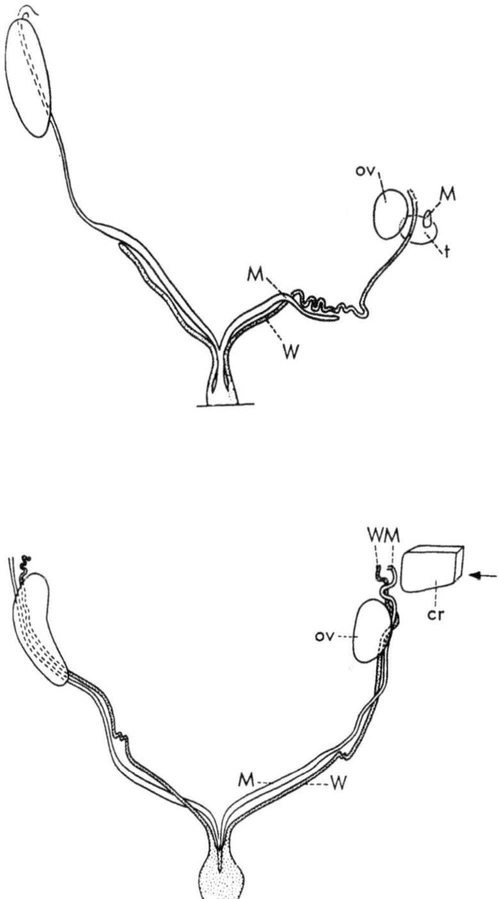

Fig. 19. Comparizon of the effect of a foetal testis and of a crystal of testosterone propionate on the genital tract of female rabbit foetuses: *Top,* a foetal testis was grafted near the ovary on day 20. Partial inhibition of the Müllerian duct and partial maintenance of the Wolffian ducts were observed on day 28. [After Jost: Arch. Anat. Microscop. Morphol. Exptl. **36**, 271—315 (1947)]. *Bottom:* a crystal of testosterone propionate was implanted near the ovary on day 20. There was maintenance of the Wolffian ducts and no inhibition of the Müllerian ducts. cr, crystal of testosterone propionate; M, Müllerian ducts; ov, ovary; t, testicle; W, Wolffian duct (stippled). (After Jost, pp. 160—180 in IIIe Reunion Endocrinol. Langue Française, Masson et Cie,. 1955)

other parts of the genital tract. A foetal testis grafted on a female rabbit foetus *in utero* (or a rat testis placed near parts of the genital tract *in vitro*, Picon, 1969), exerts these actions; testosterone has only the masculinizing effect but it does not inhibit the Müllerian ducts (Fig. 19). This may suggest that the foetal testis produces both a Müllerian inhibitor and a masculinizing hormone (Fig. 20).

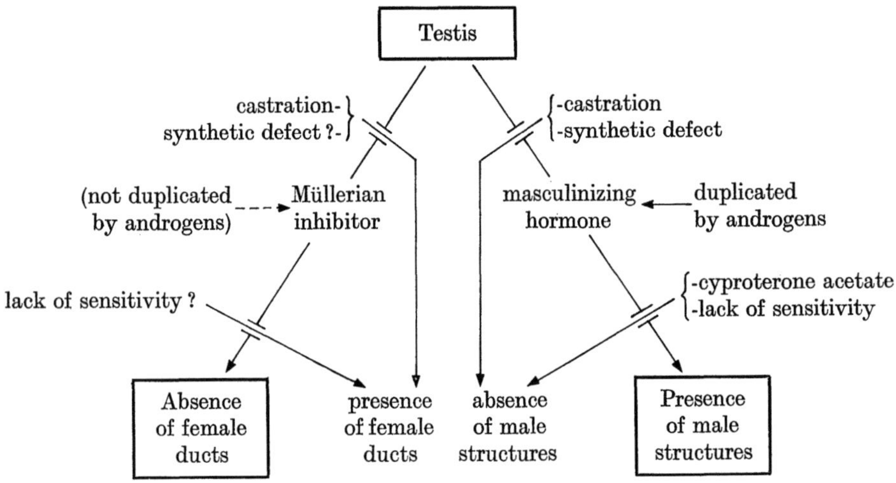

Fig. 20. Scheme summarizing the testicular control of the differentiation of the sex ducts and sex characters. Some conditions capable of interfering with the normal testicular activity are indicated (crossing arrows). The data concerning cyproterone acetate refer to experiments made on rabbit foetuses (in rats the effect is not the same)

This assumption permits a satisfactory explanation of several genital abnormalities in human beings, for instance the persistance of a uterus in an otherwise normal male. This condition could be accounted for if the foetal testis failed to produce the Müllerian inhibitor but secreted normally the masculinizing hormone. Reciprocally lack of secretion of (or lack of sensitivity of the tissues to) the masculinizing hormone could explain those cases in which no masculine character develops but the Müllerian ducts are inhibited (syndrome of testicular feminization). In beautiful experiments on rabbits which I could confirm (Jost, 1966), Dr. W. Elger (1966) produced such a condition under the influence of the antiandrogen cyproterone acetate which counteracted the androgenic hormone without preventing the Müllerian inhibition (see Fig. 20).

The role of the testicular hormone in the sexual differentiation of other parts of the body, especially some neural structures, has recently been explored with great success (see for instance HARRIS

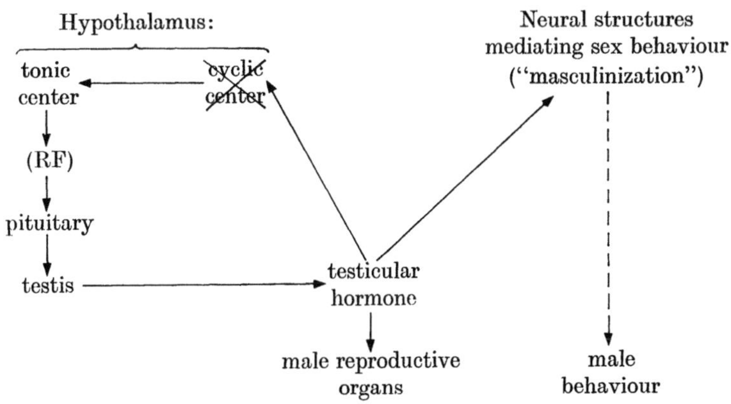

Fig. 21. Effects of the testicular hormone on neural structures during the *neonatal period* in the rat. In the hypothalamus androgens impair the further functioning of a cyclic center which is active in females. Androgens also permanently modify the neural structures mediating adult sex behaviour. In neonatally castrated males both these structures acquire the feminine pattern of functioning

and LEVINE, 1965). As has already been mentioned earlier, in the adult rat the hypothalamus imposes a continuous gonadostimulating activity on the hypophysis in the male, and a cyclical activity in the female. The male pattern of hypothalamic functioning results from the influence of the testicular hormone on the hypothalamus at a definite developmental stage (5th postnatal day) (Fig. 21). The male pattern does not appear in males castrated before that stage; in these early castrated males a cyclical hypothalamic functioning will prevail. Contrariwise the male pattern can be impressed on the hypothalamus of a young female by an injection of testosterone on the fifth day of life.

Similar neonatal (or prenatal) hormonal influences on sex behaviour of adult animals have been recognized in certain species, for instance in the rat. Female receptivity disappears when an adult female rat is castrated and it is restored by an appropriate treatment with progesterone and oestradiol. The same treatment fails to elicit female behaviour in male animals castrated as adults or in adult females which were injected with androgens during the neonatal period. But the feminizing treatment is successful if the male rats were castrated at birth.

Thus both at the level of the hypothalamic centers controlling the gonadostimulating activity and at the level of the neural structures permitting sex behaviour, the feminine pattern results essentially from the absence of the testicular influence, whereas it is suppressed by the testicular hormone, which has permanent effects (Fig. 21). These data, although they are known only in a few animal species so far, obey the type of developmental control depicted in Fig. 18 for body sex characters.

Differentiation of the Gonads

The sex type of all the sex characteristics considered so far, depends upon an early influence of testicular hormone. In order to fully understand sexual differentiation of the individual the main problem thus becomes the differentiation of the sex glands themselves.

The factors governing the development of a testis or of an ovary from the early undifferentiated primordium are still unknown. Even on purely descriptive grounds, the problem has been obscured by theoretical views and by a confusing terminology.

When one studies gonadal organogenesis in a series of foetuses of increasing gestational age, from a certain stage up it becomes possible to distinguish the gonads of male and female foetuses in histological sections. This stage is usually referred to as stage of gonadal sex differentiation. Such a wording is somewhat misleading since it is often taken to mean that the gonads of both sexes differentiate at that stage. This is not the case; at the stage under question the testis only becomes organized while the presumptive ovary still remains undifferentiated. Testicular organization becomes established, when the future seminiferous tubules, separated by intertubular cells, and the superficial albuginea testis are delineated (Fig. 22). Testicular organogenesis is a very rapid process which becomes obvious in less than 24 hours in the rabbit foetus between days 14 and 15 of pregnancy. During the same period of time nothing much happens in the gonads of the female foetuses, which remain similar to the undifferentiated primordium. As was recognized by most of the students of gonadal differentiation, at early stages ovaries are recognizable only because they are not testes.

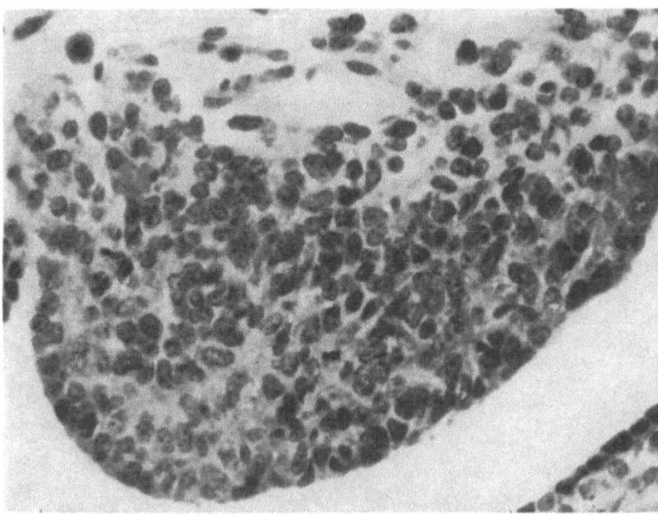

a

Fig. 22 a—e. Histological sections through gonads of rabbit foetuses ($\times$ 210). a: a 13 day-old foetus, apparently still undifferentiated

Later on, in presumptive ovaries the germ cells show the first premeiotic changes, a process which does not occur in testes before prepuberty.

Ovarian organogenesis is a slow process (Table 1). A true ovarian organization, characterized by the development of primary follicles separated by a connective stroma, is completed only in the

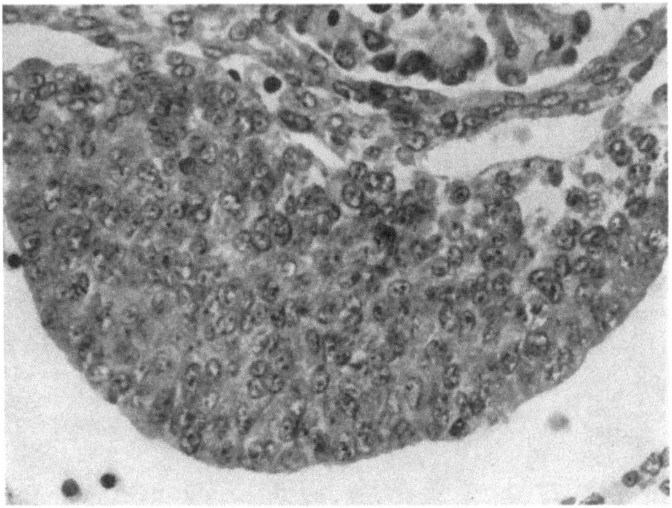

b

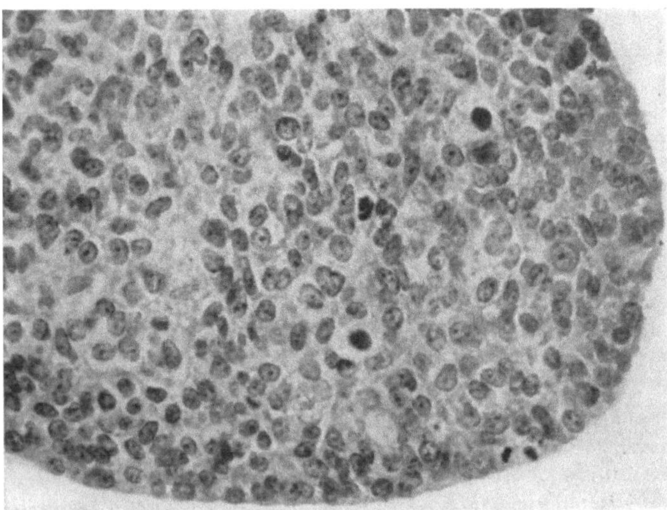

d

Fig. 22. b and c: two 14 day-old foetuses; the sex of the gonad still is histologically uncertain. Tentatively the gonad c is considered as a possible presumptive testis since irregularities suggesting cellular movements are discernible

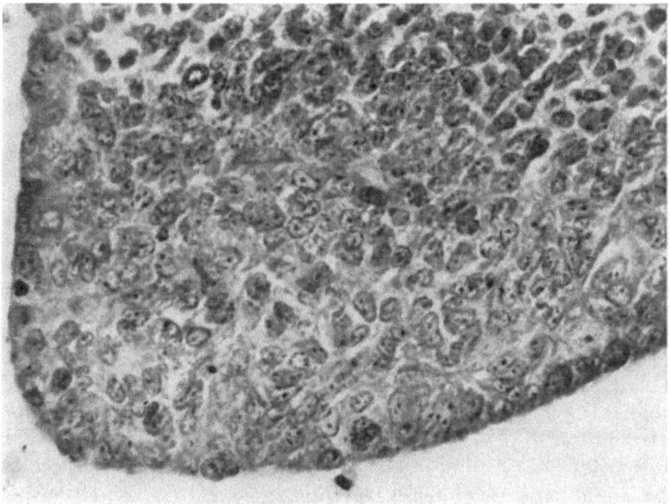

c

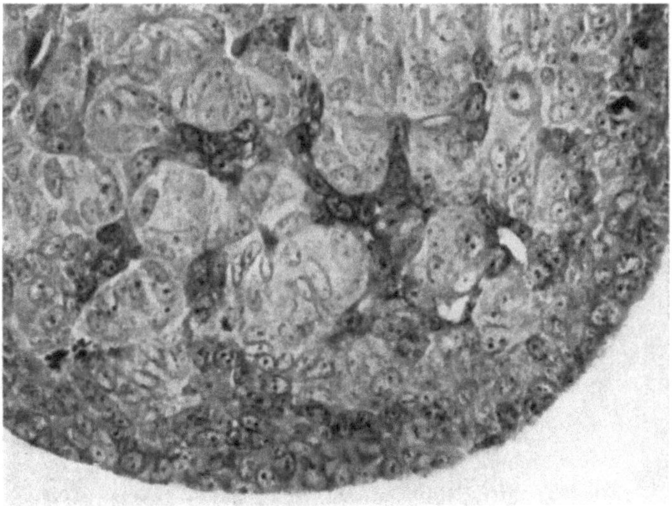

e

Fig. 22. d and e: 15 day-old foetuses. In e the testis is now well differentiated. In the presumptive female (d) the gonad is still undifferentiated second half of pregnancy in humans or after birth in many small mammals.

Table 1. *Chronology of gonadal development in three species: testicular organogenesis (seminiferous tubules, Leydig cells) precedes ovarian organogenesis by far* (post-nat. = post-natally)

	Rat (days)	Rabbit (days)	Human (weeks)
Seminiferous tubules	13—14	14—15	6— 7
Leydig cells	16	19	8
First premeiotic ovocytes	18	post-nat.	11—12
Ovarian follicles and stroma	post-nat.	post-nat.	18—28

On a purely chronological basis there is a profound disymmetry in testicular and in ovarian organogenesis; the testes diverge very soon and very rapidly from the undifferentiated condition, long before ovarian organogenesis begins. Testicular organogenesis seems to be actively triggered by a mechanism depending on the

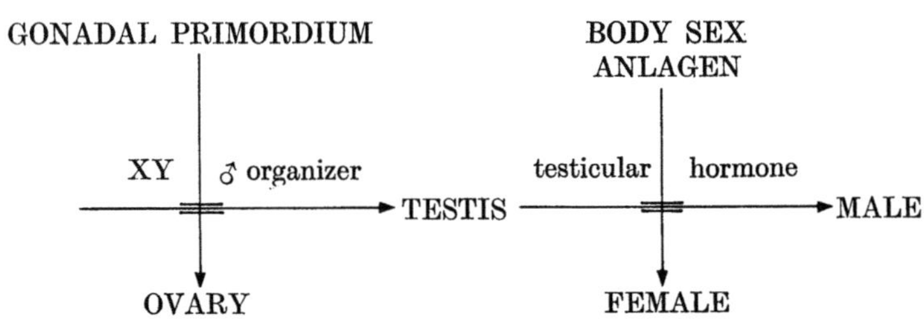

Fig. 23. Scheme illustrating the concept of completely asymmetrical processes in sex differentiation: the gonads as well as the body sex would become feminine if not diverted from doing so by a masculinizing mechanism, (compare with Fig. 16 and 18)

presence of a Y chromosome in males, in an embryonic primordium which in the absence of this trigger would slowly become an ovary (Jost, 1965, 1970). This new working hypothesis proposes a parallel between all sex structures: all would become feminine if not diverted from doing so by a masculinizing mechanism (Fig. 23).

This concept of dissymmetric developmental processes during sex differentiation, requires experimental verification, but it seems worth consideration in a field in which only theories have been presented so far, whereas the intimate mechanisms still remain to be elucidated.

* * *

It is customary to end a lecture with a conclusion. In the present case, this would be out of place. We shall now have the privilege of listening at detailed and elaborate presentations given by experts on many facets of reproductive physiology. Before doing so, I should like to still evoke a few other images in reproductive physiology, lactation and maternal care, love for the newborn baby and teaching how to become a standing, a speaking, and eventually a thinking human being.

References

CORNER, G. W.: Les hormones dans la reproduction sexuelle. Corréa, Paris. Translated by S. DANYSZ and A. BUSSARD, from "Hormones in Human Reproduction", Princeton Univ. Press, N. J. (1942).

ELGER, W.: Die Rolle der fetalen Androgene in der Sexualdifferenzierung des Kaninchens und ihre Abgrenzung gegen andere hormonale und somatische Faktoren durch Anwendung eines starken Antiandrogens. Arch. Anat. micr. Morphol. expér. **55**, 657—743 (1966).

EVERETT, J. W., SAWYER, CH. H.: A 24-hour periodicity in the "LH-release apparatus" of female rats, disclosed by barbiturate sedation. Endocrinology **47**, 198—218 (1950).

FARRIS, E. J., GRIFFITH, J. Q., JR. (eds.): The rat in laboratory investigation. J. B. Lippincott Co., 2nd ed. (1949).

GREENE, R. R.: Embryology of sexual structure and hermaphroditism. J. clin. Endocrinol. **4**, 335—348 (1944).

HARPER, M. J. K.: Ovulation in the rabbit: the time of follicular rupture and expulsion of the eggs, in relation to injection of luteinizing hormone. J. Endocrin. **26**, 307—316 (1963).

HARRIS, G. W.: Elaboration et excrétion des hormones gonadostimulantes. Introduction. Arch. Anat. micr. Morphol. expér. **56**, 385—389 (1967).

— LEVINE, S.: Sexual differentiation of the brain and its experimental control. J. Physiol. **181**, 379—400 (1965).

JOST, A.: Recherches sur la différenciation sexuelle de l'embryon de Lapin. III. Rôle des gonades foetales dans la différenciation somatique. Arch. Anat. micr. Morphol. expér. **36**, 271—316 (1947).

JOST, A.: Recherches sur le contrôle de l'organogenèse sexuelle du Lapin et remarques sur certaines malformations de l'appareil génital humain. Gynéc. et Obst. (Paris) **49**, 44—60 (1950).
— Problems of fetal endocrinology: the gonadal and hypophyseal hormones. Recent Progr. Hormone Res. **8**, 379—418 (1953).
— Modalities in the action of gonadal and gonadostimulating hormones in the foetus. Mem. Soc. Endocrinol. No. 4, 237—248 (1955).
— Biologie des androgènes chez l'embryon. IIIe Réunion des Endocrinologistes de langue française pp. 160—180. Masson et Cie. (1955).
— Gonadal hormones in the sex differentiation of the mammalian fetus. In: "Organogenesis" (R. L. DEHAAN and H. URSPRUNG, eds.). Holt, Rinehart and Winston, Inc. (N. Y.), pp. 611—628 (1965).
— Steroids and sex differentiation of the mammalian foetus. Proc. IInd. Internat. Congr. Hormonal Steroids, Milano. Excerpta Med. Int. Congr. Ser. No. 132, pp. 74—81, Amsterdam. (1966).
— Hormonal factors in the sex differentiation of the mammalian fetus. In: "A Discussion on Determination of Sex". Phil. Trans. Roy. Soc. London, **259**, 119—130 (1970).
PICON, R.: Action du testicule foetal sur le développement *in vitro* des canaux de Müller chez le Rat. Arch. Anat. micr. Morphol. expér. **58**, 1—19 (1969).
TURNER, C. D.: General endocrinology, 2nd edition. W. B. Saunders, Philadelphia & London. (1955).
WITSCHI, E.: The inductor theory of sex differentiation. Jour. Fac. Sci. Hokkaido Univ. Ser. VI., Zool. **13**, 428—439 (1957).
— Biochemistry of sex differentiation in vertebrate embryos. In: "The biochemistry of animal development" (R. WEBER, ed.), Vol. 2, chapter 4, pp. 193—225. Academic Press, New York. (1967).

The Significance of Hormones in Mammalian Sex Differentiation as Evidenced by Experiments with Synthetic Androgens and Antiandrogens

W. Elger, F. Neumann, H. Steinbeck, and J. D. Hahn
Experimentelle Forschung Pharma, Schering AG
Abteilung für Endokrinologie

With 9 Figures

Since it became possible to do experiments with synthetic androgens, the role of androgens in the differentiation of the male external genital organs could be elucidated (Dantchakoff, 1936; Jost, 1947, 1955).

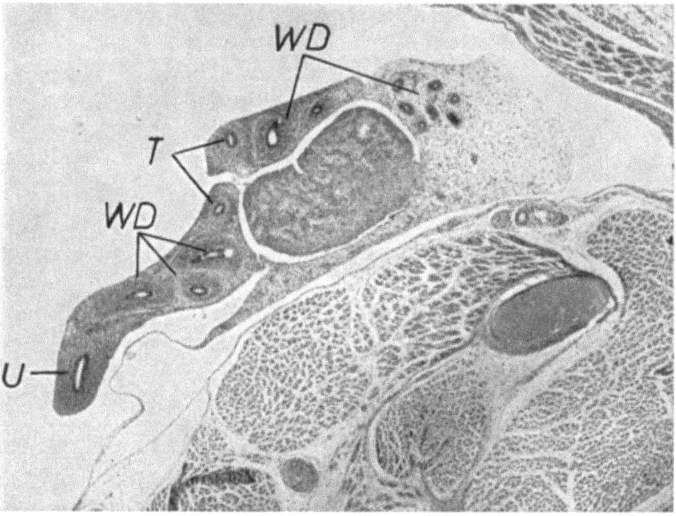

Fig. 1. Female rat fetus at the 22nd day of pregnancy. Mother treated with 1.0 mg methyltestosterone / d from d 15 to d 21 of pregnancy subcutaneously (d 1 = sperms in the vagina). Epididymal structures as well as deferent ducts are established but in spite of this ovaries and uterine horns are almost normally developed. *T* = fallopian tube, *U* = uterus, *WD* = epididymal structures, ductus deferens

It remained somewhat less clear, however, to what extent androgens are involved in the differentiation of the internal sexual organs, i.e. differentiation of gonads and gonoducts. Whereas the whole fetal testis exerts a destructive influence on the Müllerian structures, as can be seen in normal male sex differentiation or in free-martins, there is no substantial effect of androgens alone on

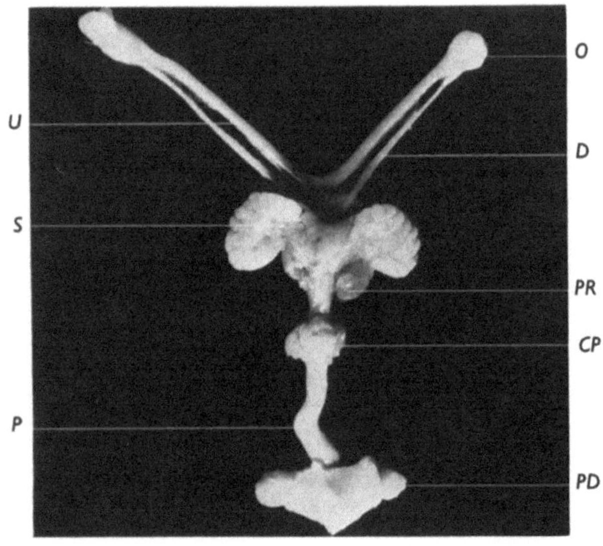

Fig. 2. Genital tract of an adult virilized female rat. Mother treated with 10.0 mg methyltestosterone / d from d 15 to d 22 of pregnancy subcutaneously. The adult virilized offspring has been treated for 15 days with 0.3 mg testosterone propionate / d subcutaneously. Along with ovaries and uterus, male sexual ducts and accessory sexual glands do exist. *U* = uterus, *S* = seminal vesicle, *P* = penis, *O* = ovary, *D* = deferent duct, *PR* = prostate, *CP* = cowper's gland, *PD* = preputial gland

ovaries and Müllerian ducts in female mammalian fetuses, although complete masculine transformation of the external genitals can be achieved. As concerns the internal genitals, intersexual structures result from androgen treatment in female fetuses (Fig. 1).

After the end of the hormone-sensitive phase in the differentiation of a given sexual structure, any modification is irreversibly fixed. The structures once established can also be made functional later (Fig. 2).

Besides gonadal differentiation, the retrogression of the Müllerian ducts seems to be one of the most interesting problems in sexual

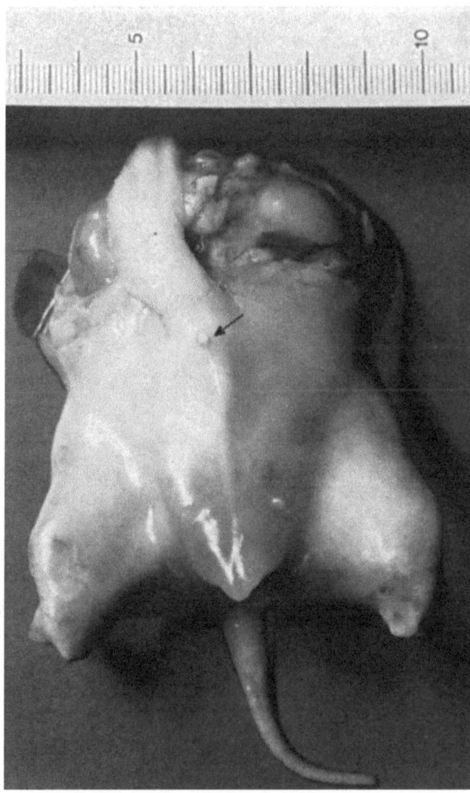

Fig. 3a—d. Effects of CPA on the differentiation of the internal and external sexual organs of various species. a) External genitals of a male sheep fetus (d 80 of pregnancy). Urethral orifice located umbilically (arrow)

differentiation. It became possible to reinvestigate the significance of androgens in gonadal and Müllerian duct development after the strong antiandrogenic properties of cyproterone acetate – in short

CPA – had been discovered by Neumann (Hamada et al., 1963; Neumann, 1964; Neumann and Hamada, 1964).

Besides modification of various androgen target structures, e.g. the external genitals and the Wolffian ducts, we never observed

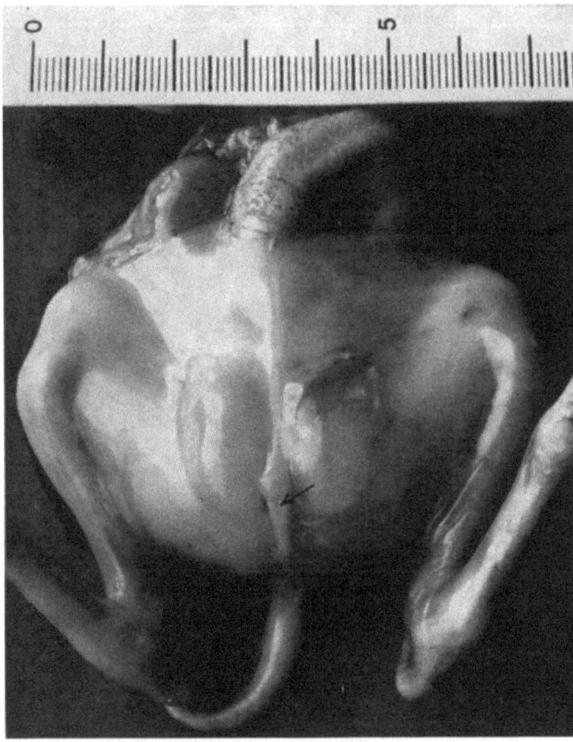

Fig. 3b. External genitals of a male sheep fetus (d 80 of pregnancy), mother treated with 10.0 mg CPA/kg daily i.m. from d 21—79 of pregnancy. Hypospadia (arrow)

any influence of this androgen-antagonist on gonadal differentiation or Müllerian duct retrogression. This is the most important result of our experiments on male fetuses of various mammalian species. Very typical findings can be demonstrated, for example, in rabbits (Elger, 1966) and other species (Fig. 3a, b, c).

Along with essentially normal testes, neither Wolffian nor Müllerian ducts do exist, as shown in Fig. 3d. Thus the gonads lost their cannular connections with the urinary tract. By this result,

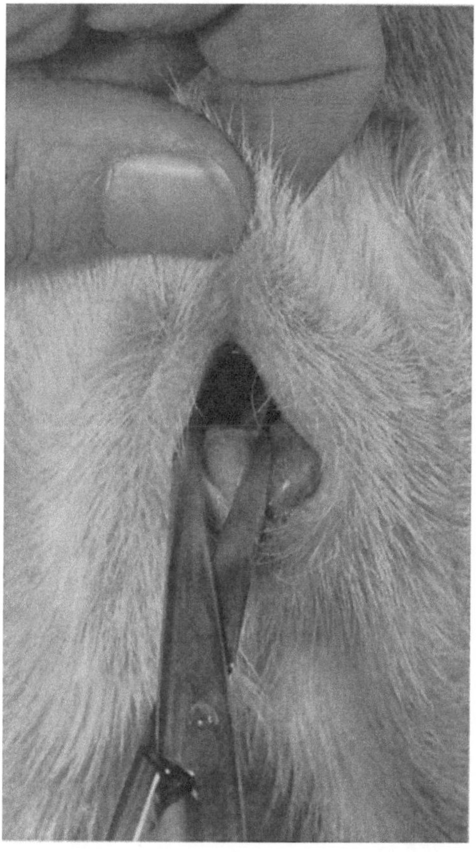

Fig. 3c. External genitals of an adult male dog, mother treated with 10.0 mg CPA/kg daily i.m. from d 22—62. Well developed vagina

Jost's theory is confirmed-that the fetal testis does not only produce androgens but also a special hormone that causes Müllerian duct suppression.

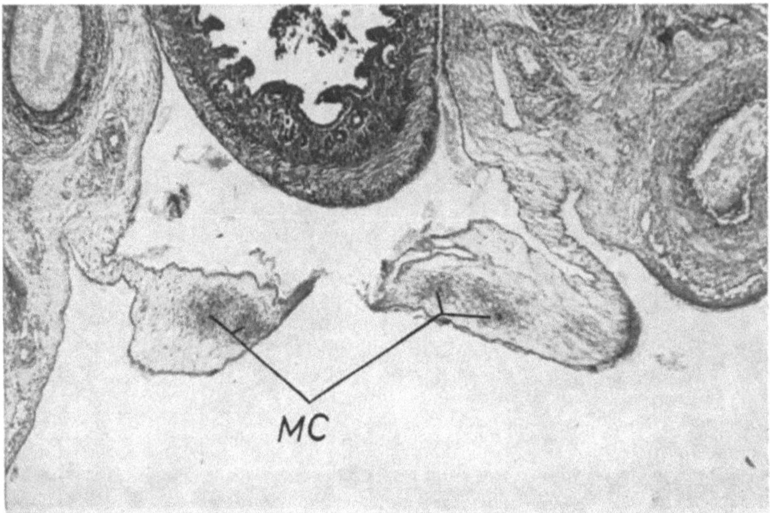

Fig. 3d. Internal genitals of a male rabbit fetus (d 30 of pregnancy), mother treated with 5.0 mg CPA/kg daily i.m. from d 13—24 of pregnancy. Mesogenital folds lacking both duct systems, former location of Wolffian and Müllerian ducts marked by mesenchymal cell condensations (*MC*)

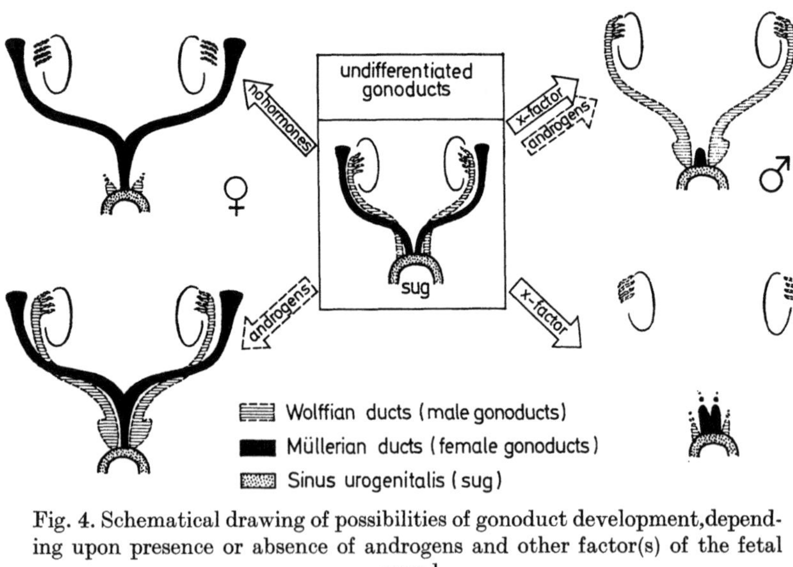

Fig. 4. Schematical drawing of possibilities of gonoduct development, depending upon presence or absence of androgens and other factor(s) of the fetal gonad

Two basic possibilities of intersexuality at the level of the gonoducts, and also normal sexual differentiation, can be explained by this theory.

For the clinician, it might be interesting to see that the syndrome of "testicular feminization" in genetically male individuals, which is characterized by a female phenotype with testis but no uterus, can be mimicked by the inhibitory effect of CPA on androgen action. To illustrate in more detail how the Wolffian ducts react on CPA treatment, the rabbit can be used as example. It is possible to titrate, so to say, the fetal androgens against CPA. Beyond a critical antiandrogen dose – in this case 5.0 mg/kg – a typical all-or-none reaction of Wolffian duct retrogression is achieved (ELGER, 1966; ELGER et al., 1970).

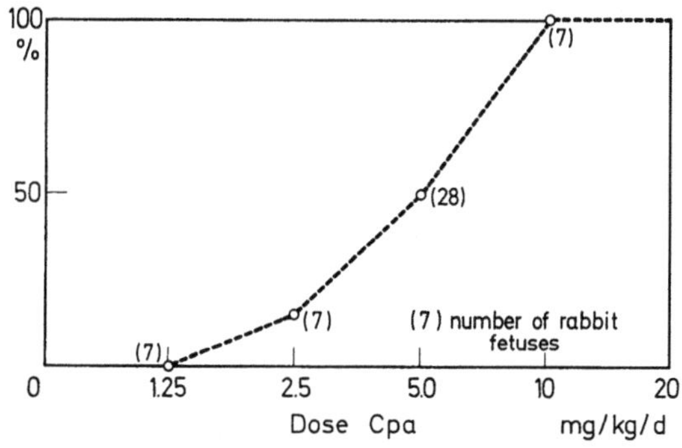

Fig. 5. Wolffian duct retrogression in male rabbit fetuses dependent on the CPA dose ('all-or-none' reaction)

Pilot studies in other species led us to assume that similar conditions are widely distributed over the mammalian kingdom. For example, we found the syndrome of gonoductal absence after CPA treatment also in dogs (STEINBECK et al., 1970), hamsters (ELGER, unpubl.) and sheep (ELGER et al., unpubl.).

A noteworthy exception to this rule is the behavior of Wolffian ducts in rats. Even under excessive CPA doses, we never found retrogressed Wolffian ducts in male rat fetuses. On the contrary, these ducts were invariably well developed and differentiated, and only the seminal vesicles were found smaller than normal. However,

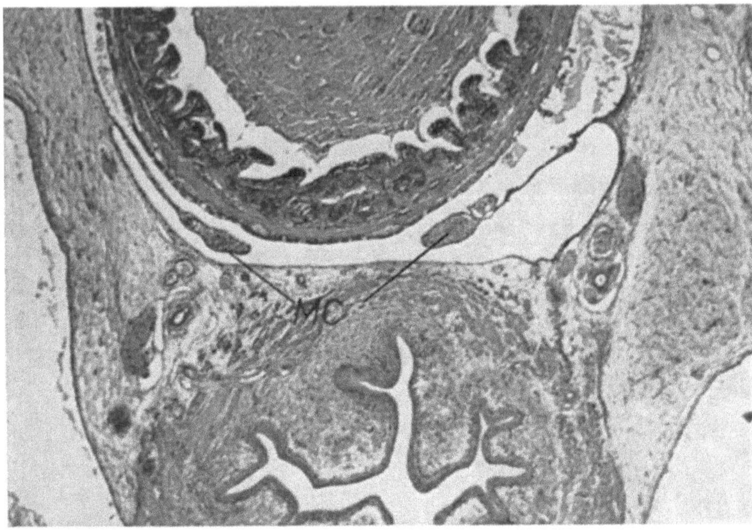

Fig. 6. Internal genitals of a male dog fetus, mother treated with 10.0 mg CPA/kg daily from d 23—44 of pregnancy. Mesogenital folds lacking both duct systems, former location of Wolffian and Müllerian ducts marked by mesenchymal cell condensations (*MC*)

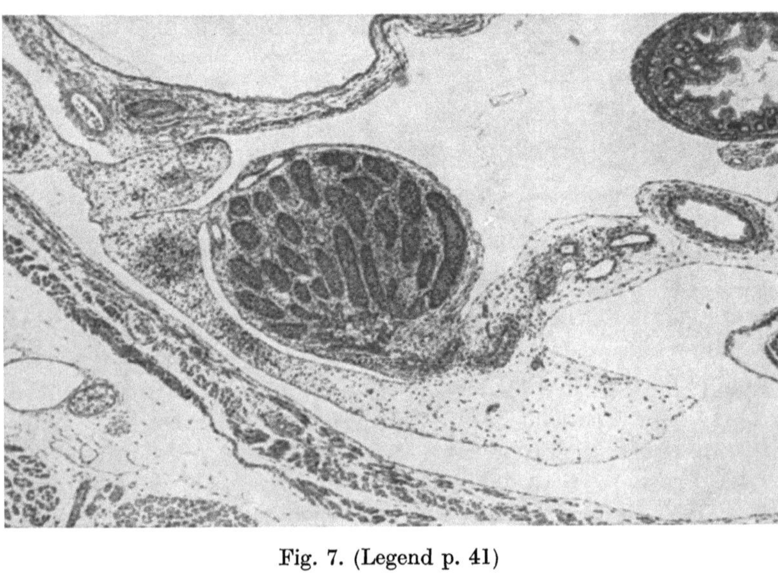

Fig. 7. (Legend p. 41)

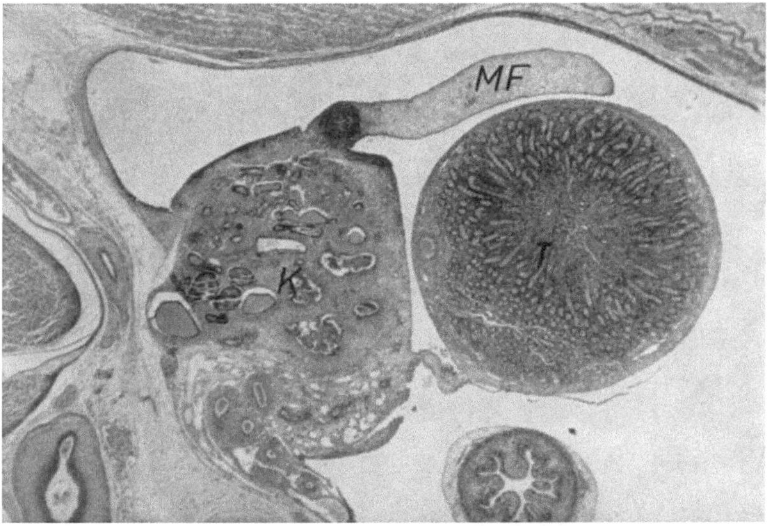

Fig. 8. Internal genitals of a male sheep fetus (d 65 of pregnancy), mother treated with 60.0 mg/kg CPA daily from d 21—64 of pregnancy. Absence of gonoducts next to a normal testis. *K* = mesonephros, *T* = testis, *MF* = mesogenital fold

the external genital organs were invariably strongly feminized (Fig. 9).

In order to investigate this particular resistance of the Wolffian ducts in male rat fetuses, we did an experiment on strongly virilized female rat fetuses.

With doses of either 1.0, 3.0 or 10.0 mg of methyltestosterone the Wolffian ducts were maintained and further differentiated. Much to our surprise we found that the effect of methyltestosterone on the Wolffian ducts was readily abolished by concomitant CPA treatment in a dose-dependent manner. In a dose ratio of 1.0 or 3.0 mg of methyltestosterone and 30.0 mg of CPA, the majority of

---

Fig. 7. Internal genitals of a male hamster fetus, mother treated with 10.0 mg CPA/animal daily from d 8—15 of pregnancy. Absence of gonoducts next to a normal testis

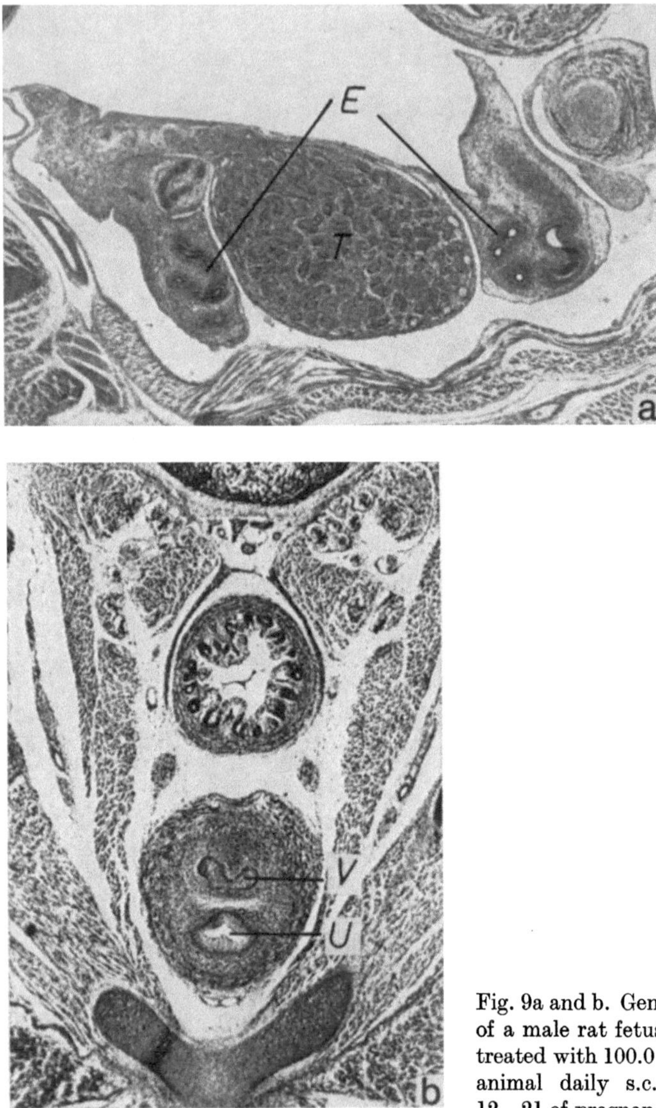

Fig. 9a and b. Genital tract of a male rat fetus, mother treated with 100.0 mg CPA/animal daily s.c. from d 12—21 of pregnancy. a) No interference of CPA treatment with Wolffian duct differentiation (epididymis) at the gonadal level. b) Formation of a vagina at the level of the urogenital sinus. This process is normally inhibited by fetal androgens. E = epididymis, T = testis, U = urethra, V = vagina

Table. *Effects of methyltestosterone alone or in combination with CPA on Wolffian duct stabilization and differentiation in female rat fetuses*

Treatment of the pregnant rats (d 15—d 21 p.c.)	♀ foetuses (d 22) n	♀ Wolffian ducts — deferent ducts (parauterine and more posterior portions)			
		Absence % ( ) = *n* of ducts	Partial regression (blind ending rudiment) % ( ) = n of ducts	Complete with signs of degeneration % ( ) = n of ducts	Complete well maintained duct % ( ) = n of ducts
MT 1.0 mg/d s.c.	11	0 (0)	0 (0)	0 (0)	100 (22)
MT 1.0 mg/d CPA 30.0 mg/d s.c.	12	66.7 (16)	20.8 (5)	12.5 (3)	0 (0)
MT 3.0 mg/d s.c.	8	0 (0)	0 (0)	0 (0)	100 (16)
MT 3.0 mg/d CPA 30.0 mg/d s.c.	12	45.8 (11)	25 (6)	29.2 (7)	0 (0)
MT 10.0 mg/d s.c.	9	0 (0)	0 (0)	0 (0)	100 (18)
MT 10.0 mg/d CPA 30.0 mg/d s.c.	10	5 (1)	10 (2)	75 (15)	10 (2)

Wolffian ducts retrogressed. Even the effect of 10.0 mg of methyltestosterone was inhibited by only 30.0 mg of CPA. In this group most of the Wolffian ducts survived but underwent no farther differentiation.

This experiment demonstrates that CPA has a strong affinity to androgen receptors also in rats. Therefore, it is unclear even more now why the Wolffian ducts of rat fetuses do not respond to CPA treatment whereas rabbit, dog, hamster and sheep fetuses do.

Our results show also the limitations of a biological experiment. Apart from this, we shall gain more detailed knowledge when the structure and the mechanism of action of the embryonal inductors will be clarified.

References

DANTCHAKOFF, V.: L'hormone male adulte dans l'histogenese sexuelle du mammifere. C. R. Soc. Biol. (Paris) **123**, 873 (1936).

ELGER, W.: Die Rolle der fetalen Androgene in der Sexualdifferenzierung des Kaninchens und ihre Abgrenzung gegen andere hormonale und somatische Faktoren durch Anwendung eines starken Antiandrogens. Arch. Anat. micr. Morph. exp. **55**, 658—738 (1966).

— unpubliziert (1970).

— HAHN, J. D., HORST, P.: unpubliziert (1970).

— NEUMANN, F., BERSWORDT-WALLRABE, R. VON: The influence of androgen antagonists and progestogens on the sex differentiation of different mammalian species. In: HAMBURGH, M. (ed.): Hormone Regulation. New York: Appleton-Century-Crofts 1970. In press.

HAMADA, H., NEUMANN, F., JUNKMANN, K.: Intrauterine antimaskuline Beeinflussung von Rattenfeten durch ein stark gestagen wirksames Steroid. Acta endocr. (Kbh.) **44**, 380—388 (1963).

JOST, A.: Recherches sur la differenciation sexuelle de l'embryon de lapin . II. Action des androgenes de synthese sur l'histogenese genitale. Arch Anat. micr. Morph. exp. **36**, 242—270 (1947).

— Modalities in the action of gonadal and gonadstimulating hormones in the foetus. Mem. Soc. Endocr. **4**, 237 (1955).

NEUMANN, F.: Nachweis intrauteriner Virilisierung durch Sagittalschnitte von Rattenfeten. 10. Symp. Ges. f. Endokrinologie, Wien 1963, S. 297 bis 300. Berlin-Heidelberg-New York: 1964.

— HAMADA, H.: Intrauterine Feminisierung männlicher Rattenfeten durch das stark gestagen wirksame 6-Chlor-Δ^6-1,2α-methylen-17α-hydroxyprogesteronacetat. 10. Symp. Dtsch. Ges. f. Endokrinologie, Wien 1963, S. 301—304. Berlin-Heidelberg-New York: 1964.

STEINBECK, H., NEUMANN, F., ELGER, W.: Effect of an antiandrogen on the differentiation of the internal genital organs in dogs. J. Reprod. Fertil. (im Druck) (1970).

Hypothalamic LH-Releasing Hormone: Chemistry, Physiology and Effect in Humans

A. V. Schally, A. Arimura, A. J. Kastin, J. J. Reeves, C. Y. Bowers and Y. Baba

Endocrine and Polypeptide Laboratories and Endocrinology Section of the Medical Service, Veterans Administration Hospital and Department of Medicine, Tulane University School of Medicine, New Orleans, La., USA

W. F. White

Scientific Divisions, Abbott Laboratories, North Chicago, Illinois, USA

With 15 Figures

I. Introduction

There is now good evidence that the secretion of LH and FSH from the anterior pituitary gland is controlled principally by the central nervous system and a feedback system, involving sex steroids [1, 2]. The hypothalamus stimulates the secretion of follicle stimulating hormone (FSH), luteinizing hormone (LH) and other pituitary hormones by releasing regulatory substances into portal blood flowing from the median eminence region into the pars distalis [3, 4]. Table 1 lists the known hypothalamic hormones and the new nomenclature we have proposed for them [4]. In this report we shall be concerned only with LH-releasing hormone (LH-RH) and FSH-releasing hormone (FSH-RH). Their function is to augment the secretion of LH and FSH.

With much simplification it can be said that the sex steroids derived from the ovary and the testes exert mainly an inhibitory influence called a "negative feedback" by acting principally on the hypothalamus [5—7]. These complex relationships are shown in Fig. 1. Solid arrows indicate stimulatory effects; interrupted arrows inhibitory influences. It should be mentioned here that estrogen can, under certain conditions, stimulate LH release [8]. This would create a situation of "positive feedback" [9]. Also it should be

Table 1. *The nomenclature of some hypothalamic neurohumors*

Present name		Proposed name	
Hypothalamic factor	Abbreviation	Hypothalamic hormone	Abbreviation
Corticotropin-releasing factor	CRF	Corticotropin-releasing hormone	CRH
Luteinizing hormone-releasing factor	LRF	Luteinizing hormone-releasing hormone	LH-RH
Follicle-stimulating hormone-releasing factor	FSH-RF	Follicle-stimulating hormone-releasing hormone	FSH-RH
Thyrotropin-releasing factor	TRF	Thyrotropin-releasing hormone	TRH
Growth hormone releasing factor or somatotropin-releasing-factor	GRF or SRF	Growth hormone-releasing hormone	GH-RH
Prolactin inhibiting factor (mammals)	PIF	Prolactin release inhibiting hormone	PRIH
Melanocyte stimulating hormone (MSH) release-inhibiting factor	MIF	MSH-release-inhibiting	MRIH
Growth hormone-inhibiting factor	GIF		

Modified from SCHALLY et al. [4]. Courtesy of Academic Press.

mentioned that there is evidence suggesting that estrogen and progesterone can exert some direct effect on the pituitary [9—11]. Another type of feedback can also apparently exist in which the inhibitory impulses are provided by the pituitary hormones themselves [12, 13]. For this system the name of "short" or "internal" feedback mechanism has been proposed [12], but it will not be discussed at this time.

In this review we propose to cover our latest studies on LH-releasing hormone (or LH-RH) and FSH releasing hormone (or FSH-RH). We will also present our work on some compounds affecting fertility, by reporting some recent results concerning the actions of clomiphene and the oral contraceptive steroids on the hypothalamo-pituitary axis in rats. In addition we will briefly mention the action of LH-RH in sheep and describe the effects of administration of LH-RH to humans.

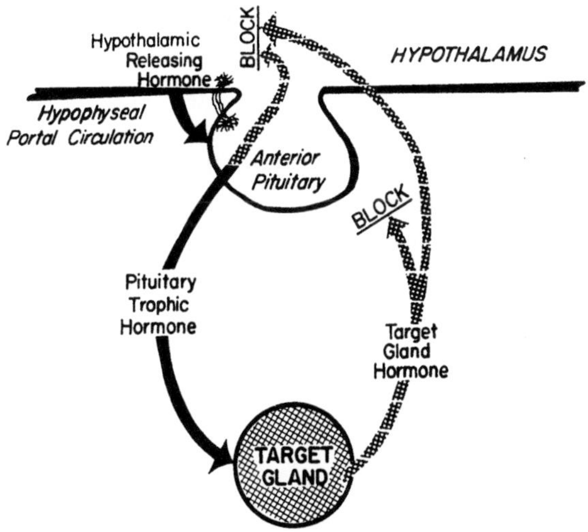

Fig. 1. Hypothalamo-hypophysial-target organ relationships. Solid arrows denote stimulatory effects; interrupted arrows, inhibitory effects. There are three negative feedback circuits discernible: (*1*) the long-established loop between target gland and anterior pituitary, (*2*) "long" feedback loop between target gland and hypothalamus and (*3*) "short" "internal" or "auto" feedback loop between the anterior pituitary and the hypothalamus. (From "Some Aspects of Neuroendocrinology" by SCHALLY, A. V., LOCKE, W., KASTIN, A. J. BOWERS, C. Y. in Year Book of Endocrinology, 1967—1968. Copyright 1967, Year Book Medical Publishers, Inc. Used by permission)

II. Purification of Porcine LH-RH

LH-RH was prepared from fragments comprising the pituitary stalk and the median eminence region as follows:

1. Pulverization of pig hypothalami on dry ice and defatting with acetone and petroleum ether [4, 14].

2. Extraction of the defatted powder with 2 N acetic acid and lyophilization of the extracts [4, 14].

3. Concentration of the lyophilized extracts by re-extraction with glacial acetic acid and lyophilization of the supernatant [4, 14].

4. Gel filtration on a preparative column (15.5 $\times$ 180 cm) of Sephadex G-25 in 1 M acetic acid [4, 14, 15].

5. Extraction of the LH-RH active fractions with phenol followed by recovery [14, 15].

6. Chromatography on carboxymethylcellulose (CMC), using ammonium acetate buffers [14, 15].

7. Rechromatography on CMC [4, 14].

8. Free flow-electrophoresis in pyridine acetate buffer at pH 6.3 [16].

9. Countercurrent distribution (CCD) in a system of 0.1 % acetic acid: 1-butanol : pyridine = 11 : 5 : 3 (*v*/*v*) [16].

10. Partition chromatography in a system consisting of 1-butanol: acetic acid:water:benzene = 4 : 1 : 5 : 0.33 [16].

TLC of Highly Purified Pig LH-RH

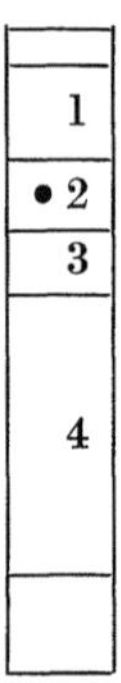

	Zone	LH-RH activity plasma LH ng/ml mean ± SE	*P*
Solvent front			
	1	8.5 ± 2.93	NS
	2	88.2 ± 8.6	< 0.001
	3	9.5 ± 2.6	NS
	4	11.0 ± 2.9	NS
	Saline	12.6 ± 2.5	——

Fig. 2. Thin layer chromatography of porcine LH-RH in 1-butanol: acetic acid : water = 4 : 1 : 5

Thin-layer chromatography (TLC) of LH-RH purified according to this procedure, revealed only one spot in the system 1-butanol: acetic acid: water = 4 : 1 : 5 positive to chlorine/o-tolidine. This spot was associated with the LH-RH activity [17]. There was no LH-RH activity in other areas of the TLC plate (Fig. 2).

The amino acid content of the LH-RH purified by CCD was 30 % of the dry, weight, but after purification by chromatography the amino acid content fell to 7.8 %.

Recent results indicate that LH-RH activity is abolished following incubation with chymotrypsin, subtilisin, amino peptidase M and papain [17]. This supports the concept that LH-RH may be a polypeptide. The molecular weight of LH-RH is of the order of several hundred (WHITE, W. F., unpublished). The biological activity of materials at various purification stages is described in the sections which follow.

Table 2. *Human LH-RH. LH-RH activity in fractions from CMC measured by elevation of plasma LH in ovariectomized rats treated with estrogen and progesterone*

Fraction[a] No.	RIA of Rat LH mμg LH-RH/ml plasma[b] (Mean of 3 rats) ± S.E.		Stimulation P vs control
	before i.v. injection	after i.v. injection	
Control saline	5.5 ± 0.26	3.9 ± 0.17	—
140—150	4.25 ± 0.25	3.5 ± 0.5	NS
151—160	5.5 ± 0.7	4.3 ± 1.1	NS
161—170	6.2 ± 0.7	5.1 ± 0.2	NS
171—180	5.5 ± 0.26	107.5 ± 9.0	0.001
181—190	7.4 ± 1.5	16.5 ± 1.7	0.01
191—200	6.6 ± 0.46	5.2 ± 0.26	NS
201—210	6.9 ± 0.3	4.7 ± 0.7	NS
211—220	5.7 ± 0.3	4.1 ± 0.26	NS

[a] 30 μl of each tube/rat = equivalent 0.4 SME/rat or approx. 4 μg dry wt/rat for active fractions.

[b] As NIH-LH-S-14 read directly on the standard curve.

From SCHALLY et al. [18]. Courtesy of J. Clinical Endocrinology and Metabolism and J. B. Lippincott Co.

Purification of human LH-RH.

Acetic acid extracts from 339 human hypothalami were prepared in the same way as described for porcine tissues. Due to the small amounts of material available, these extracts were purified only by gel filtration on Sephadex G-25, phenol extraction and then by chromatography on CMC [18]. The pattern of separation of human LH-RH on CMC is given in Fig. 3 and the biological activity is given in Table 2.

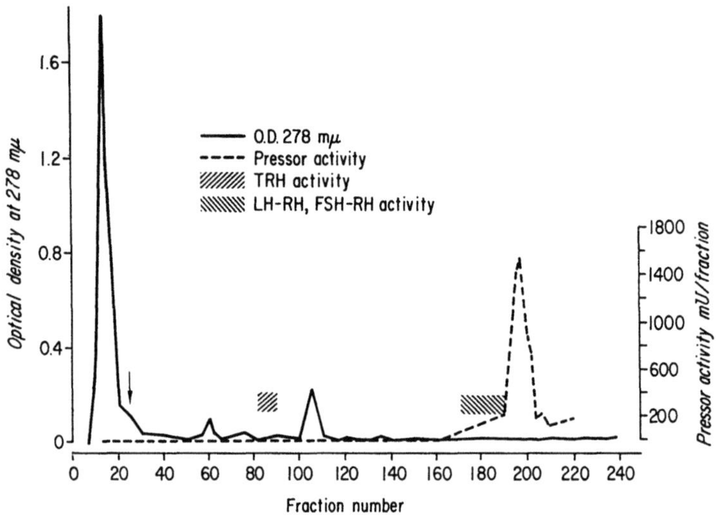

Fig. 3. Ion exchange chromatography of 334 mg of phenol concentrate of human hypothalami on CMC column 2.8 × 60 cm, equilibrated with 0.002 M, pH 4.6 ammonium acetate buffer. Gradient to 0.1 M pH 7 buffer through 2000 ml mixing flask applied at tube No. 25. Flow rate 250 ml/hr. Fraction size 25 ml. Aliquots of each tube used for TRH tests were 18 μl/mouse, for LH-RH bioassays 30 μl/rat, and for FSH-RH 500 μl/beaker. From Schally et al. [18]. Courtesy of J. Clinical Endocrinology and Metabolism and J. B. Lippincott and Co.

III. Determination of Biological Activity of LH-RH

1. Studies in Rats

From our studies, and those of others, the following actions of LH-RH preparations of various degrees of purity are apparent:

A. LH-RH will elevate plasma LH in ovariectomized rats treated with

(a) estrogen and progesterone [19, 14],
(b) testosterone [20],
(c) oral contraceptive steroids [21, 22].

B. LH-RH will raise plasma LH in
(a) castrated male rats treated with testosterone [20, 4],
(b) intact rats at varous stages of the estrous cycle [23],
(c) intact rats treated with progesterone.

These studies have been reviewed previously [2, 4]. Recently the results obtained by bioassay for LH [24] were confirmed [22—28] by the use of specific radioimmunoassays for rat LH [29, 30]. PARLOW et al. [26, 27] and SCHALLY et al. [2, 22, 31] obtained up to 20-fold elevations of serum LH after intravenous administration of nanogram doses of highly purified LH-RH to ovariectomized, estrogen and progesterone treated rats. Some of these results are shown in Table 3. They demonstrate significant stimulation of LH release as measured by bioassay or radioimmunoassay. Other recent studies on rats are described in Section V, A, B, C, F.

Table 3. *Effect of highly purified porcine LH-RH on LH release in vivo*

Group	LRH dose nanog dry wt. per rat	Plasma LH activity OAA change μg/100 ± S.E.	P group	P	Radioimmuno-assay for rat LH serum LH[a] mμg/ml
1. Saline	—	— 3.0 ± 0.5	—	—	<131
2. LRH	0.5	— 8.2 ± 0.9	2 vs 1	0.001	243
3. LRH	5	—15.6 ± 1.2	3 vs 1	0.001	1090
			3 vs 2	0.01	
4. LRH	25	—22.0 ± 1.7	4 vs 1	0.001	2460
			4 vs 3	0.05	

[a] Expressed in terms of NIAMD Rat-LH-RP-1.

Modified from SCHALLY et al. [2, 31] and PARLOW et al. [27]. Courtesy of Academic Press, Endocrinology, J. B. Lippincott Co., and Williams and Wilkins Co.

C. Purified LH-RH will induce ovulation in rats [32, 33]. As little as 1 μg injected i.v. or i.c. into Nembutal blocked 4 day cyclic rats on the day of pro-estrus induced ovulation. In contrast, LH-RH was ineffective in hypophysectomized rats [32].

Table 4. *Stimulation by porcine LH-RH of the release of LH in vitro from pituitaries of rats ovariectomized and pre-treated with estrogen and progesterone*

Group of pituitaries[a]	LH-RH added to medium µg/pituitary	LH released expressed in terms of the LH standard[b]			LH released by stimulated pituitaries assayed against control pituitaries		F
		LH released µg	µg LH/mg pit/hr	95% fiducial limits	ratio of LH release	95% confidence limits	
Control	—	12.2	0.23	0.089—0.603	1.0	—	
stimulated	0.022	23.1	0.44	0.195—1.01	2.45	1.10—5.47[c]	8.02
Control	—	8.8	0.22	0.06 —0.79	1.0	—	
stimulated	0.07	33.0	0.78	0.39 —1.5	4.81	1.35—13.2[c]	15.7[c]
Control	—	6.14	0.106	0.034—0.332	1.0	—	
stimulated	0.11	21.7	0.388	0.214—0.703	2.47	1.54— 3.96[c]	23.1[c]

[a] 7—9 pituitary halves/flask.

[b] Assay in terms of standard NIH-LH-S-1.

[c] Significant difference between samples. Significant stimulation shown by lower fiducial limit greater than 1. as well as by variance ratio *F* for samples.

From Schally et al. [14]. Courtesy of Endocrinology and J. B. Lippincott Co.

D. LH-RH is very effective in stimulating LH release *in vitro* as shown by SCHALLY and BOWERS [34]; SCHALLY et al. [4, 14]; PIACSEK and MEITES [35] and JUTISZ et al. [36]. Some of these results are shown in Table 4. *In vitro* tests offer strong proof that LH-RH acts directly on the pituitary tissue. The time course of the release of LH *in vitro* is identical with that of FSH (WHITE, W. F., unpublished). The main effect of LH-RH is probably on LH release, but evidence that LH-RH also can exert some direct or indirect action on LH synthesis is presented in Section V, B.

2. Studies in Rabbits

Recent results of HILLIARD, SCHALLY and SAWYER [37] indicate that highly purified LH-RH will induce ovulation in estrous rabbits when infused in doses of 0.1—1 μg. This method was popularized primarily by HARRIS and his co-workers who induced ovulation in rabbits after intrapituitary infusion or after intravenous administration of crude extracts or purified materials [38—40]. Other results in rabbits are briefly mentioned in Section V, E.

3. Studies in Nutria

Porcine LH-RH will raise plasma LH levels in nutria (*myocastor coypus*) untreated or treated with sex steroids [41].

Studies on the effects of LH-RH in sheep and humans are described in Sections V, C and V, G.

IV. Follicle Stimulating Hormone-Releasing Hormone (FSH-RH)

The preparations of highly purified porcine LH-RH and purified human LH-RH contained FSH-RH activity. This FSH-RH activity may have been inherent to LH-RH or due to contamination with FSH-RH.

FSH-RH activity was determined on the basis of:

1. Stimulation of FSH release *in vitro* from pituitaries of

(a) normal intact male rats [42],

(b) ovariectomized rats pretreated with estrogen and progesterone [43—45],

(c) castrated rats pretreated with testosterone propionate [44].

2. Elevation of plasma FSH in rats castrated and pretreated with steroids as measured by a radioimmunoassay for rat FSH [26, 46].

3. Elevation of plasma FSH in humans ([47]) as measured by radioimmunoassay for human FSH [48] (see Section V, G).

Among assays used previously for the measurement of FSH-RH activity and subsequently abandoned was the depletion of pituitary FSH content in normal male rats [49, 50] or in castrated male rats pretreated with testosterone propionate [51]. The use of the assays based on pituitary FSH depletion was discontinued due to the apparent lack of their specificity. The evidence for this was as follows: histamine, putrescine, spermine and spermidine were isolated from porcine hypothalamic extracts and were shown to deplete the pituitary of FSH *in vivo* [52]. However in rats pretreated with antihistamines FSH-RH depleted pituitary FSH content while histamine did not do so [43]. Subsequent *in vitro* work

Table 5. *Effect of porcine FSH-RH-preparations at various stages of purity on FSH release in vitro from rat pituitaries*

Group of pituitaries	Dose μg/ml	FSH content of medium. Ovarian wt mg ± SE	P vs control
Control, normal male pits	—	59.6 ± 4.4	—
Exp. normal male pits	0.0006	69.2 ± 7.4	N.S.
Exp. normal male pits	0.0025	87.4 ± 6.2	0.01
Exp. normal male pits	0.01	105.2 ± 17.6	0.05
Exp. normal male pits	0.04	196.2 ± 17.9	0.001
Control normal male pits	—	47.8 ± 3.8	—
Exp. normal male pits	0.001	61.6 ± 2.8	0.02
Exp. normal male pits	0.003	82.4 ± 5.1	0.001
Control, O.E.P. pits[a]	—	74.8 ± 11.6	—
Exp. O.E.P. pits	0.01	152.0 ± 17.3	0.01
Control Castrated T.P. pits[b]	—	59.5 ± 4.6	—
Exp. Castrated T.P. pits	0.075	67.8 ± 7.2	N.S.
Exp. Castrated T.P. pits	0.37	92.0 ± 7.8	0.01

[a] Pituitaries from ovariectomized, estrogen and progesterone treated rats.
[b] Pituitaries from castrated, testosterone propionate treated male rats.

Modified from Schally et al. [44]. Courtesy of Endocrinology and J. B. Lippincott Co.

shown in Tables 5 and 6, clearly demonstrated that purified porcine FSH-RH devoid of free polyamines and in doses as low as 1 ng/ml released FSH from pituitaries of normal male rats, castrated, testosterone propionate-pretreated rats or ovariectomized estrogen and progesterone pre-treated rats (Table 5), while putrescine, cadaverine, spermine, spermidine and histamine did not do so even in doses one hundred to one thousand times greater than those of FSH-RH (Table 6) [44]. Moreover, *in vivo* results show that putrescine after i.c. administration or intrapituitary infusion did not increase plasma FSH or LH as measured by RIA [53].

Table 6. *Lack of effect of some polyamines histamine on FSH release in vitro from pituarities of normal male rats*

Exp. No.	Group of pituitaries	Addition	Dose µg/ml	FSH content of Medium, ovarian wt mg ± SE	P vs control
1	Control	—	—	65.3 ± 4.1	—
	Experimental	Putrescine	0.01	77.9 ± 6.7	N.S.
2	Control	—	—	65.2 ± 5.4	—
	Experimental	Putrescine	0.01	68.8 ± 4.8	N.S.
3	Control	—	—	49.0 ± 4.2	—
	Experimental	Spermidine	1.0	60.2 ± 6.5	N.S.
	Experimental	Cadaverine	0.2	54.8 ± 7.1	N.S.
	Experimental	Putrescine	0.2	53.4 ± 5.4	N.S.
	Experimental	Spermine	1.0	67.0 ± 8.8	N.S.
4	Control	—	—	61.6 ± 4.2	—
	Experimental	Sperimine	1.0	53.4 ± 4.0	N.S.
	Experimental	Cadaverine	0.2	58.0 ± 4.1	N.S.
5	Control	—	—	68.0 ± 6.1	—
	Experimental	Agmatine	1.5	73.6 ± 7.2	N.S.
6	Control[a]	—	—	73.5 ± 5.9	—
	Experimental[a]	Histamine	0.2	71.1 ± 6.0	N.S.

[a] Pituitaries from ovariectomized rats pre-treated with estrogen and progesterone.

Modified from SCHALLY et al. [43, 44]. Courtesy of Endocrinology and J. B. Lippincott Co.

In contrast, intravenous administration of 72 ng of highly purified LH-RH/FSH-RH preparation of porcine origin increased plasma FSH in ovariectomized, estrogen and progesterone (OEP) treated, urethane anaesthetized rats as measured by RIA for rat FSH (Fig. 4) [27]. A 10 nanog dose (given i.v.) induced a smaller rise. In castrated male rats pretreated with testosterone propionate, 72 μg of the same LH-RH/FSH-RH preparation administered i.v. under urethane anaesthesia doubled plasma FSH levels at 30 min, with the FSH levels remaining elevated when measured 1 hour later (Fig. 5).

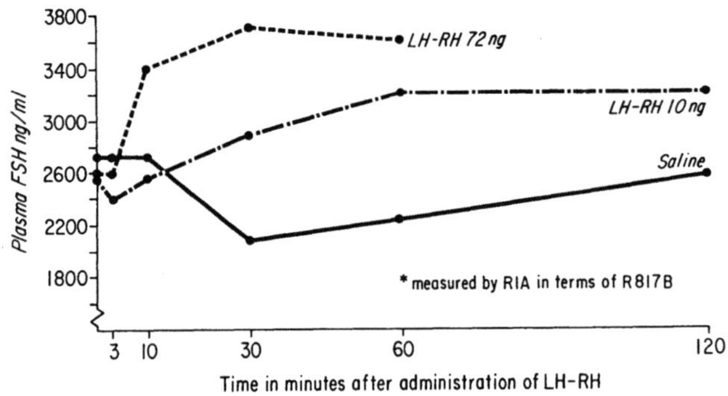

Fig. 4. Time study of the elevation of plasma FSH in ovariectomized, estrogen progesterone-treated urethane anesthetized rats after administration of 10 and 72 ng of porcine LH-RH. From [27]. Courtesy of Endocrinology and J. B. Lippincott and Co.

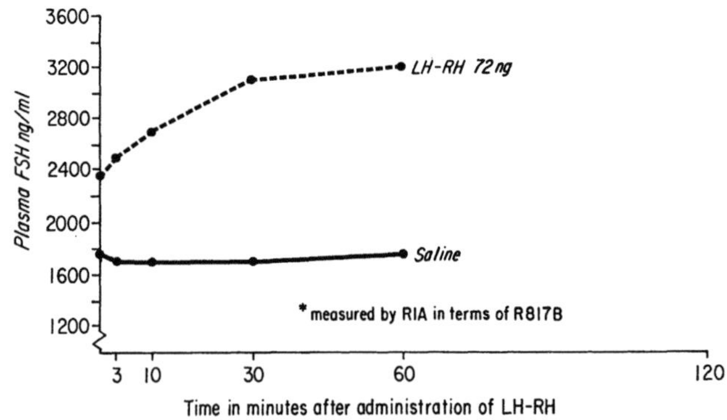

Fig. 5. Time study of the elevation of plasma FSH in castrated, testosterone treated male rats urethane anesthetized after administration of 10 and 72 ng of porcine LH-RH. From [27]. Courtesy of Endocrinology and J. B. Lippincott and Co.

Since the polyamine *per se* do not release FSH substances *in vitro*, in assays thought to be the most direct and specific [44, 54] they do not represent the primary physiological FSH-RH, i.e. a substance originating in the hypothalamus by means of which the hypothalamus controls the secretion of FSH from the anterior pituitary gland by a direct and specific physiological action on the FSH-producing cells of the pituitary [4, 44]. Furthermore it seems that the early claim [43] of separation of FSH-RH activity from LH-RH activity of porcine hypothalamic preparations by column electrophoresis may not be valid [43]. Instead, it is probable that we have separated LH-RH from polyamines since the effect of the fractions containing these latter materials was measured only by depletion of pituitary FSH content *in vivo* [44]. In our recent work, the technique of free flow electrophoresis separated free polyamines (high cathodic mobility) from materials with *in vitro* FSH-RH activity (moderate positive charge) [16]. FSH-RH fractions thus prepared still had considerable LH-RH activity [16]. Further purification by countercurrent distribution and column chromatography did not separate these activities. While it cannot be excluded that the presence of both these activities in one fraction is due to mutual cross-contamination, it is more probable, at least in the case of materials of porcine origin, that FSH-RH activity may be intrinsic to LH-RH [16, 44, 47]. It should also be mentioned that the time course of the release of FSH *in vitro* is identical with that of LH (White, W. F., unpublished). The attempts to inactivate LH-RH by various chemical reactions have always led to the destruction of FSH-RH activity (Kennedy, J. F. and W. F. White, unpublished).

V. Recent Studies with LH-RH

Since some of our early studies with LH-RH based on determination of LH by the OAAD method have been reviewed on several occasions [2, 4, 55], we will restrict ourselves to the review of the most recent results which utilize the measurement of LH by specific radioimmunoassays.

A. Time Responses in Vivo

The results shown in Fig. 6 indicate that 3 min after i.v. administration of 10 or 72 ng of porcine LH-RH to ovariectomized, estrogen and progesterone pretreated (OEP), urethane anaesthetized rats, plasma LH levels rose [27]. The rise was more than 3 fold in the case of the higher dose. The LH levels continued to rise 10 min after LH-RH administration and reached a peak at 30 min, when the elevation was 11 fold in the case of the high dose, as compared with the corresponding control. At 1 hour the LH levels with either dose were still elevated. In contrast, saline injection induced no rise in LH levels [27].

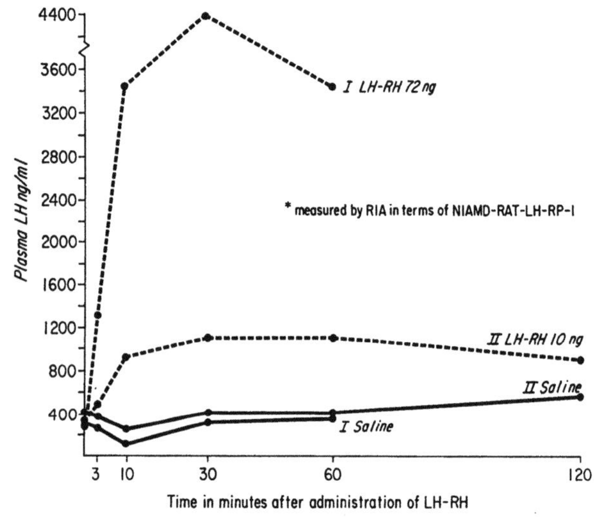

Fig. 6. Time study of the elevation of plasma LH in ovariectomized, estrogen progesterone-treated urethane anesthetized rats after administration of 10 and 72 ng of porcine LH-RH. From [27]. Courtesy of Endocrinology and J. B. Lippincott and Co.

In castrated male rats pretreated with testosterone propionate and anesthetized with urethane, LH-RH also induced a marked rise in plasma LH (Fig. 7). The time course of this elevation was similar to that of the OEP rats in that plasma LH started to rise at 3 min, continued rising at 10 min and peaked at 30 min, but the rise was 5—6 fold compared with the control value. The LH levels at 2 hrs in the case of lower dose (10 ng) had essentially returned to control values [27]. When LH-RH was injected i.c. the magnitude of the response was twice as great as that induced by i.v. administration.

Normal rats under urethane anesthesia also responded to LH-RH but their sensitivity was much smaller than that of castrated, steroid-treated rats (Fig. 8). As before, the peak of LH was achieved in 20—30 min, but the maximal rise in plasma LH was less than 100 % [27].

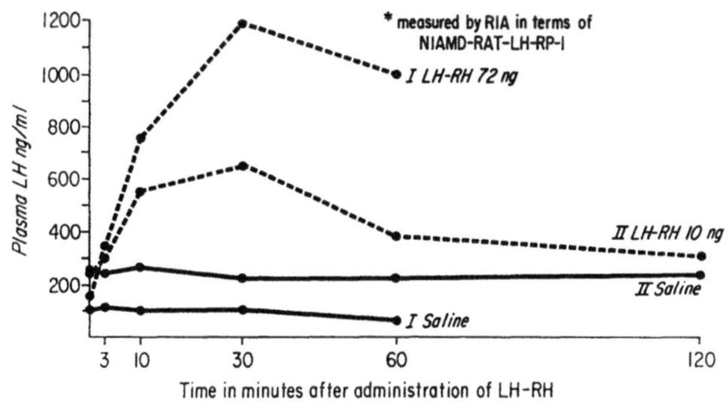

Fig. 7. Time study of the elevation of plasma LH in castrated, testosterone-treated male urethane anesthetized rats after administration of 10 and 72 ng of porcine LH-RH. From [27]. Courtesy of Endocrinology and J. B. Lippincott and Co.

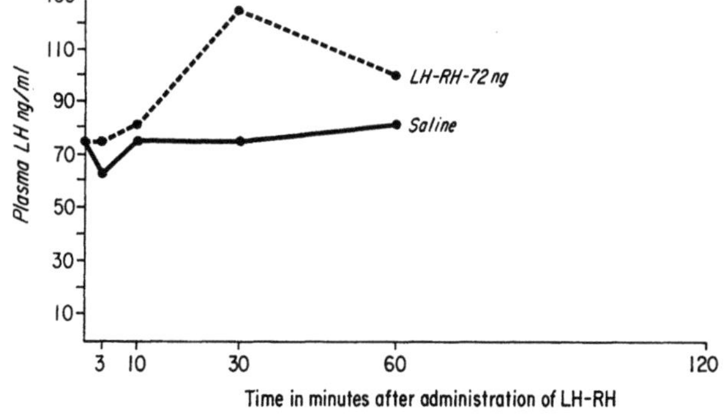

Fig. 8. Time study of the elevation of plasma LH in normal male urethane anesthetized after administration of 10 and 72 ng of porcine LH-RH. From [27]. Courtesy of J. B. Lippincott and Co. and Endocrinology

B. Effect of LH-RH on Release and Synthesis of LH and FSH in Tissue Cultures

In contrast to the well established effect of hypothalamic hormones on the release of gonadotropins, the evidence for the stimulatory effect of LH-RH on synthesis of LH and FSH remains scanty. The recent results of MITTLER, ARIMURA and SCHALLY [56], utilizing tissue cultures, offer some evidence that porcine LH-RH can stimulate synthesis as well as release of LH and FSH. Tissue cultures of rat anterior pituitaries were carried out in T. C. medium 199 for 3 days. The LH content of tissues and medium was estimated by radioimmunoassay [30] and FSH content both by bioassay [57] and radioimmunoassay [26, 46]. It can be seen in Table 7 that

Table 7. *Effects of porcine LH-RH preparations on medium and tissue LH contents in rat pituitary cultures*

Treatment	Total of 6 doses of LH-RH per pituitary in μg	LH content as mμg NIH-LH-S_{14} per pituitary by radioimmunoassay			
		Medium	Tissue	Total	Increase
Control	—	846	1444	2290	
Stimulated	0.46	2166	468	2634	344
Control	—	682	1006	1688	
Stimulated	1.25	1925	333	2258	570
Control	—	472	1027	1499	
Stimulated	0.25	1208	445	1653	154
Control	—	440	748	1188	
Stimulated	6.25	1705	165	1870	682
Control	—	600	2152	2752	
Stimulated	0.31	2625	315	2940	188
Tissue cultured	—	—	730	—	—
for 2 days only	—	—	560	—	—

From MITTLER, ARIMURA and SCHALLY [56]. Courtesy of the Proceedings of the society for Experimental Biology and Medicine and Academic Press.

addition of microgram amounts of highly purified porcine LH-RH twice daily to cultures of pituitaries significantly increased the quantities of LH and FSH released [56]. In each experiment in which LH-RH was added, pituitary LH content was depleted and averaged only 29 % of the controls after 3 days. Experimental medium from pituitaries stimulated with LH-RH averaged 324 % of the controls and contained more LH than could be accounted for by the reduction in the tissue content. The mean increase in the total medium and tissue content was equivalent to 388 mμg of NIH-LH-S-14 per pituitary. FSH release and synthesis are shown in Table 8. In 4 out of 5 experiments radioimmunoassayable FSH

Table 8. *Effects of porcine LH-RH preparations on medium and tissue FSH contents in rat pituitary cultures*

Exp. No.	Treatment	Total of 6 doses of LH-RH per pituitary in μg	FSH by radioimmunoassay (as μg NIAMD rat FSH RP-1 per pituitary)			
			Medium	Tissue	Total	Increase
1	Control	—	2.03	2.49	4.52	
	Stimulated	0.46	6.23	0.77	7.00	2.48
2	Control	—	1.76	1.73	3.49	
	Stimulated	1.25	10.53	0.87	11.40	7.91
3	Control	—	3.94	1.20	5.14	
	Stimulated	0.25	10.08	1.39	11.47	6.33

From MITTLER, ARIMURA and SCHALLY [56]. Courtesy of Proceedings of the Society for Experimental Biology and Medicine and Academic Press.

in the pituitaries was depleted, while the experimental medium averaged 326 % of the controls. There was an apparent synthesis of 4.3 μg of rat FSH per pituitary. Hypothalamic LH-RH also significantly stimulated the release of bioassayable FSH in 2 experiments. This stimulation was probably even greater since we encountered unexpected instability of bioassayable FSH when the pituitaries were stimulated with LH-RH. This may be due to extrusion of proteolytic enzymes from certain pituitary storage granules. In conclusion, LH-RH seems to stimulate synthesis as well as release of LH and FSH in tissue cultures [56].

C. Effect of LH-RH in Sheep

Another animal in which the response to LH-RH can be conveniently studied due to the availability of a reliable radioimmunoassay for LH [58] is the sheep. Intracarotid administration of 1 μg purified LH-RH into the conscious intact male sheep (ram) and female sheep (ewe) raised serum LH 2—5 ng/ml [59]. Maximal response in the ewes and the ram were obtained with 3 μg LH-RH which increased serum LH levels 6—18 ng/ml 2.5—10 min after i.c. administration [59]. Higher levels of LH could not be obtained by increasing the doses of LH-RH. On the other hand, in the castrated male sheep (wether), a 20 ng/ml elevation of serum LH could be obtained with 1 μg LH-RH and by increasing the dose of LH-RH to 27 μg serum LH levels as high as 160 ng/ml could be obtained [59]. A typical elevation of serum LH induced by 9.0 μg of purified LH-RH as measured by radioimmunoassay in the ewe, ram and the wether is shown in Fig. 9. This demonstrates that the pituitary is more responsive to LH-RH in the castrated than in the intact sheep. The peak of serum LH levels was observed about 5 min after i.c. injection of LH-RH in both normal and castrated sheep. The serum LH started decreasing 10 min after injection. In Nembutal anesthetized sheep, serum LH levels remained high 60 min after injection, suggesting a different metabolic rate of LH clearance between conscious and anesthetized sheep. The elevation of serum LH levels was obtained in the absence of any lowering of serum prolactin levels [60]. This indicates that LH-RH and prolactin release inhibiting hormone (PRIH) are separate.

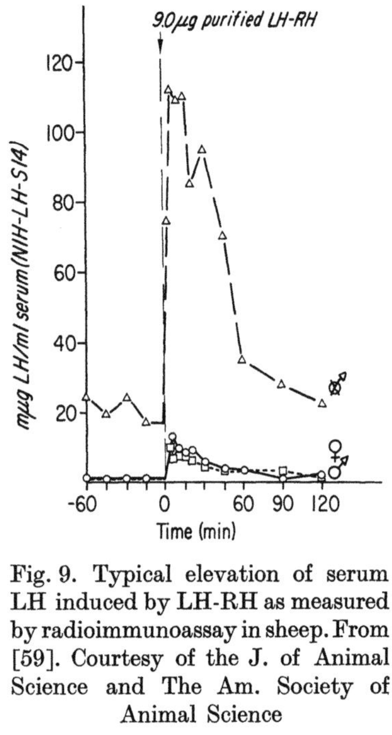

Fig. 9. Typical elevation of serum LH induced by LH-RH as measured by radioimmunoassay in sheep. From [59]. Courtesy of the J. of Animal Science and The Am. Society of Animal Science

In an experiment involving a male and a female fetal sheep, the *in utero* infusion of 30 μg of a purified porcine hypothalamic LH-RH fraction produced a marked rise in serum LH and FSH in the male, but not in the female fetus [61]. Both LH and FSH were measured by radioimmunoassays [58, 62]. The LH and FSH peaks for the male occurred within eight minutes of infusion and were nearly superimposable. Additional studies are under way to establish the reason for the apparent resistance of the female fetus to the stimulation of gonadotropin release.

D. Effect of LH-RH in the Chimpanzee

In repeated experiments separated by several days, the i.v. infusion of 15 and 30 μg per kg body weight of a purified LH-RH fraction of porcine origin into an immature male chimpanzee resulted in marked rises in serum levels of both LH and FSH with almost superimposable peaks [63]. Both the LH and the FSH peaks occurred within 10 minutes of sample infusion and both showed an approximate doubling of base-line levels [63].

E. Studies with Oral Contraceptive Steroids

The development of new methods of birth control may be one of the most pressing tasks facing humanity [64]. Of the methods presently available, perhaps the best one is based on oral administration of contraceptive preparations composed of a synthetic progestogen in combination with a small dose of synthetic estrogen [65—67]. Since we have LH-RH readily available, we thought that it would be of interest to evaluate the effect of pretreatment with ovulation inhibiting steroids on plasma FSH and LH levels in rats and on the subsequent response to LH-RH [21, 22]. Plasma LH levels were measured by ovarian ascorbic acid depletion (OAAD) assay [24] and by radioimmunoassay (RIA) [29]. Plasma FSH levels were also measured by RIA. The results are shown in Tables 9—11. These results clearly indicate that:

(1) Large doses of various progestogens alone and their combinations with estrogens exert inhibitory effects on LH secretion since they depressed the high plasma levels of LH found in ovariectomized rats to between 1/5—1/3 of its basal concentration [22]. In most cases, the combination of progestins with ethinyl estradiol or mestranol was more effective in lowering plasma LH than estrogen alone. It is not clear whether this is due to a potentiation or a summation of effects [21].

Table 9. *Effect of norethynodrel, ethynodiol diacetate, norethindrone and mestranol and their combination on plasma LH levels and response to LH-RH in ovariectomized rats*

Group	Treatment[a]	Dose μg	LRH dose μg	Plasma LH activity OAA change μg/100 mg ± SE	P group	P DUNCAN's test
Exp. 1						
1	Control	20000	—	—17.2 ± 0.9	—	—
2	Norethynodrel	20000	—	—10.8 ± 2.6	2 vs 1	0.01
3	Norethynodrel	20000	0.4	—32.0 ± 0.9	3 vs 2	0.01
					3 vs 1	0.01
4	Ethynodiol diacetate	2000	—	— 9.2 ± 1.2	4 vs 1	0.01
5	Ethynodiol diacetate	2000	0.4	—27.4 ± 1.8	5 vs 4	0.01
					5 vs 1	0.01
6	Norethindrone	5000	—	— 9.4 ± 1.7	6 vs 1	0.01
7	Norethindrone	5000	0.4	—20.2 ± 2.0	7 vs 6	0.01
					7 vs 1	NS
Exp. 2						
1	Control	—	—	—19.0 ± 0.4	—	—
2	Norethindrone	5000	—	— 9.0 ± 1.5	2 vs 1	0.01
3	Norethindrone	5000	0.4	—20.2 ± 2.3	3 vs 2	0.01
4	Mestranol	200	—	—10.6 ± 2.8	4 vs 1	0.01
5	Mestranol	200	0.4	—22.2 ± 1.8	5 vs 4	0.01
6	Ortho-Novum	4000	—	— 5.0 ± 1.4	6 vs 1	0.01
	(norethindrone	4000			6 vs 2	NS
	+ mestranol)	200			6 vs 4	0.05
7	Ortho-Novum	4000				
	(norethinodrone	200	0.4	—18.0 ± 1.7	7 vs 6	0.01
	+ mestranol)					
8	Enovid (norethynodrel + mestranol)	10000	—	— 5.4 ± 1.2	8 vs 1	0.01
					8 vs 4	0.05
9	Enovid (norethynodrel + mestranol)	10000	0.4	—20.8 ± 1.4	9 vs 8	0.01
10	Ovulen (ethynodiol diacetate + mestranol)	2200	—	— 4.0 ± 1.8	10 vs 1	0.01
					10 vs 4	0.05
11	Ovulen (ethynodiol diacetate + mestranol)	2200	0.4	—23.0 ± 1.6	11 vs 10	0.01
					11 vs 1	NS

[a] Daily for 5 days; doses given sc in 0.2 ml sesame oil.

From SCHALLY et al. 1968 [21]. Courtesy of J. Clin. Endocr. Metab. and J. B. Lippincott Co.

(2) Most progestogens in combination with a synthetic estrogen will also lower plasma FSH levels in the same animals although the extent of suppression is, in most cases, much less than that for LH [21].

(3) None of these steroids given singly or in combination blocked the response to LH-RH. In fact, minute doses of LH-RH caused 6—13 fold elevation of plasma LH in ovariectomized rats pretreated with contraceptive steroids and induced LH levels higher than those found in ovariectomized rats [21, 22].

It may be important to note that blockade of LH release as shown by suppression of plasma LH can be obtained with doses of oral contraceptive combinations some 200 times smaller than those which were used (Tables 9—11) in an attempt to block the response to LH-RH [21, 22]. However, neither estrogen, progesterone, nor a variety of contraceptive steroids can block the stimulating effect of LH-RH on the release of LH in ovariectomized rats. Other studies by SCHALLY et al. [22], shown in Table 12, indicate that oral contraceptives do not depress the sensitivity of pituitary tissue of ovariectomized rats to exogenous LH-RH. It can be seen that LH-RH administered to ovariectomized rats induced only a 74 % increase in plasma LH levels and no change in FSH levels. In the same rats, pretreatment with Ovulen lowered plasma LH to less than 1/5 of its basal concentration. A subsequent injection of LH-RH induced a 760 % increase in the plasma LH level. Although the absolute plasma LH value in the steroid suppressed rats after LH-RH was slightly smaller than in the untreated ovariectomized rats, this presumably resulted from the lower basal LH concentration in the former group; the increase after LH-RH was 460 mμg/ml in the untreated group and 765 mμg/ml in the steroid treated group. One may surmise that other contraceptive steroids would show the same effects in ovariectomized rats.

However, progesterone and possibly other progestogens may exert some direct effect on the pituitary. This is indicated by results in rats and rabbits [28, 37]. In normal rats the administration of 25 mg progesterone 48 hrs before the intracarotid injection or intrapituitary infusion of LH-RH decreased the responses to small doses of LH-RH [28]. These results are shown in Table 13. Similarly, administration of progesterone induced a partial blockade of ovulation in estrous rabbits in response to intrapituitary infusion of

Table 10. *Effect of sc administration of large doses of various sex-steroids on plasma LH and FSH levels and the response to LH-RH in ovariectomized (OVX) rats*

Group	Treatment[a]	Dose mg	LH-RH Dose μg	Radioimmunoassays of Plasma mean LH[b] n g/ml	Radioimmunoassays of Plasma mean FSH[b] n g/ml	Plasma LH activity OAA change g/100 mg ± SE	P group	P Duncan's test
Exp. 1								
1	OVX control	—	—	531	2800	—13.4 ± 1.0	—	—
2	Megestrol acetate	10	—	213	3000	— 1.8 ± 4.2	2 vs 1	0.01
3	Ovex (megestrol acetate + ethinyl estradiol)	10 0.2	—	119	1800	— 1.0 ± 2.8	3 vs 1 3 vs 2	0.01 NS
4	Ovex (megestrol acetate + ethinyl estradiol)	10 0.2	0.16	963	—	—12.0 ± 2.9	4 vs 3	0.01
5	Lynestrenol	10	—	119	1500	— 0.2 ± 1.6	5 vs 1	0.01
6	Lyndiol (lynestrenol + mestranol)	10 0.3	—	119	1600	— 0.0 ± 2.6	6 vs 1 6 vs 5	0.01 NS
7	Lyndiol (lynestrenol + mestranol)	10 0.3	0.16	969	1300	—16.6 ± 1.2	7 vs 6	0.01

Exp. 2								
1	OVX control	—	—	1188	6500	—20.8 ± 0.9	—	—
2	AY 11 440	20	—	219	2600	—11.6 ± 0.9	2 vs 1	0.01
3	AY 11 440 + ethinyl estradiol	20 1	—	281	2150	— 9.8 ± 1.4	3 vs 1	0.01
							3 vs 2	NS
4	AY 11 440 + ethinyl estradiol	20 1	0.11	2475	2650	—18.8 ± 1.5	4 vs 3	0.01
5	Ethinyl estradiol	1	—	381	1825	—12.0 ± 1.3	5 vs 1	0.01
6	Schering AG. SH 1040	25	—	594	3450	—11.2 ± 2.2	6 vs 1	0.01
7	Schering AG. SH 1040	25	0.11	1781	5350	—22.2 ± 2.2	7 vs 6	0.01

[a] Daily for 5 days; injections given s.c. in Sesame oil. — 3—4 Donor rats per group. — Five assay rats per group.

[b] Values are the mean of two replicate radioimmunoassays. — Reference preparations for LH and FSH in RIA were NIAMD Rat LH-RP-1 and R 817 B respectively.

From Schally et al. [22]. Courtesy of Endocrinology and J. B. Lippincott Co.

Table 11. *Effect of large doses of eight oral contraceptive preparations on plasma LH and FSH levels and on the response to LH-RH in ovariectomized rats as measured by the RIA*

Group No.	Treatment	Dose[a] mg	LH-RH dose μg	Plasma LH mμg/ml[b]	Plasma FSH mμg/ml
1	Ovariectomized control	—	—	653	3900
2	Oracon				
	(dimethisterone	20	—	153	1800
	+ ethinyl estradiol)	0.08			
3	Oracon				
	(dimethisterone	20	0.1	2338	2525
	+ ethinyl estradiol)	0.08			
4	Ovral	1			
	(norgestrel	1			
	+ ethinyl estradiol)	0.1	—	137	2775
5	Ovral				
	(norgestrel	1	0.1	905	2150
	+ ethinyl estradiol)	0.1			
6	Ortho novum				
	(norethindrone	5	—	130	1500
	+ mestranol)	0.25			
7	Ortho novum				
	(norethindrone	5	0.1	1610	2650
	+ mestranol)	0.25			
8	Ovulen				
	(ethynodiol diacetate	2	—	199	2150
	+ mestranol)				
9	Ovulen				
	(ethynodiol diacetate	2	0.1	1860	2100
	+ mestranol)				
10	Provest				
	(medroxyprogesterone	2	—	142	2200
	acetate				
	+ ethinyl estradiol)	0.01			
11	Provest				
	(medroxyprogesterone	2	0.1	1313	2750
	acetate				
	+ ethinyl estradiol)	0.01			
12	Enovid				
	(norethinodrel	10	—	156	1500
	+ mestranol				
13	Enovid				
	(norethinodrel	10	0.1	1485	2325
	+ mestranol				

Table 11 (continued)

Group No.	Treatment	Dose[a] mg	LH-RH dose µg	Plasma LH mµg/ml[b]	Plasma FSH mµg/ml
14	C-Quens (chlormadinone acetate + mestranol)	10 0.4	—	172	1850
15	C-Quens (chlormadinone acetate + mestranol)	10 0.4	0.1	1843	1850
16	Norlestrin (norethindrone acetate + ethinyl estradiol)	5 0.16	—	135	2575
17	Norlestrin (norethindrone acetate + ethinyl estradiol)	5	0.1	1778	2650

[a] Daily for 5 days.
[b] Mean of two replicate RIA.

Reference preparations for LH and FSH in RIA were NIAMD Rat LH-RP-1 and R817B respectively.

From Schally et al. [22]. Courtesy of Endocrinology and J. B. Lippincott Co.

Table 12. *Comparison of the responses to LH-RH in ovariectomized rats with and without pretreatment with ovulen*

Group	Treatment	Dose LH-RH µg	Plasma LH mµg/ml[b]	Δ	Plasma FSH mµg/ml[c]
1	Untreated	—	620		2100
				460	
2	Untreated	0.1	1080		1800
3	Ovulen 2 mg[a]	—	115		1250
				765	
4	Ovulen 2 mg[a]	0.1	880		1500

[a] Daily for 5 days.
[b] In terms of reference preparation NIAMD Rat LH-RP-1.
[c] In terms of reference preparation NIAMD Rat FSH-RP-1.

From Schally et al. [22]. Courtesy of Endocrinology and J. B. Lippincott Co.

threshold doses of porcine LH-RH [37]. Out of twenty rabbits which ovulated in response to an infusion of 0.2 μg LH-RH, only 6 ovulated when pretreated 10—19 hrs previously with 2 mg progesterone. When the dose of LH-RH was increased to 1.0 μg, half of the animals tested in which progesterone had blocked ovulation in response to 0.2 μg, ovulated in spite of progesterone [37]. In discussing these results it should be mentioned that in 1965

Table 13. *Effect of progesterone*[a] *on the response to treshold doses of luteinizing hormone-releasing hormone (LH-RH) in intact female rats*

Treatment	Group	Substance injected i.c.	Plasma LH[b] level ng/ml as NIH-LH-S-14				DUNCAN's new MRT P
			Before	After LH-RH	Δ	mean Δ	
Oil Carrier	1	Saline	0.8	0.1	—0.7		
			ND	ND	0	—0.1	
			ND	0.3	0.3		
			ND	ND	0		
Oil Carrier	2	LH-RH 10 ng	1.0	2.4	1.4		1 vs 2
			ND	1.0	1.0	1.5	<0.05
			ND	1.2	1.2		
			0.5	2.8	2.3		
Oil Carrier	3	LH-RH 50 ng	ND	3.7	3.7	3.6	1 vs 3
			ND	2.7	2.7		<0.01
			2.2	7.9	5.7		3 vs 2
			1.4	3.8	2.4		<0.01
Progesterone	4	Saline	ND	ND	0		4 vs 1
			ND	ND	0	0.05	NS
			ND	ND	0		
			ND	0.2	0.2		
Progesterone	5	LH-RH 10 ng	ND	1.0	1.0		5 vs 4
			ND	ND	0	0.5	NS
			ND	1.1	1.1		5 vs 2
			ND	ND	0		NS
Progesterone	6	LH-RH 50 ng	ND	1.4	1.4		6 vs 4
			ND	1.7	1.7	1.6	<0.05
			ND	1.6	1.6		6 vs 3
			ND	1.5	1.5		<0.01

[a] 25 mg given 48 hrs previously.
[b] Measured by radioimmunoassay (00-Rat LH-RIA).

From ARIMURA and SCHALLY [28]. Courtesy of Endocrinology and J. B. Lippincott Co.

KANEMATSU and SAWYER showed in rabbits that implants of norethindrone in the hypothalamus prevent copulation-induced ovulation but that similar implants in the pituitary did not do so [68]. However, later in the same laboratory HILLIARD et al. [69] reported that ovulation induced by intrapituitary infusion of crude hypothalamic extracts was blocked by pretreatment of rabbits with norethindrone. Recently SPIES et al. [70], in collaboration with the same group, also concluded that chlormadinone blocks ovulation in the rabbit by an action exerted on the pituitary gland.

A number of possible explanations exist for these results. The most likely explanation is that the LH-RH used by HILLIARD et al. [69] and SPIES et al. [70] was very weak and constituted only a threshold dose which could indeed be blocked by pretreatment with progestins. For example, in both these studies crude hypothalamic extracts and not purified LH-RH were used [69, 70]. It is possible that these extracts may contain, in addition to some LH-RH, factors influencing the release of LH-RH. These substances may have also backtracked around the cannulae into the hypothalamus and thus could possibly have stimulated LH-RH release in the unblocked animal [2]. In the steroid-blocked rat, this could not occur if steroids acted on the hypothalamus. Other considerations such as decrease in responsiveness of the ovary to exogenous (or endogenous) LH after treatment with chlormadinone could have also contributed to the results obtained by SPIES et al. [70].

Furthermore, recent results indicate that steroids may affect both the hypothalamus and the pituitary (see above [28, 37]). The results of SPIES et al. [70] are in agreement with the results of others. Thus DÖCKE et al. [71] found that intrapituitary implants of chlormadinone blocked ovulation in rats. DÖCKE and DÖRNER [72] put forward a complex hypothesis according to which progestogens, in the absence of estrogen, block the pituitary receptor sites and suppress their responsiveness to LH-RH. Speaking against this view is the fact that excellent responses to LH-RH can be obtained in ovariectomized rats pretreated with massive doses of various progestogens in the complex absence of estrogen as shown by SCHALLY et al. [21]. The results of HILLIARD et al. [69]; SPIES et al. [70]; and DÖCKE et al. [71] are also not in agreement with the data obtained by EXLEY et al. [73] and HARRIS and SHERRATT [74]. The observations of EXLEY et al. [73] were extended by HARRIS and SHERRATT [74] who demonstrated that administration of large doses of chlormadinone acetate failed to block the ovulatory response in conscious rabbits after electrical stimulation of the hypothalamus or intrapituitary infusion of hypothalamic extracts containing LH-RH. They also suggested that the suppression of FSH secretion by chlormadione could explain part of the discrepany between their results and those of SPIES et al. [70]. The papers by HARRIS and SHERRATT [74] and SCHALLY et al. [21, 22] provide a more complete discussion of these results.

In conclusion, the effect on the central nervous system-hypothalamus-pituitary axis may well represent the principal action of contraceptive steroids. Our results [2, 21, 22], and those of HARRIS and SHERRATT [74] and others, strongly suggest that although estrogens and progesterone exert some effects on the pituitary, the hypothalamus or another CNS center may be the main central site of action of contraceptive steroids.

F. Studies with Clomiphene in Rats

Clomiphene citrate, a synthetic non-steroidal substance, is comprised of two geometic isomers cisclomiphene and transclomiphene. In view of well established findings that in the human female clomiphene can facilitate LH release since it induces ovulation and regular ovulatory type menses in previously anovulatory, amenohorreic women, we decided to investigate its effects on the hypothalamic-pituitary axis in rats and its possible interaction with LH-RH [75].

Clomiphene citrate and its two isomers were injected into ovariectomized adult rats and the effects on the release of LH and FSH were evaluated using bioassay and radioimmunoassay technics. The 3 compounds given subcutaneously for 3 days in doses of 2500—10000 μg/rat/day lowered plasma LH levels (Tables 14—15). Transclomiphene was more effective than clomiphene in suppressing LH levels at these dose levels. Rats pre-treated with these high doses of clomiphene and its isomers showed increases in plasma LH levels after administration of LH-RH (Table 14) [75]. The high plasma LH found in ovariectomized rats appeared to increase further after administration of lower doses of the compounds (15—135 μg/rat/day for 3 days); cisclomiphene resulted in the greatest and most consistent elevation (Table 15). Results obtained

Table 14. *Effect of LRH on plasma LH levels (OAA assay) of ovariectomized rats pretreated with large doses of clomiphene, its isomers and/or estrogen plus progesterone (E.P.[a])*

Group number	Treatment	Dose[b] (μg/rat)	Dose of LRH (μg/rat)	Plasma LH activity OAA change (μg/100 mg ± S.E.)	P group	P
1	Control	—	—	—14.4 ± 0.9	—	—
2	Control	—	0.25	—20.6 ± 1.6	—	—
3	E.P. (control)	—	—	— 2.7 ± 3.3	3 vs 1	0.001
4	E.P.	—	0.25	—18.7 ± 2.1	4 vs 3	0.01
5	Clomiphene	500	—	—11.8 ± 1.1	5 vs 1	N.S.
6	Clomiphene	500	0.25	—14.0 ± 1.4	6 vs 5	N.S.
7	Clomiphene	2500	—	—11.8 ± 0.7	7 vs 1	N.S.
					7 vs 3	0.01
8	Clomiphene	2500	0.25	—18.6 ± 0.8	8 vs 7	0.01
9	Clomiphene	10000	—	—10.0 ± 1.1	9 vs 1	N.S.
10	Clomiphene	10000	0.25	—17.0 ± 1.1	10 vs 9	0.01
11	Clomiphene	10000	—	— 6.8 ± 2.2	11 vs 1	0.01
					11 vs 3	N.S.
12	Clomiphene	10000	0.25	—17.6 ± 0.8	12 vs 11	0.01

[a] Single dose of 50 μg estradiol benzoate and 25 mg progesterone per rat 72 hours before sacrifice. From SCHALLY et al. [75]. Courtesy of Am. J. Obstet. Gynec. and C. V. Mosby Co.

[b] Clomiphene or isomers given subcutaneously in oil daily for 3 days except for groups No. 11 and 12 given as single dose 72 hours before sacrifice.

by radioimmunoassay showed low doses of clomiphene and its isomers increased serum FSH levels but that high doses of trans-clomiphene decreased serum FSH. Thus high doses of clomiphene and its isomers seemed to inhibit LH and FSH release, but low doses stimulated LH and FSH release and possibly facilitated the effect of LH-RH on the pituitary. The low doses of clomiphene and cisclomiphene also appeared to enhance to a slight extent the response to LH-RH in ovariectomized estrogen-progesterone treated (Table 16) or untreated rats. The response to LH-RH was not blocked by the higher doses (Table 14) thereby providing evidence for a central blocking site which is not the pituitary but a CNS center, possibly the hypothalamus.

Table 15. *Effect of clomiphene and its isomers on plasma LH and FSH levels of ovariectomized rats determined by radioimmunoassay*

Group number	Treatment	Dose[a] (μg/rat)	Serum FSH			Serum LH		
			Assay number	(mμg/ml)[b]	mean	Assay number	(mμg/ml)[c]	mean
1	Control	—	1	2800	2700	1	1031	1056
			2	2600		2	1081	
2	Clomiphene	5	1	4200	4100	1	1125	1122
			2	4000		2	1119	
3	Clomiphene	15	1	4300	4150	1	875	897
			2	4000		2	919	
4	Clomiphene	45	1	3500	3850	1	1375	1407
			2	4200		2	1438	
5	Clomiphene	135	1	4250	3925	1	844	904
			2	3600		2	963	
6	Clomiphene	10000	1	2900	2950	1	288	304
			2	3000		2	319	
7	Cisclomiphene	5	1	3600	3500	1	1375	1238
			2	3400		2	1100	
8	Cis-clomiphene	15	1	3900	3950	1	1563	1482
			2	4000		2	1400	
9	Cis-clomiphene	45	1	5500	4750	1	1188	1276
			2	4000		2	1363	
10	Cis-clomiphene	135	1	4350	3975	1	1125	1053
			2	3600		2	981	
11	Cis-clomiphene	10000	1	4500	4150	1	531	547
			2	3800		2	563	
12	Trans-clomiphene	5	1	3600	3800	1	1219	1241
			2	4000		2	1263	
13	Trans-clomiphene	15	1	2650	2725	1	844	813
			2	2800		2	781	
14	Trans-clomiphene	45	1	3950	3875	1	844	963
			2	3800		2	1081	
15	Trans-clomiphene	135	1	4800	4200	1	813	766
			2	3600		2	719	
16	Trans-clomiphene	10000	1	1750	1775	1	181	197
			2	1800		2	213	

[a] Given subcutaneously in oil daily for 3 days.

[b] Expressed in terms of a rat FSH reference preparation 1.4 × NIH-FSH-S-1 in biologic potency.

[c] Expressed in terms of a rat LH preparation 0.03 × NIH-LH-S-1 in biologic potency.

From SCHALLY et al. [75]. Courtesy of Am. J. Obstet. Gynec. and C. V. Mosby Co.

Table 16. *Effect of LRH on plasma LH levels (OAA assay) of ovariectomized rats pretreated with small doses of cisclomiphene and estrogen plus progesterone (E.P.)*

Group number	Treatment	Dose (μg/rat	Dose of LRH (μg/rat)	Plasma LH activity OAA change (μg/100 mg ± S.E.)	P group	P
1	E.P. (control)	—	—	— 6.3 ± 0.8	—	—
2	E.P. (control)	—	0.4	—17.0 ± 0.7	2 vs 1	0.01
3	E.P. + cisclomiphene	5	—	— 0.8 ± 2.7	3 vs 1	0.05
4	E.P. + cisclomiphene	5	0.4	—17.6 ± 0.9	4 vs 3	0.01
					4 vs 2	N.S.
5	E.P. + cisclomiphene	45	—	— 5.5 ± 1.8	5 vs 1	N.S.
6	E.P. + cisclomiphene	45	0.4	—22.5 ± 0.9	6 vs 5	0.01
					6 vs 2	0.01
7	E.P. + cisclomiphene	135	—	— 6.5 ± 0.6	7 vs 1	N.S.
8	E.P. + cisclomiphene	135	0.4	—21.5 ± 1.0	8 vs 7	0.01
					8 vs 2	0.05

Given subcutaneously in oil daily for days.

From Schally et al. [75]. Courtesy of Am. J. Obstet. Gynec. and C. V. Mosby Co.

G. Studies in Human Subjects

The possible clinical application of LH-RH and the lack of species specificity were suggested by our recent studies [47, 76]. In the first of these studies [47], 1.5 mg of a highly purified preparation of LH-RH obtained from porcine hypothalami was injected into 8 subjects: 2 untreated men, 2 untreated women, 2 men pretreated with ethinyl estradiol and 2 women pretreated with an oral contraceptive preparation. A suitable statistical design was used. The design was that of a 4-factor partially nested factorial experiment in which serum LH and FSH were measured by specific radioimmunoassays and compared with the values obtained after injection of a control solution of lysine vasopressin [47]. Porcine LH-RH showed no detectable LH or FSH activity in the RIA systems. The results indicated that LH-RH induced a significant elevation of serum LH ($p < 0.01$) and of serum FSH ($p < 0.05$). The mean increase in serum LH was about 4-fold with a range of 1.9—7.5. Since this LH-RH preparation contained some FSH-RH

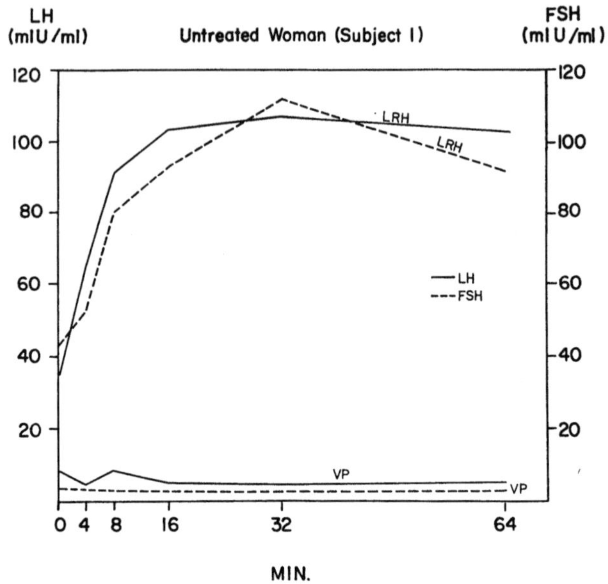

Fig. 10. Effects of administration of 1.5 mg of porcine LH-RH on the levels of LH and FSH in an untreated woman. From [47]. Courtesy of the J. Clin. Endocr. Metab. and J. B. Lippincott and Co.

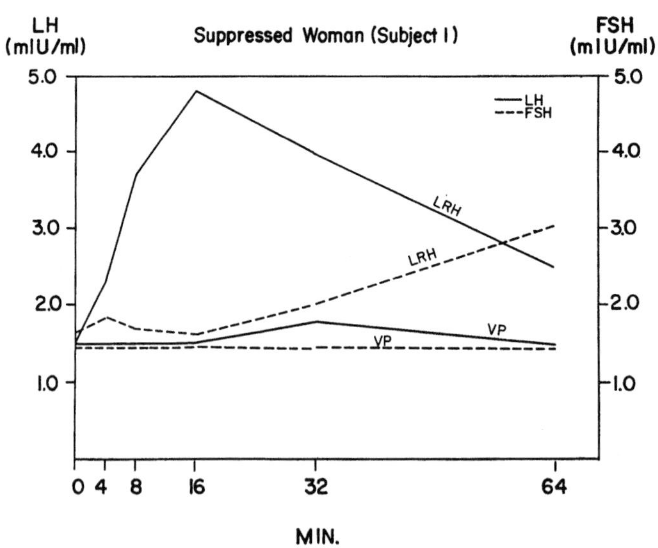

Fig. 11. Effects of administration of 1.5 mg of porcine LH-RH on the levels of LH and FSH in a woman treated with Lyndiol. From [47] Courtesy of J. Clin. Endocr. Metab. and J. B. Lippincott and Co.

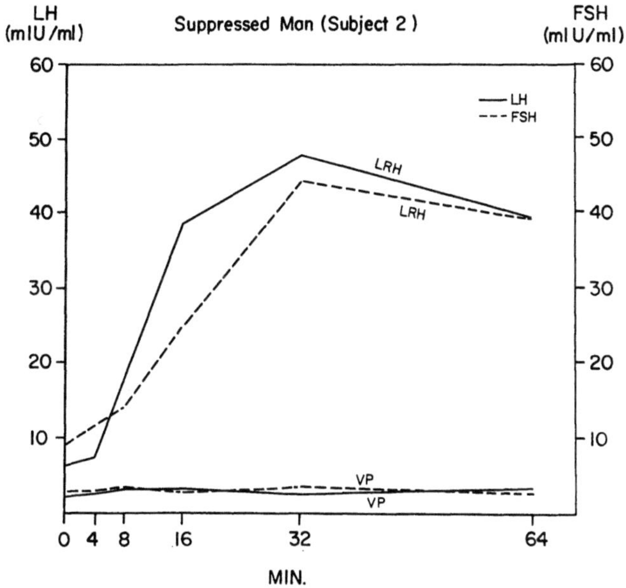

Fig. 12. Effects of administration of 1.5 mg of porcine LH-RH on the levels of LH and FSH in a man pretreated with estrogen. From [47]. Courtesy of J. Clin. Endocr. Metab. and J. B. Lippincott and Co.

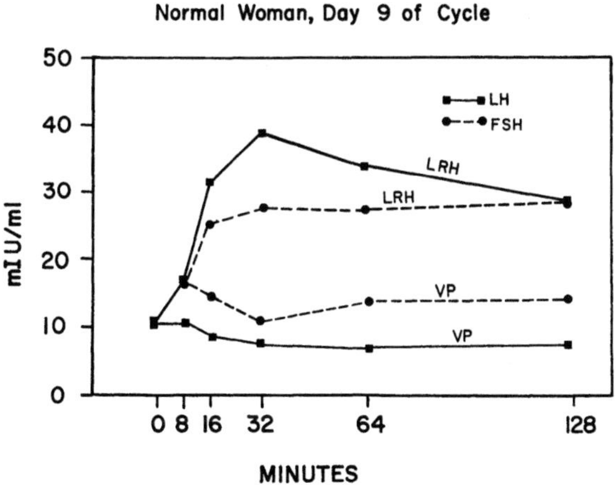

Fig. 13. Effects of administration of 0.7 mg of porcine LH-RH on the levels of LH and FSH in a normal woman, on day 9 of the menstrual cycle. From [76]. Courtesy of Am. J. Obstet. Gynec. and C. V. Mosby Co.

activity, which may be intrinsic to porcine LH-RH or due to contamination with FSH-RH, serum FSH levels were also elevated. There was no significant difference in the magnitude of response to LH-RH between normal women (Fig. 10), women pretreated with contraceptives (Fig. 11), and normal men or normal men pretreated

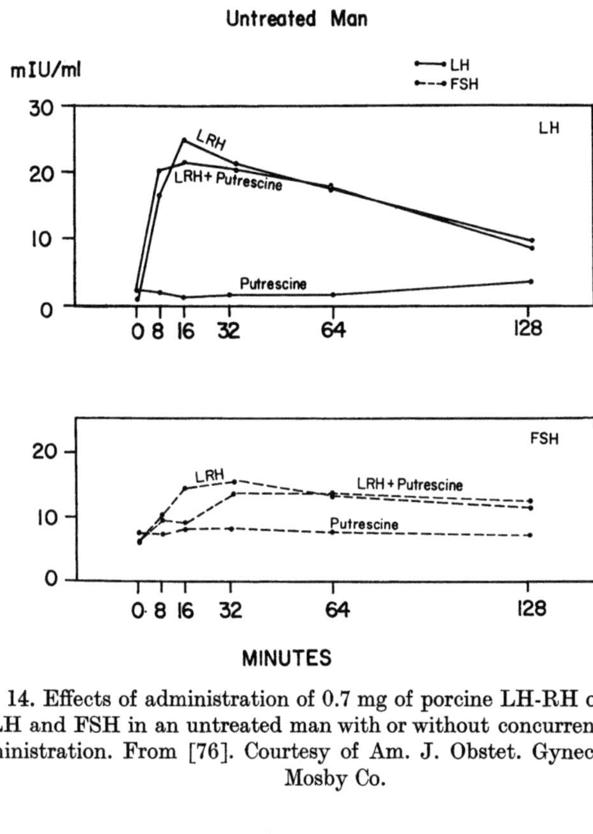

Fig. 14. Effects of administration of 0.7 mg of porcine LH-RH on the levels of LH and FSH in an untreated man with or without concurrent putrescine administration. From [76]. Courtesy of Am. J. Obstet. Gynec. and C. V. Mosby Co.

with estrogen (Fig. 12). The administration of vasopressin did not greatly change plasma LH or FSH levels. These studies were extended in 18 human subjects using 0.7 mg of a preparation of porcine LH-RH of a greater specific activity [76, 77]. The material was injected intravenously into postmenopausal women untreated as well as suppressed with an oral contraceptive preparation, normal women on the 9th day of menstrual cycle (Fig. 13), women with secondary amenorrhea and untreated males (Fig. 14). In addition, two men were injected subcutaneously with LH-RH. The mean increase in serum LH after LH-RH was about 5-fold. Four postmenopausal women were responsive regardless of whether or not their normally elevated serum LH levels were suppressed by

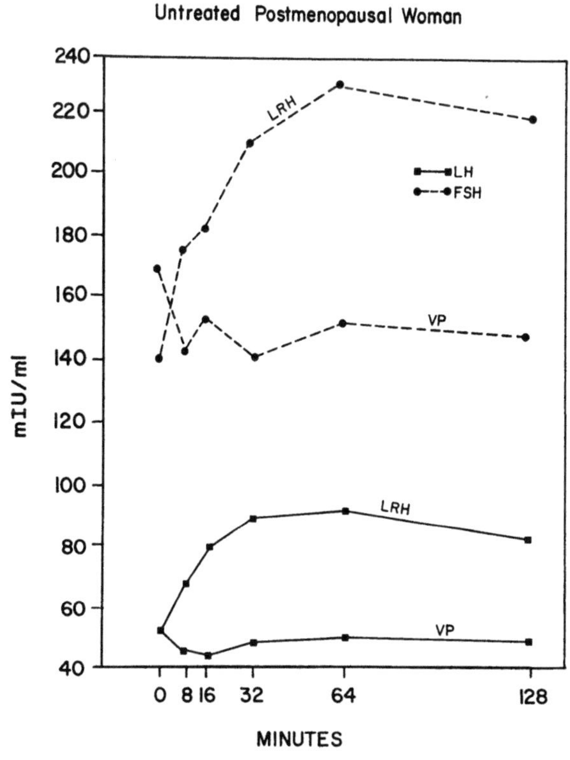

Fig. 15. Effects of administration of 0.7 mg of porcine LH-RH on the levels of LH and FSH in an untreated postmenopausal woman. From [76]. Courtesy of Am. J. Obstet. Gynec. and C. V. Mosby Co.

administration of the oral contraceptive Lyndiol (Fig. 15). Subcutaneous administration of LH-RH to men was about as effective as intravenous administration. In three men in whom plasma LH levels had been elevated by pretreatment for 8 days with clomiphene (200 mg/day), administration of porcine LH-RH caused an additional sharp increase in plasma LH [77]. Putrescine did not elevate serum LH or FSH and did not potentiate the response to LH-RH (Fig. 14).

Both these studies clearly demonstrated the effectiveness of porcine LH-RH in stimulating the release of LH in the human being and laid the foundations for future large scale clinical tests with LH-RH of natural or synthetic origin. Recently Root et al. induced some rise in serum LH levels in 3 abnormal infants with crude hypothalamic extracts using doses some 1000—2000 larger than those utilized in our studies [78]. As expected, LH-RH samples of human origin were also effective in inducing LH release in men [77].

VI. Conclusion

The *in vivo* and *in vitro* studies carried out in at least five species of animals and in man strongly support a role for LH-RH in the regulation of LH secretion. The release and synthesis of LH and FSH appears to be regulated by the interplay of hypothalamic gonadotropin releasing hormone (s) and sex steroids. These studies also demonstrate that species specificity to LH-RH is probably not a problem and suggest a possible clinical usefulness of LH-RH in the control of fertility.

Acknowledgments

Some of the original studies reviewed here were carried out in collaboration with Dr. A. F. Parlow, Department of Obstetrics and Gynecology, UCLA; Dr. Dorsey E. Holtkamp. The William S. Merrill, Co.; Drs. Hilliard and Sawyer, UCLA; Dr. A. Rees Midgley, The University of Michigan; Dr. Carlos Gual, Instituto Nacional de la Nutricion, Mexico; and Dr. J. C. Mittler in our laboratory. They will be reported in detail in original publications.

These studies were supported in part by grants from Research Service VACO, NIH grants AM-07467, AM-09094-05, AM-43308-01 and the Population Council. New York, N. Y.

References

1. Sawyer, C. H.: In: Reproduction and Sexual Behavior (M. Diamond, ed.), p. 13—23. Bloomington: Indiana University Press 1969.
2. Schally, A. V., Kastin, A. J.: In: Advances in Steroid Biochemistry (M. H. Briggs, ed.), Vol. II. New York-London: Academic Press 1970.
3. Harris, G. W.: Neural Control of the Pituitary Gland. London: E. Arnold 1955.

4. SCHALLY, A. V., ARIMURA, A., BOWERS, C. Y., KASTIN, A. J., SAWANO, S., REDDING, T. W.: Recent Progr. Hormone Res. **24**, 497—588 (1968).
5. LISK, R. D., NEWLON, M.: Science **139**, 222—224 (1963).
6. DAVIDSON, J. M.: In: Frontiers in Endocrinology (W. F. GANONG and L. MARTINI, eds.), p. 343—388. Oxford: University Press 1969.
7. — SAWYER, C. H.: Proc. Soc. exp. Biol. (N. Y.) **107**, 4—7 (1961).
8. RAMIREZ, V. D., SAWYER, C. H.: Endocrinology **76**, 1158—1168 (1965).
9. BOGDANOVE, E. M.: Vitam. and Horm. **22**, 206—260, (1964).
10. PALKA, Y. S., RAMIREZ, V. D., SAWYER, C. H.: Endocrinology **78**, 487—499 (1966).
11. BOGDANOVE, E. M.: Endocrinology **73**, 696—712 (1963).
12. MARTINI, L., FRASCHINI, F., MOTTA, M.: Recent Progr. Hormone Res. **24**, 439—496 (1968).
13. CORBIN, A., COHEN, A. I.: Endocrinology **78**, 41 (1966).
14. SCHALLY, A. V., BOWERS, C. Y., WHITE, W. F., COHEN, A. I.: Endocrinology **81**, 77—81 (1967).
15. — REDDING, T. W., BOWERS, C. Y., BARRETT, J. F.: J. biol. Chem. **244**, 4077—4088 (1969).
16. — ARIMURA, A., WHITE, W. F.: in preparation.
17. — BABA, Y., ARIMURA, A.: J. Biol. Chem. in press.
18. — ARIMURA, A., BOWERS, C. Y., WAKABAYASHI, I., KASTIN, A. J., REDDING, T. W., MITTLER, J. C., NAIR, R. M. G., PIZZOLATO, P., SEGAL, A. J.: J. clin. Endocr. Sept. (1970).
19. RAMIREZ, V. D., MCCANN, S. M.: Endocrinology **73**, 193—198 (1963).
20. SCHALLY, A. V., CARTER, W. H., ARIMURA, A., BOWERS, C. Y.: Endocrinology **81**, 1173—1176 (1967).
21. — — SAITO, M., ARIMURA, A., BOWERS, C. Y.: J. clin. Endocr. **28**, 1747—1755 (1968).
22. — PARLOW, A. F., CARTER, W. H., SAITO, M., BOWERS, C. Y., ARIMURA, A.: Endocrinology **86**, 530—541 (1970).
23. ANTUNEZ-RODRIGUES, J., DHARIWAL, A. P. S., MCCANN, S. M.: Proc. Soc. exp. Biol. (N. Y.) **122**, 1001—1004 (1966).
24. PARLOW, A. F.: In: Human Pituitary Gonadotropins (A. ALBERT, ed.), p. 300—310. Springfield: C. C. Thomas 1961.
25. GAY, V. L., REBAR, R. W., MIDGLEY, A. R.: Proc. Soc. exp. (N. Y.) **130**, 1344—1347 (1969).
26. PARLOW, A. F., DAANE, T. A., SCHALLY, A. V.: Abstracts Endocrine Society, 51st Meeting, p. 83, 1969.
27. — — ARIMURA, A., SCHALLY, A. V.: Endocrinology, in press (1970).
28. ARIMURA, A., SCHALLY, A. V.: Endocrinology, Oct. (1970).
29. MONROE, S. E., PARLOW, A. F., MIDGLEY, A. R., JR.: Endocrinology **83**, 1004—1012 (1968).
30. NISWENDER, G. D., MIDGLEY, A. R., JR., MONROE, S. E., REICHERT, L. E., JR.: Proc. Soc. exp. Biol. (N. Y.) **128**, 807—811 (1968).
31. SCHALLY, A. V.: In: J. MEITES (ed.), Proceedings of the NIH Conference on Hypothalamic Hypophysiotropic Hormones, Tucson, Arizona, p. 97. Baltimore: Williams & Wilkins Comp. (1970).

32. Arimura, A., Schally, A. V., Saito, T., Muller, E. E., Bowers, C. Y.: Endocrinology **80**, 515—520 (1967).
33. Schiavi, R., Jutisz, M., Sakiz, E., Guillemin, R.: Proc. Soc. exp. Biol. (N. Y.) **114**, 426 (1963).
34. Schally, A. V., Bowers, C. Y.: Endocrinology **75**, 312—320 (1964).
35. Piacsek, B., Meites, J.: Endocrinology **79**, 432—439 (1966).
36. Jutisz, M., Berault, A., Novella, M., Ribot, G.: Acta Endocr. **55**, 481—496 (1967).
37. Hilliard, J., Schally, A. V., Sawyer, C. H.: Abstract of the International Steroid Congress. Hamburg 1970.
38. Harris, G. W.: In: Control of Ovulation, p. 56, ed. by C. A. Villee. London: Pergamon Press Ltd. 1961.
39. Campbell, H. J., Feurer, G., Harris, G. W.: J. Physiol. **170**, 474—486 (1964).
40. Fawcett, C. P., Reed, M., Charlton, H. M., Harris, G. W.: Biochem. J. **106**, 229—236 (1968).
41. Schally, A. V., Itoh, Z., Carter, W. H., Saito, M., Sawano, S. Arimura, A., Bowers, C. Y.: Gen. Comp. Endocr. **12**, 176—179 (1969).
42. Mittler, J. C., Meites, J.: Endocrinology **78**, 500—504 (1966).
43. Schally, A. V., Saito, M., Arimura, A., Sawano, S., Bowers, C. Y., White, W. F., Cohen, A. I.: Endocrinology **81**, 882—892 (1967).
44. — Mittler, J. C., White, W. F.: Endocrinology **86**, 903—908 (1970).
45. Jutisz, M., de la Llosa, M. P.: Endocrinology **81**, 1193 (1967).
46. Parlow, A. F., Daane, T. A.: Endocrinology (in press).
47. Kastin, A. J., Schally, A. V., Gual, C., Midgley, A. R., Jr., Bowers, C. Y., Diaz-Infante, A., Jr.: J. clin. Endocr. **29**, 1046—1050 (1969).
48. Midgley, A. R., Jr.: Endocrinology **79**, 10 (1966).
49. David, A., Fraschini, F., Martini, L.: Experientia (Basel) **21**, 483—487 (1965).
50. Corbin, A., Story, J. C.: Experientia (Basel) **22**, 694—695 (1966).
51. Saito, T., Arimura, A., Muller, E. E., Bowers, C. Y., Schally, A. V.: Endocrinology **80**, 313—318 (1967).
52. White, W. F., Cohen, A. I., Rippel, R. H., Story, J. C., Schally, A. V.: Endocrinology **82**, 742—752 (1968).
53. Arimura, A., Schally, A. V.: in preparation.
54. Kamberi, I. A., McCann, S. M.: Endocrinology **85**, 815—824 (1969).
55. Schally, A. V., Arimura, A., Muller, E. E., Saito, T., Bowers, C. Y., White, W. F., Cohen, A. I., Corbin, A.: Proc. Symposia III. International Congress of Pharmacology, Pergamon Press **2**, 41—58 (1967).
56. Mittler, J. C., Arimura, A., Schally, A. V.: Proc. Soc. exp. Biol. (N. Y.), **133**, 1321—1325 (1970).
57. Steelman, S. L., Pohley, F.: Endocrinology **53**, 604—620 (1953).
58. Niswender, G. D., Reichert, L. E., Midgley, A. R., Jr., Nalbandov, A. V.: Endocrinology **84**, 1166 (1969).
59. Reeves, J. J., Arimura, A., Schally, A. V.: J. Animal Sci. in press (1970).
60. — — — in preparation.

61. FOSTER, D., NALBANDOV, A. V., WHITE, W. F.: in preparation.
62. HIRATA, A., BROWN, P., WHITE, W. F.: in preparation.
63. STEVENS, V. C., POWELL, J., WHITE, W. F.: in preparation.
64. SEGAL, S. J.: New Engl. J. Med. **279**, 364—370 (1968).
65. PINCUS, G.: Science **153**, 493—500 (1966).
66. DICZFALUSY, E.: Amer. J. Obstet. Gynec. **100**, 136—163 (1968).
67. RUDEL, H. W., KINCL, F. A.: Acta Endocr. **51**, Suppl. 105, 7—45 (1966).
68. KANEMATSU, S., SAWYER, C. H.: Endocrinology **76**, 691—699 (1965).
69. HILLIARD, J., CROXATTO, H. B., HAYWARD, J. N., SAWYER, C. H.: Endocrinology **79**, 411—419 (1966).
70. SPIES, H. G., STEVENS, K. R., HILLIARD, J., SAWYER, C. S.: Endocrinology **84**, 277 (1969).
71. DÖCKE, F., DÖRNER, G., VOIGT, K. H.: J. Endocr. **41**, 353—362 (1968).
72. — — Acta Endocr. Suppl. **119**, 163 (1967).
73. EXLEY, D., GELLERT, R. J., HARRIS, G. W., NADLER, R. D.: J. Physiol. **195**, 697—714 (1968).
74. HARRIS, G. W., SHERRATT, R. M.: J. Physiol. **203**, 59—66 (1969).
75. SCHALLY, A. V., CARTER, W. H., PARLOW, A. F., SAITO, M., ARIMURA, A., BOWERS, C. Y., HOLTKAMP, D. E.: Amer. J. Obstet. Gynec., **107**, 30—40 (1970).
76. KASTIN, A. J., SCHALLY, A. V., GUAL, C., A. R. MIDGLEY, JR., BOWERS, C. Y., GOMEZ-PEREZ, F.: Amer. J. Obstet. Gynec., in press (1970).
77. — — — — — CABEZA, A., FLORES, F.: Abstract Clinical Research, April 1970.
78. ROOT, A. W., SMITH, G. P., DHARIWAL, A. P. S., MCCANN, S. M.: Nature (Lond.) **221**, 570—572 (1969).

On the Identity of the LH- and FSH-Releasing Hormones

W. F. White

Biochemical Research, Abbott Laboratories, North Chicago, Ill./USA

With 5 Figures

In the few minutes allotted to me by the Chairman, I should like to develop the idea briefly mentioned by Dr. Schally regarding the possible identity of the LH releasing hormone and the FSH releasing hormone. We at Abbott Laboratories have been collaborating with Dr. Schally's group in an attempt to isolate and identify these substances from porcine hypothalamic extracts.

About two years ago we turned to the *in vitro* incubation technique originally introduced by Saffron and Schally, as a simple and direct means of testing for the primary or proximate releasing substances. At first we used two different types of specially prepared rat pituitary tissue for the incubations: the pituitary from the ovariectomizid, estrogen-progesterone treated female for LH release, and the pituitary from the testosterone-treated castrate male for FSH release. With these two separate test systems we quickly found that fractions which were powerful releasers of LH also released large amounts of FSH and that weakly active fractions were poor releasers of both hormones. Fig. 1 shows a typical experiment. The samples tested here were successive fractions eluted from CM-cellulose by a gentle salt gradient. It can be seen that the release of both hormones reaches a peak in Fraction 9.

After a time we reasoned that, if we were going to test the theory that the same substance released both gonadotropins, it would be desirable to study both releases in the same incubation using a common pituitary tissue for both. To this end we chose to use the pituitary from the normal adult male rat as the simplest and least constrained tissue. Such pituitaries weigh about 8 mg and contain 18 to 20 μg of LH and about 700 μg of FSH, both as the NIH standard. Fig. 2 shows some results. Here the ratio of LH release to FSH release remains fairly constant over a very large range of purity of sample.

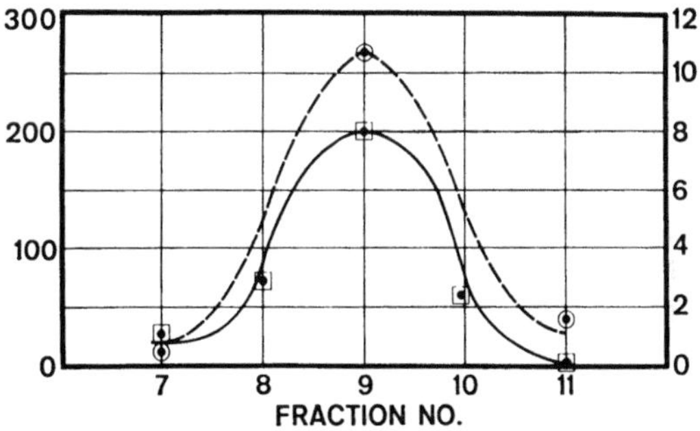

Fig. 1. Test of CM-cellulose fractions for gonadotropin release. A. FSH, using pits, from male castrate-testosterone rats. B. LH, using pits, from female, ovariectomized, E-P rats. μg FSH/PIT ⊙; μg LH/PIT ⊡

Sample	Dose per PIT	Net Release over Control AS μg NIH STD./PIT.	
		LH	FSH
Crude rat SME extract	≎ 0.125 Gland	0.51	14.7
High purity porcine fraction 3154-025-98	0.08 μg	1.55	67.7

Fig. 2. Combined test of LH and FSH release in pituitaries from normal male rats

We next reasoned that the rates of release ought to be identical if the same substance were responsible. Fig. 3 shows that when the ordinates are adjusted to compensate for the contents of the two hormones, the values for the respective releases at various time intervals fall on a smooth curve. Further, Fig. 4 shows that two types of chemical destruction are equally effective in destroying both release activities. Finally, Fig. 5 shows a typical experiment in which a highly purified releasing factor preparation is given to

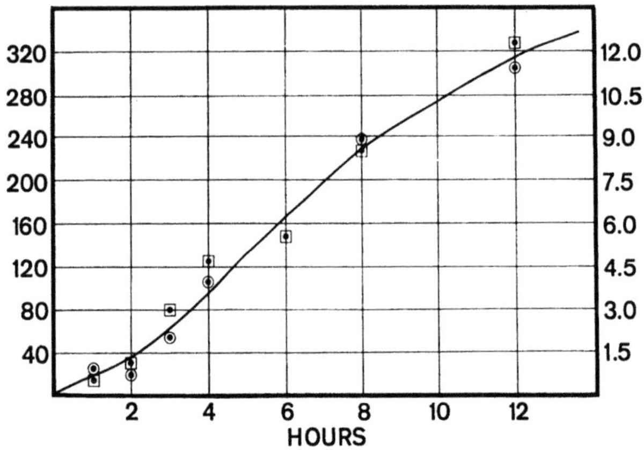

Fig. 3. Gonadotropin release in normal male pituitaries. μg FSH/PIT ⊙; μg LH/PIT ⊡

Fraction	Dose (μg/PIT)	Net Release of Gonadotropins (μg/PIT)	
		FSH	LH
1) Partially purified Porcine	188	249	4.65
2) High potency porcine	0.10	347	4.45
3) Same after 1 hr. IN HCI hydrolysis (100°)	0.20	(—8)	0.08
4) Residue after FDNB Treatment of (2)	0.40	(—7)	0.13

Incubation conditions: 6 hours at 37° in modified No. 199 medium, 16 pituitary halves per beaker.

Test methods: suitable aliquots of medium for 1) Steelman-Pohley (FSH) and 2) Parlow assays (LH).

Fig. 4. *In vitro* test for release of both FSH and LH from normal adult male pituitaries an immature male chimpanzee. It will be noted that peaks of release of the two hormones as measured by serum levels are superimposable.

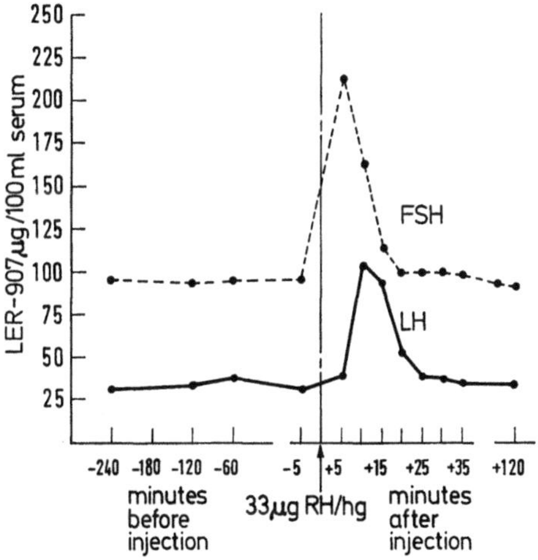

Fig. 5. Response to purified porcine gonadotropin releasing hormone. Subject: Immature 3 year-old (30 kg) Chimpanzee. Investigators: V. C. STEVENS and and J. POWELL, Columbus, Ohio. Assays: Human RIA Systems. Source of RH: W. F. WHITE, Abbott, North Chicago

In summary, our data seems to be consistent with the hypothesis of a common releasing hormone. The final proof of this theory, of course, awaits the synthesis of these substances.

Comparative Aspects of Pituitary Regulatory Pattern of Sexual Functions

H. Karg and D. Schams

Institut für Physiologie, Süddeutsche Versuchs- und Forschungsanstalt für Milchwirtschaft der Technischen Universität München, 805 Freising-Weihenstephan

With 6 Figures

In a scheme (Fig. 1) the assumed connections between the elements responsible for the regulation of the female sexual functions are summarized.

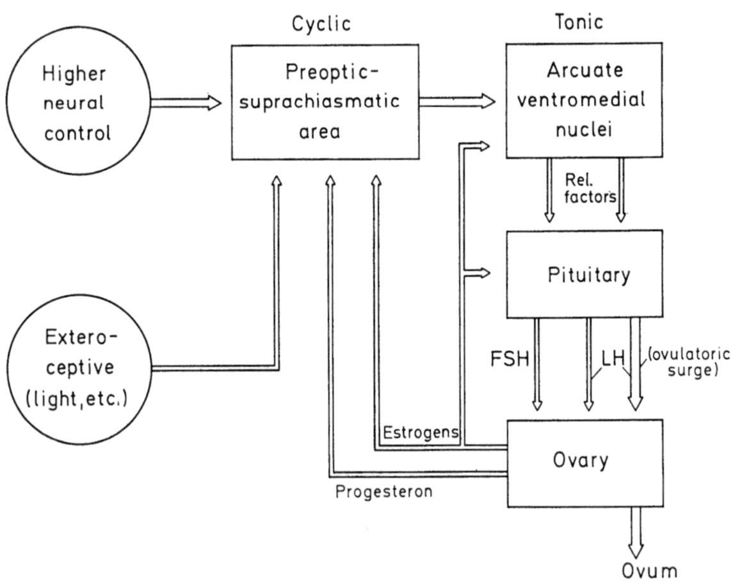

Fig. 1. Regulatoric influences on the ovarian function (mod. from C. A. Barraclough, 1966; see also F. Neumann et al., 1969)

Being invited to present a short communication I have to restrict myself to one topic and want to focus on the LH-action at the female gonad with respect of the significance of LH-appearance in peripheral blood.

Our own data were obtained by radioimmunoassay in the bovine (KARG and SCHAMS, 1968; SCHAMS and KARG, 1969a). In the comparative view this species exhibits certain attractions for this kind of research: It is possible (by palpation) to control the ovaries under physiological conditions (i.e. determination of the development of the corpus luteum and the follicle including the exact timing of ovulation); furthermore the bovine placenta is not known to secrete gonadotropins or gestagens; thus the progesterone production of the corpus luteum (and perhaps its dependancy on pituitary LH) cannot be abolished during pregnancy.

There should be discussed 3 main effects of LH:

1. The induction of ovulation,

2. the luteinization (formation of the corpus luteum),

3. the luteotropic effect (maintenance of the corpus luteum function).

Ad 1: LH as Ovulation Inducing Hormone

For this general accepted property following evidences are for instance available:

a) Injection of LH-like material induces ovulation (HANSEL and TRIMBERGER, 1951); this effect is known since long in several species and common also in therapeutic approaches in women (GEMZELL, 1965).

b) Ovulation can be inhibited by injection of an antiserum for LH (QUADRI, HARBERS, and SPIES, 1966; FÜLLER, HANSEL, and SAATMAN, 1968).

c) Intrafollicular injections of LH in rabbits have shown that the individual treated follicles can be made to ovulate by as little as 10 μg of LH; the ovulation occurs 10 hours after the injection (JONES and NALBANDOV, 1969; NALBANDOV, 1970).

d) There is the distinct time relationship between the endogenous LH-peak and the occurrance of ovulation, as has been repeatably shown (for example see Fig. 2: from SCHAMS and KARG, 1969b).

Characteristic for the release pattern is the short appearance of this very distinct elevation of LH in the peripheral blood, which reminds on a real "outpouring" of this hormone from the pituitary. (A half life time t/2 of about 35 minutes has been determined for exogenous administered LH: SCHAMS and KARG, 1969b).

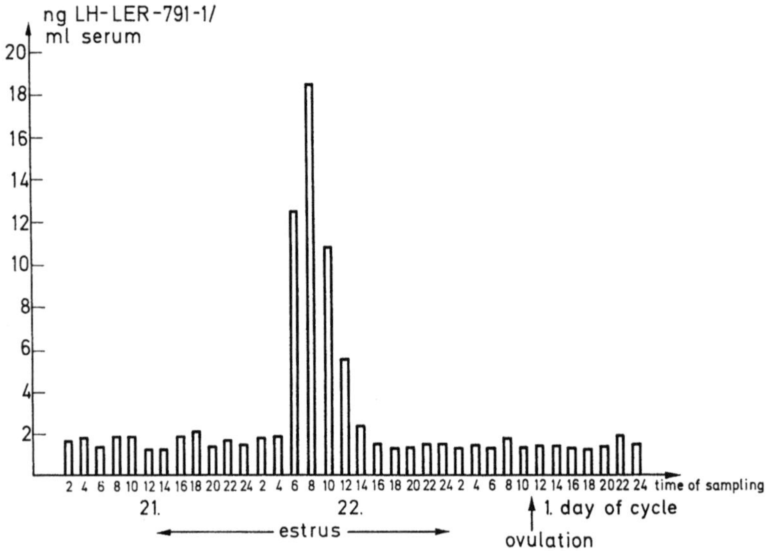

Fig. 2. Blood serum level of LH during the days around estrus in a cow (blood samples taken in 2 hours interval)

Ad 2: LH as Luteinizing Hormone

For its luteinizing action the name of the hormone originally derives from, but it is rather doubtful if LH really immediately exhibits this property: Most provokative recent experiments with rabbits and pigs (EL-FOULY and COOK, 1969; see also NALBANDOV, 1970) indicate, that removal of the ova from mature follicles immediately luteinizes the granulosa layer, which forms structures histologically indistinguishable from corpora lutea. So it is rather possible, that LH acts not directly on luteinization, but gives with the induction of ovulation just the start signal for the autonomous further performances of the deserted follicle.

Ad 3: LH as Luteotropic Hormone

There is overwhelming evidence, that LH is the main luteotropic factor at last in guinea pigs, rabbits, dogs, cats, sheep, pigs, cows and women (NALBANDOV, 1970). But also for rats and mice, which are the classical experimental animals for the definition of LTH (prolactin) as luteotropic hormone, recent experiments have shown, that – at least during pregnancy – LH has to be considered as essential in this property (MOUGDAL, 1969; LOEWIT, BADAWY, and LAURANCE, 1969). Several experiments with in vitro incubation-, glandular perfusion- or transplantation-technique, which are summarized elsewhere (MCCRACKEN et al., 1969; HERLYN, 1970), emphasize the potency of LH to stimulate the steroid secretion especially in the corpus luteum.

But there is still a controversy, what endogenous regulatory pattern of LH may be responsible for this luteotropic action. To approach these problems it is necessary to go strictly the species specific way. According to NALBANDOV (1970) in ewes the life-span of corpora lutea can be almost indefinitely prolonged by a steady infusion of minimal doses of LH. LH was also effective for the maintenance of the corpora lutea in (non pregnant or pregnant) hypophysectomized sheep (KALTENBACH, 1968). In this view the basic level of LH during the cycle is deemed to be important and the requirement of a contineous low dosis LH secretion pattern is emphasized.

Earlier experiments in the pig have shown however, that – once ovulation had occured – the formation and function of the corpus luteum is independent of hypophyseal support (BRINKLEY, NORTON, and NALBANDOV, 1964; NALBANDOV, 1970). In the bovine the concept of rather a demand of a short time strong LH-stimulation for a certain life span of the corpus luteum is supported by the findings of DONALDSON and HANSEL (1965) and WILTBANK et al. (1961); these authors were able to prolong the cycle length by single injections of LH-like material. But according to ANDERSON (1968) the stalk section on day 2 in the bovine cycle is followed by lower progesterone secretion and progesterone concentrations in the corpus luteum during the luteal phase. In this connection our results should be discussed, that in the bovine – besides the typical preovulatoric LH-peak – in some cases one other distinct LH-

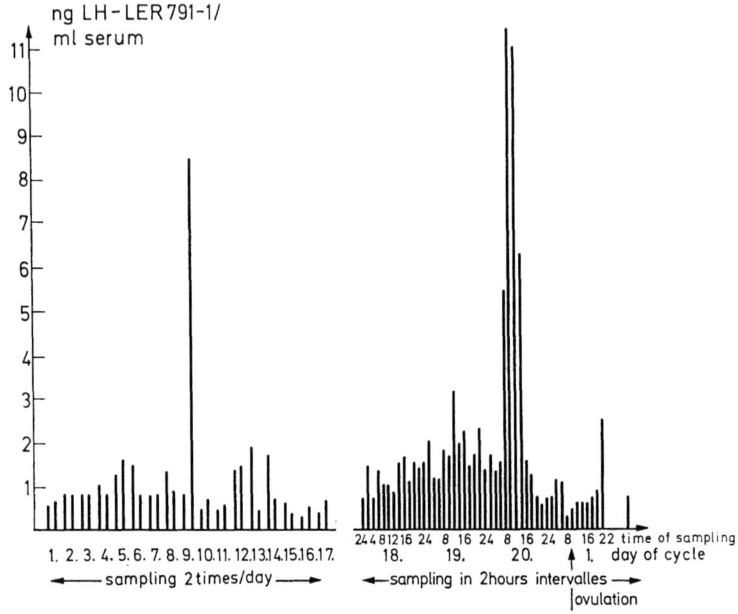

Fig. 3. Blood serum level of LH during the cycle of the cow B

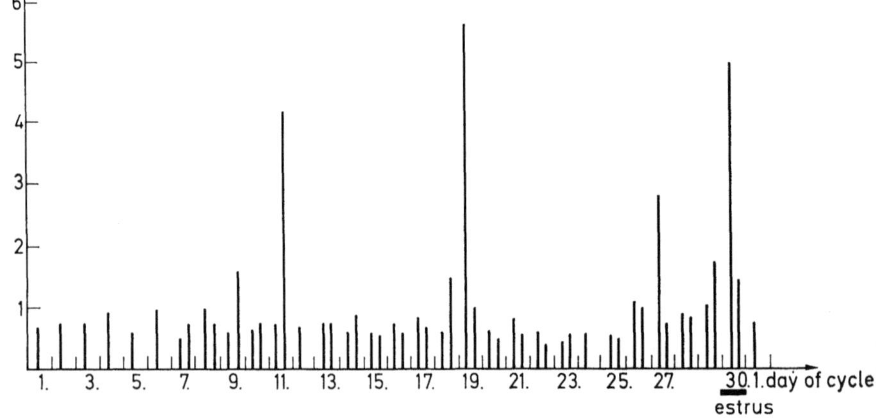

Fig. 4. Blood serum level of LH during the cycle of the cow N

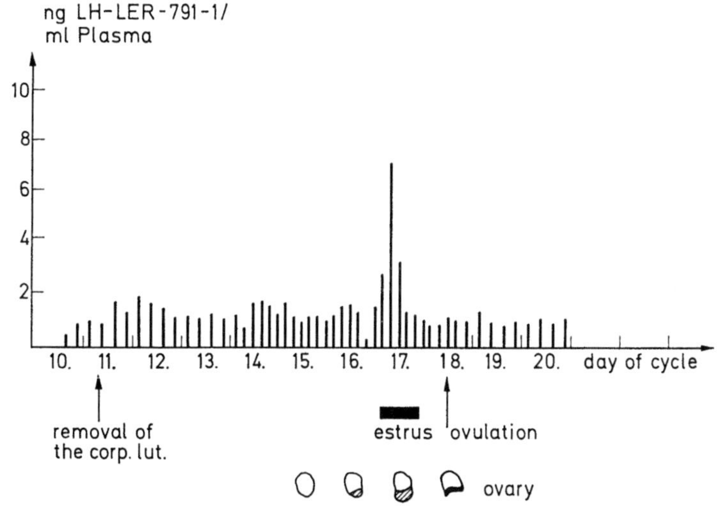

Fig. 5. Blood plasma level of LH in a cow: Shortening of the cycle after removal of the corpus luterum

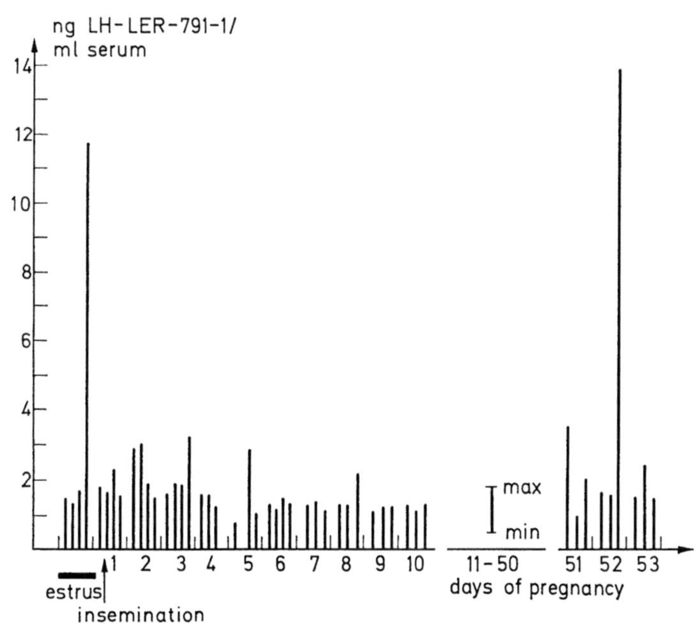

Fig. 6. Blood level of LH in cow A at time of insemination and during the first weeks of pregnancy

elevation was observable (Karg, Aust, and Böhm, 1967; Karg and Schams, 1968; Schams and Karg, 1969a), as shown in Fig. 3.

The appearance of this extra peak coincides with the increasing steroidogenical activity of the corpus luteum (Karg, Hoffmann, and Schams, 1969). At the present situation of these studies we cannot say, if the lack of detection of these extra LH-peaks in some cases is due to the short appearance time in connection with not frequent enough collecting the samples or due to individual differences. But that extra LH-peaks during the cycle (besides the one in connection with ovulation) may be discussed in this species to be luteotropic, was indicated in the results of one of our animals, which exhibits an odd but regularly prolonged cycle (30 instead of 21 days): in this individual case we observed more endogenous LH-elevations during the estrus period (Fig. 4).

This endogenous LH-release pattern resembles the findings mentioned above of artificial prolongation of the cycle after single injections of LH-like material.

Concerning the concept of the negative feedback mechanism it is interesting to observe these examples of periodic LH-release during the corpus luteum phase. Normally the preovulatory LH-peak occurs a few days after regression (or artificial removal, see Fig. 5) of the corpus luteum in the cycle.

In the bovine species, where the picture of LH appearance is probably not concealed by cross reacting compounds of placental source, the occurence of such LH-peaks during pregnancy in the presence of the corpus luteum may be characteristic for this rhythmic pattern of LH outpouring (Schams, 1969). It remains to be further evaluated, whether such peaks during pregnancy, as shown in Fig. 6, exhibit luteotropic properties.

These examples may be indicative for the importance of species specific research in endocrinology, if we want to know it exactly: the comparative synopsis than may be helpful to understand the pattern of biological regulation in its general variability.

Our experiments were supported by the Deutsche Forschungsgemeinschaft.

References

Anderson, L. L.: Hypophysial influences on ovarian function in the cow. Résumés, VIe Cong. Intern. Reprod. Anim. Insem. Artif., Paris, Vol. 1, p. 645 (1968).

BARRACLOUGH, C. A.: Modifications in the CNS regulation of reproduction after exposure of prepubertal rats to steroid hormones. Rec. Prog. Hormone Res. **22**, 503 (1966).

BRINKLEY, H. J., NORTON, H. W., NALBANDOV, A. V.: Experimental alteration of the human ovarian cycle by estrogen. Endocrinology **74**, 453 (1964).

DONALDSON, L. E., HANSEL, W.: Prolongation of life span of the bovine corpus luteum by single injections of bovine luteizining hormone. J. Dairy Sci. **38**, 903 (1965).

EL-FOULY, M., COOK, B.: A controlling role of the oocyte on the function of granulosa cells. 51st Proc. Endocrine Soc. p. 54, New York (1969).

FÜLLER, G. B., HANSEL, W., SAATMAN, R. R.: Regression of sheep corpora lutea with antibovine luteinizing hormone. J. Anim. Sci. Abstr. **27**, 1191 (1968).

GEMZELL, C.: Induction of ovulation with human gonadotropins. Rec. Prog. Hormone Res. **21**, 179 (1965).

HANSEL, W., TRIMBERGER, G. W.: Atropine blockage of ovulation in the cow and its possible significance. J. Anim. Sci. **10**, 719 (1951).

HERLYN, D.: Versuche zur analytischen Erfassung des Luteinisierungshormons im Blut weiblicher Rinder. Diss. med. vet. München 1970.

JONES, E. E., NALBANDOV, A. V.: Induction of ovulation in the rabbit by intrafollicular injection of gonadotrophic hormones. Ann. Meeting Soc. Study Reprod. 2nd, Abstr. 24 (1969).

KALTENBACH, C. C.: Pituitary control of luteal function in the ewe. J. Anim. Sci. Suppl. **27**, 204 (1968).

KARG, H., AUST, D., BÖHM, S.: Versuche zur Bestimmung des Luteinisierungshormons (LH) im Blut von Kühen unter Berücksichtigung des Zyklus. Zuchthygiene **2**, 55 (1967).

— HOFFMANN, B., SCHAMS, D.: Verlauf der Blutspiegel an Progesteron, Luteinisierungshormon und Prolaktin während des Zyklus bei einer Kuh. Zuchthygiene **4**, 149 (1969).

— SCHAMS, D.: Attempts to determine LH in bovine blood. Excerpta Medica Foundation, ICS 161, Part 2, 415 (1968) and discussion, ICS 161, Part 3, 718 (1968).

LOEWIT, K., BADAWY, S., LAURENCE, K.: Alteration of corpus luteum function in the pregnant rat by anti-luteinizing serum. Endocrinology **84**, 244 (1969).

MCCRACKEN, J. A., GLEW, M. E., LEVY, L. K.: Regulation of corpus luteum function by gonadotropins and related compounds. Advanc. Biosciences **4**, 377 (1969).

MOUGDAL, N. R.: Effect of ICSH on early pregnancy in hypophysectomized pregnant rats. Nature (Lond.) **222**, 298 (1969).

NALBANDOV, A. V.: First annual Carl G. Hartman Lecture. Comparative aspects of corpus luteum function. Biol. Reproduction **2**, 7 (1970).

NEUMANN, F., STEINBECK, H., HAHN, J. D.: Hormones and brain differentiation. Workshop conf. on integration of endocrine and non endocrine mechanism in the hypothalamus, Stresa 1969. London-New York: Pergamon press (in press).

QUADRI, S. K., HARBERS, L. H., SPIES, H. G.: Inhibition of spermatogenesis and ovulation in rabbits with antiovine LH rabbit serum. Proc. Soc. exp. Biol. (N. Y.) **123**, 809 (1966).

SCHAMS, D.: Radioimmunobiologische Bestimmung des Luteinisierungshormons (LH) im Blutserum von Kühen in den ersten zwei Monaten der Trächtigkeit. Dtsch. tierärztl. Wschr. **76**, 561 (1969).

— KARG, H.: Radioimmunologische LH-Bestimmung im Blutserum vom Rind unter besonderer Berücksichtigung des Brunstzyklus. Acta Endocr. **61**, 96 (1969a).

— — Zeitlicher Verlauf und analytische Erfaßbarkeit des endogen bzw. exogen erhöhten Blutspiegels an Luteinsisierungshormon beim Rind. Zuchthygiene **4**, 61 (1969b).

WILTBANK, J. N., ROTHLISBERGER, J. A., ZIMMERMAN, D. R.: Effect of human chorionic gonadotropin on maintenance of the corpus luteum and embryonic survival in the cow. J. Anim. Sci. **20**, 827 (1961).

Cybernetics of Mammalian Reproduction

NEENA B. SCHWARTZ

University of Illinois College of Medicine, P.O. Box 6998, Chicago, Ill. 60680/ U.S.A.

With 11 Figures

Introduction

The study of reproduction by basic scientists and clinical scientists encompasses many organs, processes and concepts. One may restrict one's efforts to the mechanism of action of a single gonadotrophic hormone on a morphological or secretory response of the gonad; one may similarly study the mechanism of action of a single gonadal steroid on accessory sex tissue, or on the nervous system, etc.; one may be concerned with gametogenesis or embryogenesis or placentation; on may examine various neurohumoral aspects of the control of the anterior pituitary by the hypothalamus, etc. Each of these problems, and a virtual infinity of others, can legitimately be called "research in reproductive biology."

We have elected to use still another approach in our search for the explanation(s) of the regulation of reproductive cyclicity in the mammalian organism. This approach has been to combine the methodologies of cybernetics, or systems analysis [4], with that of reproductive endocrinology in an attempt to include the totality of possible research methodologies. "The approach encompasses a number of stages, which are carried out simultaneously: [1] identify from extant experimental data the essential variables and connecting linkages among the system components; [2] make a model of the system; [3] simulate the system by computer and/or mathematical equations; [4] perform "experiments" on the model to verify its resemblance to the real system and to predict the results of new experiments on the real system; [5] perform these experiments and modify the model to conform to the behavior of the real system, etc." (quoted from Ref. [8]).

In the present paper we will present an overview of the approach by discussing first a general comparative model of the system, and then a more specific model of the rat estrous cycle. Finally, we will present a beginning attempt at computer simulation and try to illustrate how such simulation efforts lead us back to the biological organism for further research.

A General Model of Mammalian Reproduction Cycles

This model is seen in Fig. 1, and has been discussed in detail in Ref. [6]. The major areas of the model are as follows. On the right the ovary is represented as two microorgans: the follicle and the corpus luteum. In the middle of the diagram the anterior pituitary

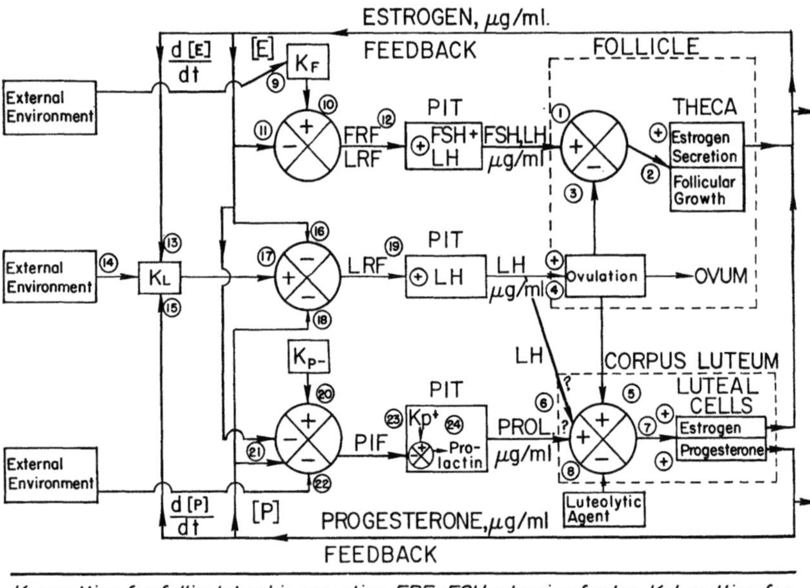

K_F = setting for folliculotrophin secretion rate
K_L = setting for luteinizing hormone secretion rate
P = progesterone E = estrogen
FRF = FSH releasing factor
LRF = LH releasing factor
PIF = prolactin inhibiting factor
Kp^+ = setting for prolactin secretion rate
Kp^- = setting for PIF secretion rate

Fig. 1. A model for the regulation of ovarian function and cyclicity in a variety of species. Symbols for the distribution volumes and utilization rates of hormones have been eliminated from the diagram. This figure is reproduced from Ref. [6], by permission is seen as three functionally separate systems: the "tonically" secreting mechanism for FSH and LH, the "ovulatory surge" system for LH (in the middle) and the system for prolactin secretion at the bottom. To the left of the diagram is seen the system of environmental inputs and steroid feedbacks which ultimately result in releasing or inhibiting factors being delivered to the pituitary from the hypothalamus.

What is postulated in this model is that a combination of FSH and LH ("folliculotrophin") induces follicular growth and estrogen secretion. This estrogen, in addition to exerting the well-known effects on accessory sex tissue and secondary sex characters (including behavior), exerts a negative feedback on folliculotrophin secretion and a positive facilitatory effect on release of the "ovulatory surge" of LH. This surge causes ovulation, changes the follicle into a corpus luteum, and terminates estrogen secretion (while increasing progesterone secretion in many species). Prolactin, in some species, and LH in other species is luteotrophic for the corpus luteum, maintaining morphological and secretory integrity.

The environmental inputs shown are derived from comparative data [10]. Seasonal breeders, such as sheep and ferrets, require optimum ambient light-dark ratios in order to show follicle growth. The rabbit and other induced ovulators require the stimulus of coitus for the release of the ovulatory surge. Even in "spontaneous" ovulators, such as the rat, mouse or hamster, the lighting environment sets the time when the ovulatory surge of LH is released [10]. As for prolactin, the rodents appear to require a mating stimulus to release this hormone for luteal maintenance during pseudopregnancy and pregnancy. Most, if not all, species require a suckling stimulus to maintain prolactin secretion for lactation.

The Events of the Rat Estrous Cycle

The temporal events of the rat estrous cycle are summarized in Figs. 2 and 3 [8, 10, 11]. In an environment of alternating light and dark, rats run predominantly either four-day or five-day cycles terminated by ovulation (which can occur in the absence of the male) at four or five day intervals. Ballooning of the uterus, from distension with intralumenal fluid, and vaginal cornification are classic signs of precedent estrogen secretion. The termination of uterine ballooning and the abrupt appearance of mating behavior are signs of estrogen secretion followed by progesterone secretion (Fig. 3). Blood levels of LH and FSH secreted by the anterior pituitary are low or undetectable except for a few hours starting at 14.00 hours on the afternoon of the day called *proestrus*; the tremendous surge of the LH (and also perhaps the FSH) seen at that

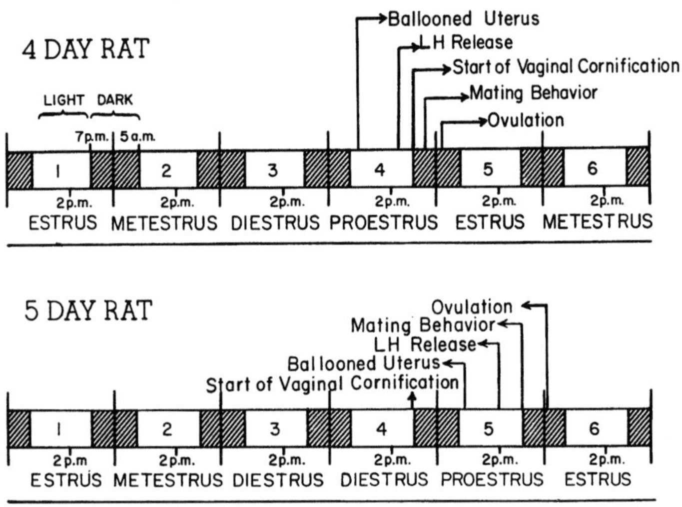

Fig. 2. The timing of the manifest events of the rat estrous cycle. Reproduced from Ref. [9], by permission

time is responsible for progesterone secretion and the ovulation occuring after midnight (Fig. 2 and 3). There is a great deal of evidence that estrogen secretion, which occurs on the day before proestrus as the result of a small amount of LH secretion (see [8, 11]), causes not only the uterine, vaginal and behavioral events, but also the "surge" secretion of LH and FSH.

There is another signal which is necessary for the surge release of the LH and FSH in the rat. This is presumably a neural function, related to the time of day: administration of neural blocking agents (such as barbital) on the day of proestrus before 14.00 hours, but not after 16.00, prevents LH release and ovulation for 24 hours. The "critical period" for LH release lies between about 14.00 to

16.00 hours. Readministration of the blocking agent at 14.00 of the next day (which would have been the day of estrus, Fig. 2) again blocks ovulation [8]. Thus, it appears that both estrogen and a "time of day" biological clock signal are necessary for LH release and ovulation. Reversal of light and dark in the environment reverses the time of LH release and ovulation.

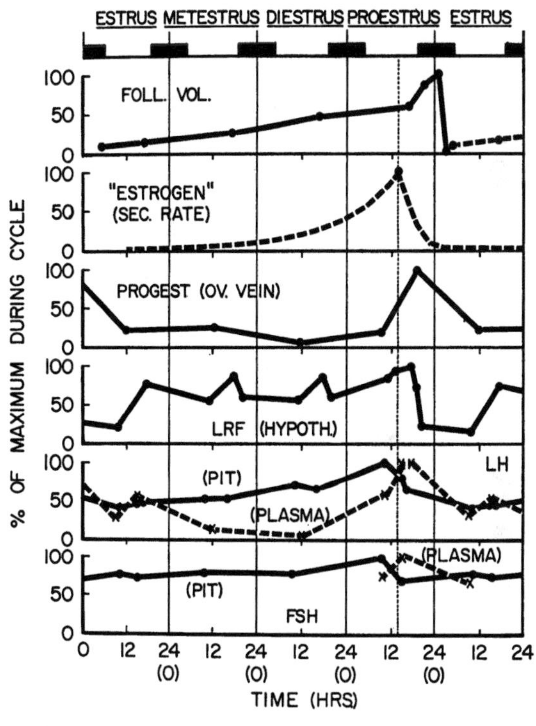

Fig. 3. A number of variables seen on the time scale of a four-day cycle. The vertical dotted line seen on the day of proestrus denotes 14.00 hours. The figure is reproduced by permission from Ref. [9], which lists all of the original references on which the data points are based

That there is a *negative* feedback of estrogen on the hypothalamus and/or pituitary to suppress LH and FSH secretion has also been well documented [8]. Ovariectomy inreases plasma LH and FSH, and chronic injection of estrogen depresses this secretion.

Model of the Regulation of the Cycle in the Female Rat

A simplified model of the rat estrous cycle has been described for endocrinological and engineering audiences [7, 9, 12]. A more detailed model is presented in Fig. 4, and further described in Fig. 5 and Tables 1–4. This model and the accompanying material were developed in detail in Ref. [8].

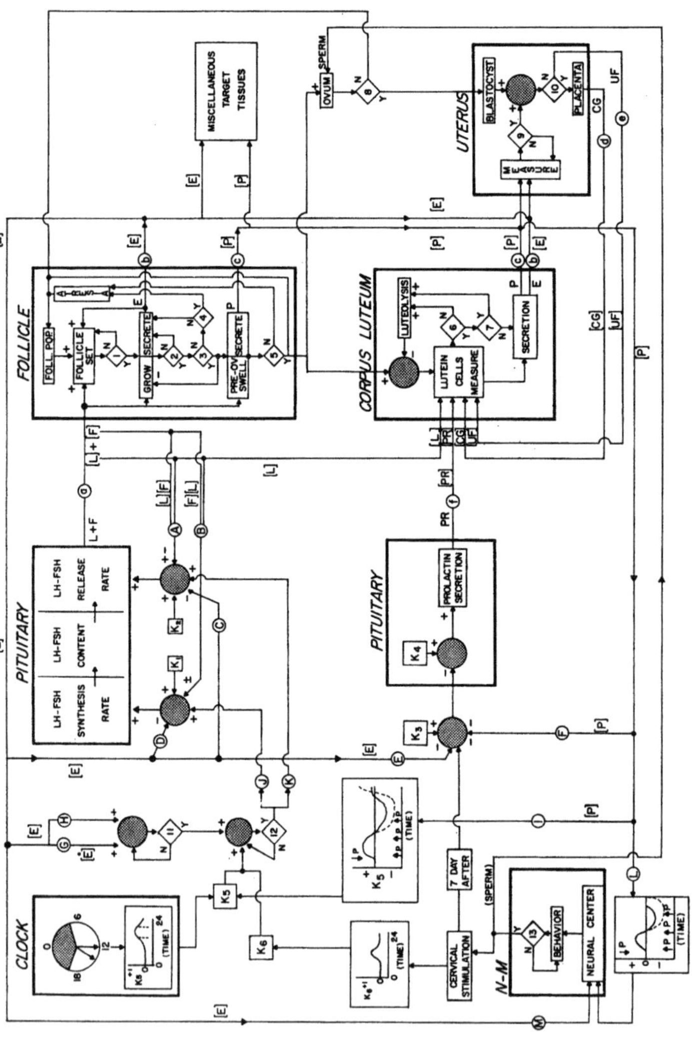

Fig. 4. A model for the control of the rat estrous cycle. See Fig. 5 and Tables 1 to 4 for a detailed explanation of the specific decision functions, settings, distribution and loss symbols, and detectors and multipliers in this figure. Reproduced by permission from Ref. [8]

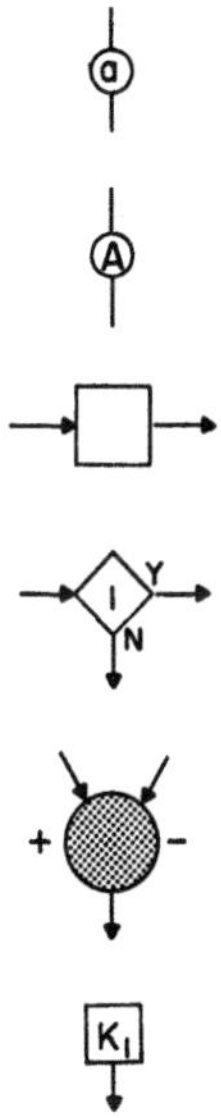

Fig. 5. Explanation of symbols in Fig. 4. Reproduced from Ref. [8], by permission

Table 1. *Explanation of decision functions (Fig. 4)*

1. Are LH and FSH both present?
2. Are follicles "ready" to ovulate?
3. Is (LH) rising rapidly?
4. Have 3 or more days passed since No. 2 above was "yes"?
5. Has ovulation occurred?
6. Are prolactin and/or chorionic gonadotropin present?
7. Is LH or "uterine factor" (UF) present?
8. Has fertilization taken place?
9. Is the uterus "prepared" for implantation?
10. Did the blastocyst implant?
11. Are (E) and (E) in proper range?
12. Are K_5 and/or K_6 at threshold level?
13. Did mating occur?

Reproduced by permission from Ref. [8].

Table 2. *Explanation of settings (Inputs) (Fig. 4)*

K_1	Setting(s) for LH and FSH rate(s) of synthesis
K_2	Setting(s) for LH and FSH rate(s) of release
K_3	Setting for factor inhibiting prolactin release
K_4	Setting (pituitary) for prolactin release
K_5	Clock input (modifiable by progesterone) for LH surge release
K_6	Cervical stimulation in put for LH surge release

Reproduced by permission from Ref. [8].

Table 3. *Explanation of distribution and loss symbols (Fig. 4)*

(a)	Volume of distribution and time constants for LH and FSH
(b)	Volume of distribution and time constant for estrogen
(c)	Volume of distribution and time constant for progesterone
(d)	Volume of distribution and time constant for chorionic gonadotropin
(e)	Volume of distribution (direct channel uterus to ovary ?) and time constant for uterine luteolytic factor
(f)	Volume of distribution and time constant for prolactin

Reproduced by permission from Ref. [8].

Table 4. *Explanation of detectors and multipliers (Fig. 4)*

(A)	"Internal feedback" receptors for LH and FSH on release rates of LH and FSH
(B)	As in (A), but for LH and FSH synthesis rates
(C)	Negative feedback receptor and feedback constant for estrogen levels on LH and FSH release
(D)	As in (C), but for LH and FSH synthesis
(E)	Feedback receptors for estrogen on prolactin release
(F)	Feedback receptors for progesterone on prolactin release
(G)	Receptor and differentiator for estrogen level for surge system
(H)	Receptor for estrogen level for surge system
(I)	Receptor for progesterone for effect on clock signal (K_5)
(J)	Multiplier for surge system signal for LH and FSH synthesis rates
(K)	Multiplier for surge system signal for LH and FSH release rates
(L)	Receptor for progesterone for mating behavior
(M)	Receptor for estrogen for mating behavior

Reproduced by permission from Ref. [8].

The model in Fig. 4 attempts to represent a number of experimental situations, in addition to the normal estrous cycle: pregnancy, pseudopregnancy, ovariectomy and hypophysectomy; injection of barbiturates, anti-estrogens, anti-gonadotrophins, estrogen and progesterone. With respect to our description of the estrous cycle (Figs. 2 and 3) the model postulates the following.

As estrogen levels drop at the end of the cycle (as a result of the action of the LH surge on the follicles), the negative feedback of the steroid on LH and FSH synthesis and secretion rates falls and the blood levels of the gonadotrophic hormones rise. This recruits a new set of follicles, which begin to grow and secrete estrogen. Every day a biological clock function (K_5) assumes a sequential set of values, becoming positive for a few hours around 14.00, thus representing the "critical period" for LH (and FSH) release. On the day when estrogen levels reach a threshold level (or a threshold rate of change) decision function # 11 signals that estrogen levels are correct, and when 14.00 hours is reached and K_5 enters the proper range, decision function # 12 fires (Fig. 4, Table 1). This becomes the day of proestrus, since the subsequent triggering of gonadotrophins causes ovulation, cessation of estrogen secretion and the rapid secretion of progesterone.

Mathematical and Computer Representations of the Estrous Cycle

The underlying problem seen by all of those who have attempted to simulate or mathematically represent the relationships between the ovary and the pituitary has been that of obtaining cyclicity. It has been recognized on theoretical grounds by a number of investigators in this area that it would be difficult to achieve a cyclic solution to values of estrogen and LH with just a simple "push-pull" negative feedback system [1, 3, 5]. More recent analysis of the problem introduces a relaxation oscillation which can lead to cyclicity [1, 14].

Our own approach also has been to treat the system as an irreducible set of two relaxation oscillations [9, 12, 13]. In the model seen in Fig. 4 each decision function (Table 1) represents a potential relaxation oscillation. The two critical ones for yielding cyclicity are # 5, which converts the follicle into a corpus luteum, with different secretory characteristics, and # 12, which causes the "dump" of LH which leads to the aforementioned event. We have labelled this irreducible system as a "push-pull-click-click" system [9].

Although there is a scarcity of real data for parameter values for construction of computer simulations, attempts in this direction have been made with the collaboration of Paul Waltz, of the IBM Data Center. One such simulation using the IBM CSMP [2] has recently been published [13]. Another program and resultant simulation are described in Figs. 6–11. Only a small fraction of the elements and functions seen in Fig. 4 are included in the program.

```
BETA - LOSS RATE OF LH
A2- NEGATIVE FEEDBACK OF ES ON LH PRODUCTION.
A0- SET POINT SECRETION RATE OF LH
ALPHA2- LOSS RATE OF ES
M- POSITIVE EFFECT OF ES ON ITS RATE OF PRODUCTION.
A3- GAIN RATE OF ES IN THE PRESENCE OF LH
LHCP- LH CONCENTRATION IN THE CRITICAL PERIOD
ESCP- ES CONCENTRATION IN THE CRITICAL PERIOD
THES- THRESHOLD VALUE OF ES FOR ALLOWING SURGE TO OCCUR
CT- TIME OF THE CRITICAL PERIOD AS MEASURED ON THE 24-HOUR CLOCK
CLDATA- DATA POINTS FOR THE SURGE AS A FUNCTION OF THE CLOCK
CCON- A CONVERSION CONSTANT FOR EXPERIMENTING WITH CLOCK VALUES
```

Fig. 6. Explanation of parameters for program for rat cycle, utilizing IBM CSMP (Schwartz and Waltz, unpublished observations)

```
INITIAL
   PARAMETER  BETA=1.5,            A0=0.156,  ...
              A2 = 75.,  ...
   M = (966., 12* .025), ...
              ALPHA2=1.38, A3=0.0078, ...
              LHCP=0.09, ESCP=0.0028, THES=0.00138, ...
              CT=16.0
   FUNCTION  CLDATA= (0.,0.), (6.,0.), (10.,0.), (11.,0.), ...
              (12.,0.2), (13.,0.5), (14.,1.0), (14.5, 1.0), ...
              (15.,1.0), (16.,1.0),(17.,1.0), (18.,0.5), ...
              (19.,0.2), (20.0,0.0), (22.0,0.0), (24.0,0.0)
   PARAMETER  CCON=0.1
```

Fig. 7. INITIAL segment of program (see legend for Fig. 6)

Fig. 6 describes the parameters to be used. In the INITIAL segment of the program, seen in Fig. 7, numerical values are presented. Some of these values are estimated from data in the literature [8, 13], and some have been "invented". In the initial segment of the program "zero time" for the simulation is set at 16.00 hours on the afternoon of proestrus (Figs. 2 and 3), and the initial state of the system is defined (LHCP, ESCP). M was permitted in successive simulations to take values between 966.000 and 966.300.

```
*      THE SECOND SECTION CONSISTS OF THE DYNAMIC RELATIONS BETWEEN
*SYSTEM VARIABLES.  AN EXPLANATION OF THESE VARIABLES IS AS FOLLOWS-
*     LHDOT- TIME DERIVATIVE OF THE LH CONCENTRATION
*     LH- LH CONCENTRATION
**     UNITS  MICROGRAMS OF LH PER MILLILITER OF BLOOD PLASMA (UG/ML)
*     SURGE- ADDITION TO LHDOT WHEN SURGE SYSTEM IS ACTIVATED
*     ESDOT- TIME DERIVATIVE OF LH
*     ES- ESTROGEN CONCENTRATION
**     UNITS  MICROGRAMS OF ES PER MILLILITER OF BLOOD PLASMA (UG/ML)
*     SIGNAL- A VARIABLE =0. WHEN SURGE IS ON, 1. OTHERWISE
*     LH1 AND ES1- THE (POSSIBLY NEGATIVE ) INTEGRALS OF LHDOT AND ESDOT
*     CLOCK- THE ACTUAL VALUE OF THE SURGE AS A CLOCK FUNCTION ONLY
*     BLH- LH PRODUCTION RATE ( BETA-LH)
```

Fig. 8. Explanation of variables used in program (see legend for Fig. 6)

```
DYNAMIC
   BLH = -A2*ES + A0 + SURGE
   LHDOT = -BETA*LH + BLH
   ESDOT = -ALPHA2 * ES + A3*LH + M*ES*ES*SIGNAL
   LH1= INTGRL(LHCP,LHDOT)
   ES1= INTGRL( ESCP,ESDOT)
   LH= AMAX1(0.,LH1)
   ES = AMAX1(0.,ES1)
  PROCED  SURGE,SIGNAL = SS(ES,THES,CLOCK)
   IF(SURGE)  10,10,2
10 IF( ES - THES ) 1,2,2
 1 SURGE = 0.
   SIGNAL = 1.
   GO TO 3
 2 SURGE = CLOCK
   IF ( ( SIGNAL .LE. 0.0 )  .OR.  ( SURGE .GE. .1 ) )  SIGNAL = 0.
 3 CONTINUE
  ENDPRO
   CLOCK = CCON *  AFGEN(CLDATA,AMOD(TIME + CT,24.))
   FINISH  ES =  .0050
```

Fig. 9. DYNAMIC segment of program (see legend for Fig. 6)

The DYNAMIC segment of the program, which consists of the dynamic relations between variables (Fig. 8), is seen in Fig. 9. It first describes differential equations for estrogen and LH, which are simulated *continuously*; estrogen and LH levels are limited to zero and above. The release of the LH surge is simulated *logically* by the use of a PROCED section (Fig. 9). This section permits the SURGE to assume values defined by the CLDATA parameter (Fig. 7) (simulating the 24 hour clock), thus to "turn on" the LH surge whenever estrogen levels have reached threshold (THES) and the time of day is "right".

```
*     THE FINAL SECTION CONSISTS OF SIMULATION RUN AND OUTPUT CONTROL
*STATEMENTS.  THESE TELL HOW LONG TO RUN THE SIMULATION, WHEN TO
*PRINT VALUES, ETC.
   TERMINAL
      TIMER  FINTIM=240.,PRDEL=1.0,OUTDEL=1.0
      METHOD   SIMP
*  720 HOURS  = 30 DAYS.  PRDEL, OUTDEL - INTERVALS OF 1 HOURS
      PRINT  ES,LH,SURGE,SIGNAL, BLH
      LABEL   ESTROGEN CONCENTRATION VS. TIME
      PRTPLT  ES
      LABEL  LH CONCENTRATION VS. TIME
      PRTPLT  LH
END
STOP
```

Fig. 10. TERMINAL segment of program (see legend for Fig. 6)

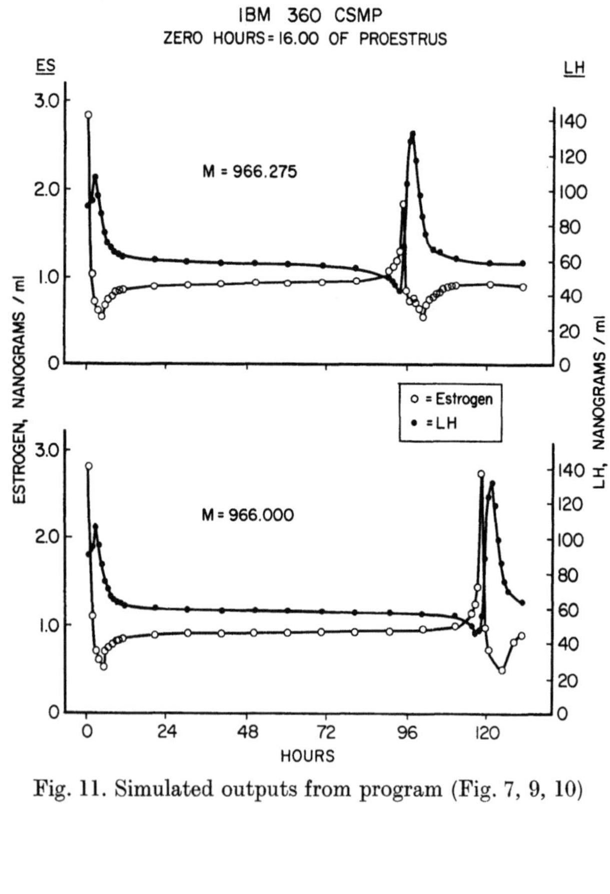

Fig. 11. Simulated outputs from program (Fig. 7, 9, 10)

The TERMINAL segment of the program is seen in Fig. 10. It defines the length of the simulation and the outputs. Two simulations, at two different M values, are seen in Fig. 11; one yields a four-day cycle, the other a five-day cycle. It does a reasonably faithful job of reproducing cyclic levels of estrogen and LH (Fig. 3, [8]) and fits with the concept that five-day cycles result from a slower secretion rate of estrogen [11]. The program also will simulate "persistent estrus" if "ovulation" is not permitted to take place (CCON set to zero). However, when intermediary values of M from those seen in Fig. 11 are used, ES values go out of bounds, which is certainly not the case under normal conditions.

There are many oversimplifications in the program shown, and some of the parameters may be orders of magnitudes off. Furthermore, the hypotheses which underlie the program (and the model in Fig. 4) most certainly are not wholly correct. Has there been a gain for the biologist in attempting the modelling and computer simulations?

Heuristic Value of the Cybernetic Approach

The general value of the approach described in this paper is first the mnemonic function of the models (Figs. 1, 4) and the explicit nature of the hypotheses involved both in modelling and programming for computer simulation [8]. These methods provide a tool for summarizing thought processes which appears unexcelled by other more usual techniques. Fuzzy thinking is revealed rather quickly in the course of preparing model and computer programs.

Furthermore, the approach is tremendously productive of experiments. First, in trying to give numerical values to parameters of the system, the biologist immediately realizes how many such critical parameters have not been measured. This is not only partly the result of lack of methodology, but also partly the result of lack of definition of the problem. Second, once a computer program is available, "experiments" can easily be performed on the computer which provide outputs which can be compared to actual experiments or can be used to pretest real experiments. For example, the behavior of the estrogen levels in the program seen in Fig. 7, 9, and 10, when M is permitted to vary, clearly indicates that a postulated positive feedback of estrogen on its own rate of production in the ovary cannot be of this particular nature or range of values. In subsequent simulations [13] this role of estrogen is postulated to act through follicular size as an intervening variable.

Finally, the cybernetic approach itself can be viewed as "explanatory". This may be more controversial than simply viewing it as "descriptive". MESAROVIC has enunciated this issue in the following statement [4]: "The fundamental question for the community of biologists is whether an explanation on the systems theoretic basis is acceptable as a true scientific explanation in the biologic inquiry."

Acknowledgements

The published and unpublished work described in this paper has been supported in part by PHS Research Grant HD-00440. I would like to express my gratitude to Mr. PAUL WALTZ, of the IBM Data Center, Chicago, who has been a most stimulating colleague in a continuing collaborative effort using the IBM CSMP technique of system simulation.

References

1. DANZIGER, L., ELMERGREEN, G. L.: Mathematical models of endocrine systems. Bull. math. Biophys. **19**, 9—18 (1957).
2. *I.B.M.Application program:* System/360 continuous system modelling program (360-CX-16X) User's manual No. H20-0367-2.
3. LAMPORT, H.: Periodic changes in blood estrogen. Endocrinology **27**, 673—680 (1940).
4. MESAROVIC, M. D.: Systems theory and biology. Berlin-Heidelberg-New York: Springer 1968.
5. RAPOPORT, A.: Periodicities of open linear systems with positive steady states. Bull. math. Biophysics **14**, 171—183 (1952).
6. SCHWARTZ, N. B.: Newer concepts of gonadotropin and steroid feedback control mechanisms. In: GOLD, J. J. (Ed.): Textbook of gynecologic endocrinology, pp. 33—50. New York: Hoeber Medical Div. Harper & Row 1968.
7. — A model for the control of the rat estrous cycle. J. Basic Eng. 321—324 June (1969).
8. — A model for the regulation of ovulation in the rat. Rec. Prog. Hormone Res. **25**, 1—55 (1969).
9. — Modelling and control in gonadal function. In: STEAR, E. B., KADISH, A. H. (Eds.). Hormonal control systems, supp. 1. Mathematical biosciences, pp. 229—255. New York: Elsevier Co. 1969.
10. — Control of rhythmic secretion of gonadotrophins. In: MOTTA, M., FRASCHINI, F., MARTINI, L. (Eds.): NATO workshop conference on integration of endocrine and non-endocrine mechanisms in the hypothalamus. London-New York: Academic Press (in press).

11. SCHWARTZ, N. B., ELY, C. A.: Comparison of effects of hypophysectomy, antiserum to ovine LH, and ovariectomy on estrogen secretion during the rat estrous cycle. Endocrinology **86**, 1420—1435 (1970).
12. — HOFFMANN, J. C.: A model for the control of the mammalian reproductive cycle. Excerpta med. Int. Cong. **132**, 997—1003 (1967).
13. — WALTZ, P.: Role of ovulation in the regulation of the estrous cycle. Fed. Proc. (in press).
14. THOMPSON, H. E., HORGAN, J. D., DELFS, E.: A simplified mathematical model and simulations of the hypophysisovarian endocrine control system. Biophys. J. **9**, 278—291 (1969).

Biogenesis of Androgens

EMIL STEINBERGER

Albert Einstein Medical Center, Research Laboratories
Division of Endocrinology and Reproduction, Philadelphia, Pa 19141

With 4 Figures

The observation that castration produces systemic effects dates back to antiquity [1]. The classical studies of BERTHOLD [2], which furnished physiologic evidence that testes are producing a hormonal substance, provided an impetus for a concerted effort for the investigation of the endocrine functions of the testes. Although testosterone was isolated from testicular tissue by the 1930's [3], it wasn't until SRERE'S [4] and Brady's [5] demonstration that testicular tissue is able to incorporate radiolabeled acetate into testosterone that the capacity of testicular tissue for *de novo* testosterone synthesis has been conclusively demonstrated.

These observations have been subsequently confirmed by a number of investigators [6—10] and extended to demonstrate secretion of the *de novo* synthesized radiolabeled testosterone by the testes [6] and isolation of testosterone from the venous effluent of human testis [11].

In the past two decades, considerable effort has been expended to elucidate the biogenetic pathways leading to formation of testosterone. A large body of information has been generated concerning the enzymatic relationships relevant to the testosterone synthesis. The available data permit us to conclude that the process of androgen biogenesis in the testis can be viewed as a chain of biosynthetic pathways: Acetate → squalene → cholesterol → pregnenolone → testosterone.

Testes of some mammals utilize extragonadally formed cholesterol, but the bulk of evidence strongly suggests that in most

This work has been supported by Grant No. HD 04178 from the United States Public Health Service and by a grant from: The Ford Foundation.

mammalian species cholesterol is synthesized *de novo* in the testes [12, 13]. Although only a small fraction of total testicular cholesterol is utilized for androgen biogenesis [14], it still serves as the major precursor for biologically important steroids [14, 15, 16].

The biosynthetic pathways leading to formation of cholesterol have been the topic of numerous investigations. The key steps (acetate → mevalonic acid → squalene → lanosterol → cholesterol) have been elucidated, but a number of details still await further study. The biogenesis of squalene from acetate proceeds, most likely, via mevalonic acid, farnesyl and nerolidyl (Fig. 1). Evidence

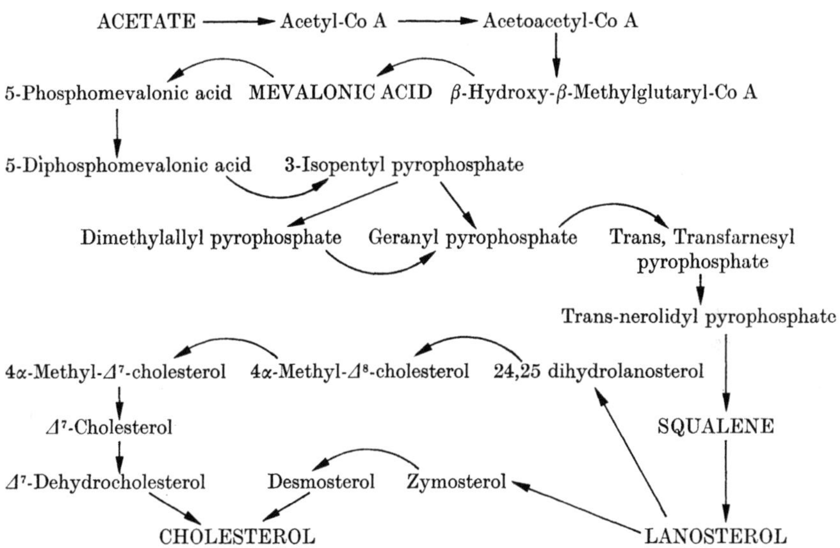

Fig. 1. Biosynthetic pathways leading to formation of cholesterol from acetate

for this pathway has been provided by a number of investigators [17–22]. Both carboxyl and methyl carbons of acetate can be utilized [23]. The pathways by which squalene is metabolized to cholesterol (Fig. 1) are still tentative.

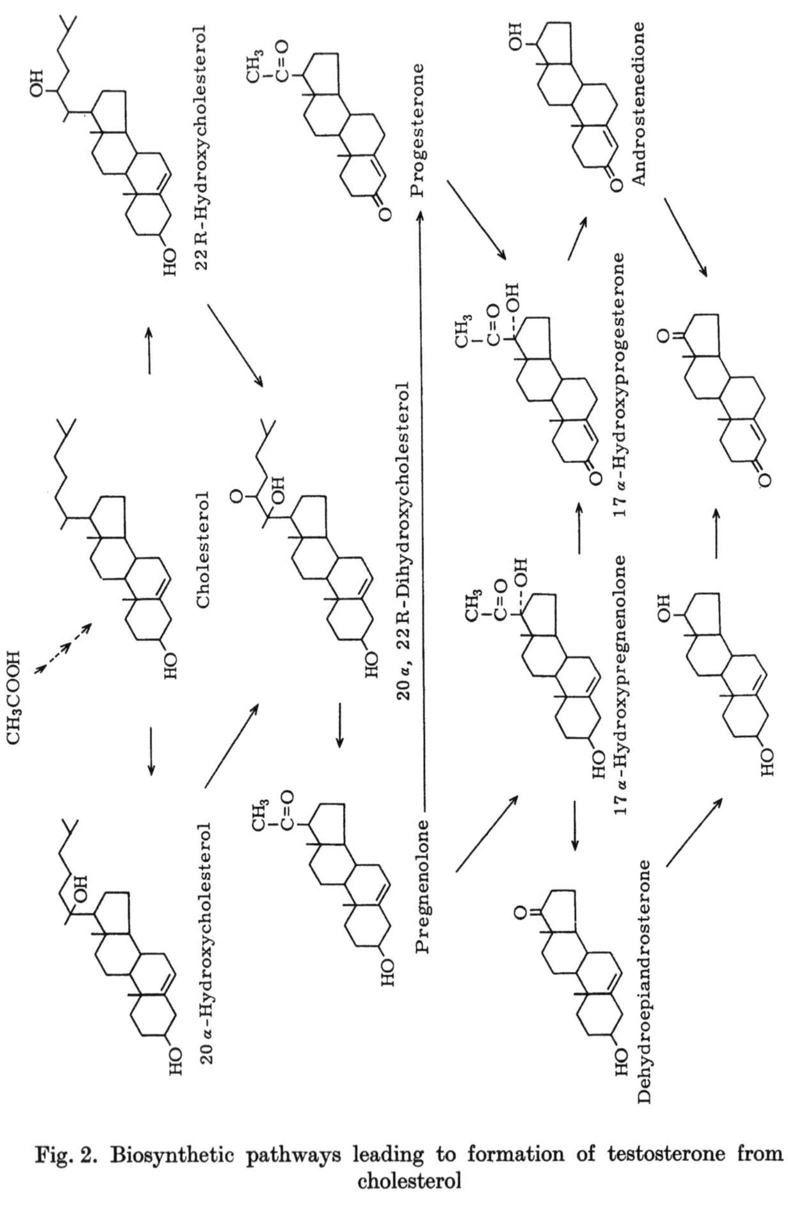

Fig. 2. Biosynthetic pathways leading to formation of testosterone from cholesterol

A number of investigators demonstrated utilization of lanosterol for biogenesis of cholesterol [24, 25]. The details of this pathway are also not entirely clear [18, 26–29]. The transformation of lanosterol to cholesterol can occur via the 24,25-dihydrolanosterol and Δ^7-dehydrocholesterol pathway [18, 26, 28] or via desmosterol [28]. The preferred pathway remains to be determined.

The transformation of cholesterol to pregnenolone involves essentially splitting of a carbon-carbon bond of the cholesterol side-chain [30]. The reactions proceed, most likely, in the manner illustrated in Fig. 2. However, the details of this series of transformations still remain to be elucidated [31].

The conversion of pregnenolone to androgens has been the most extensively investigated segment of the androgen biosynthetic pathway. Since most of the studies dealt with *in vivo* systems, the postulated pathways (Fig. 2) represent only the *capacity* of the tissue to affect the observed transformations. Even experiments utilizing *in vitro* or *in vivo* techniques of perfusion with radiolabeled precursors probably do not reflect the normal physiologic state, since the experimental conditions (anesthesia, anticoagulation, etc.) affect the metabolic capacity of the testes [32, 33]. Consequently, it is proper to define several possible pathways, with the understanding that the physiologically preferred pathway still remains to be determined.

Inspection of the proposed pathways (Fig. 2) reveals a number of possible combinations. However, there are two major pathways – one via the Δ^5, the other via the Δ^4 compounds – through which the biosynthetic process can proceed. It is the first step in the metabolic transformations of pregnenolone which determines which pathway will be followed. If pregnenolone is first hydroxylated to form 17α-hydroxypregnenolone, the Δ^5 pathway is chosen; if, on the other hand, pregnenolone is oxydized at C_3 position and isomerised ($\Delta^5 \rightarrow \Delta^4$), progesterone is formed and transformations proceed via the Δ^4 pathway. It should be emphasized that at any point along the Δ^5 pathway, conversion to a Δ^4 compound can occur.

The enzymes and co-factor requirements effecting these transformations are listed in Table 1. They have been discussed in detail in several recent reviews [34–36].

Table 1. *Enzymes related to androgen formation from cholesterol*

Name of enzyme	Required cofactor	Substrate	Product
20α-Hydroxylase	NADPH, O_2	Cholesterol	20α-Hydroxy-cholesterol
		22R-Hydroxy-cholesterol	20α-22R-Dihydroxy-cholesterol
22R-Hydroxylase	NADPH, O_2	Cholesterol	22R-Hydroxy-cholesterol
		20α-Hydroxy-cholesterol	20α-22R-Dihydroxy-cholesterol
C_{20}-C_{22} Lyase	NADPH, O_2	20α-22R-Dihydroxycholesterol	Pregnenolone
Δ^5-3β-Hydroxysteroid dehydrogenase with isomerase	NAD and NADP	Pregnenolone	Progesterone
17α-Hydroxylase	NADPH, O_2	Progesterone	17α-Hydroxy-progesterone
C_{17}-C_{20} Lyase	NADPH, O_2	17α-Hydroxy-progesterone	Androstenedione
17β-Hydroxysteroid dehydrogenase	NADPH	Androstenedione	Testosterone

Hormonal Control of Androgen Biogenesis

In vivo administration of gonadotropins with ICSH-like activity increases testosterone secretion by the testes [31]. Addition of gonadotropins *in vitro* to incubates of testicular tissue promotes incorporation of acetate-^{14}C into testosterone [37], increases the mass of testosterone in testicular tissue [38], and promotes conversion of cholesterol to testosterone.

The biochemical site of ICSH action received considerable attention in the past decade. The suggestion of HALKERSTON et al. [39] and KORITZ [40] that ICSH, most likely, acts on the biosynthetic steps beyond cholesterol has been confirmed by HALL [41], who provided considerable evidence that ICSH "stimulates steroidogenesis at some step(s) beyond cholesterol and before pregnenolone", most likely by stimulating the 20α-hydroxylation of cholesterol [42].

In long-term experiments, SAMUELS and HELMREICH [43] demonstrated increased concentration of Δ^5-3β-hydroxysteroid dehydrogenase in testes of hypophysectomized rats treated with HCG. Long-term treatment of rodents with HCG also produces increased activity of C_{17}-C_{20} lyase [44—46], 17α-hydroxylase [44, 47, 48], Δ^5-3β-hydroxysteroid dehydrogenase and Δ^4-5α reductase [46].

These findings indicate that most, if not all, enzymes involved in conversion of pregnenolone to testosterone are activated by gonadotropins. However, the possibility exists that the increase of enzyme activities other than cholesterol 20α-hydroxylase may reflect simply a "trophic" activity of the gonadotropin, including changes in the Leydig cell numbers and their total metabolic function.

Studies concerned with human testicular tissue yielded somewhat controversial results. SCHOEN [49] and SLAUNWHITE et al. [9] suggested that 17β-hydroxysteroid dehydrogenase activity is suppressed in absence of gonadotropins, but C_{17}-C_{20} lyase is not affected. TAMAOKI and SHIKITA [50] failed to confirm the decrease in 17β-hydroxysteroid dehydrogenase, but demonstrated decrease in the lyase and 17α-hydroxylase. CARSTENSEN [51], in studies involving incubation of human testicular tissue *in vitro* in presence of ICSH, concluded that the gonadotropins stimulate lyase activity but have no effect on Δ^5-3β-hydroxysteroid dehydrogenase.

This lack of consistency in results obtained by the various investigators could be due to a number of factors. Not only were the utilized clinical and experimental conditions different, but also the testicular tissue used in most of these studies came from aged males undergoing orchiectomy for carcinoma of the prostate. Recently, STEINBERGER et al. [52] demonstrated marked variation in the relative activity of enzymes concerned with steroid biosynthesis in testes from different individuals of this age group and diagnostic category. It is clear that studies utilizing normal testicular tissue from young adults need to be conducted to first establish a normal pattern before physiologic studies are undertaken.

Structure-function Relationships

It has been generally accepted that Leydig cells are the primary source of testicular androgens [53—55]. Direct evidence for this has been provided by Christensen and Mason [56], who demonstrated conversion of progesterone to testosterone by Leydig cells manually separated from the seminiferous tubules, and Steinberger et al. [57, 58] who demonstrated steroid bioconversion in pure cultures of Leydig cells.

Electron microscopic studies of Leydig cells directed attention to the importance of smooth endoplasmic reticulum in the steroidogenic process [59]. Ultrastructural studies and data from experiments utilizing cell fractionation techniques permit a tentative

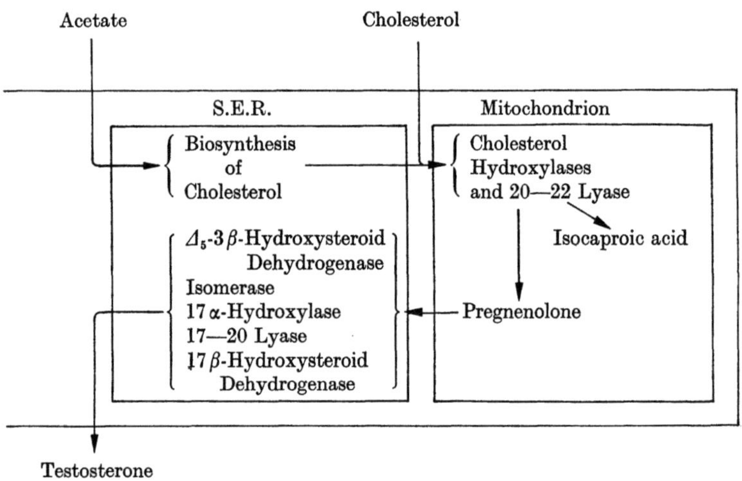

Fig. 3. Relationship of cellular organelles to testosterone biogenesis. Acetate is delivered to the smooth endoplasmic reticulum (SER), within the cell cytoplasm, where it is utilized for synthesis of cholesterol. The *de novo* formed cholesterol, or cholesterol delivered from the vascular system, is transported to the mitochondria, where it is metabolized to pregnenolone. Pregnenolone is transported back to the SER, where it is metabolized to testosterone which is secreted

formulation of the relation of cellular organelles to steroid biogenesis. The current concept is diagrammatically illustrated in Fig. 3.

The synthesis of cholesterol takes place in the smooth endoplasmic reticulum or adjacent cytoplasm [22, 24]. The side-chain cleavage of cholesterol takes place in the mitochondria [60], and the formed pregnenolone returns to the smooth endoplasmic reticulum where it is converted to testosterone [61, 62]. The mechanisms concerned with transport of the various steroid intermediates into the different organelles, as well as mechanisms involved in secretion of the final product by the Leydig cell, have not been investigated in any detail.

Differentiation of the Steroid Biosynthetic Pathways in Developing Testes

In the developing testes, the capacity for testosterone synthesis and secretion changes with the stage of the development. The changes are not manifested by a gradually increasing capacity of the testes to form testosterone. Apparently qualitative alterations occur, reflected by a change in the ratio of androstenedione to testosterone [63—65].

Studies of the steroid biosynthetic pathways and the respective enzymes in testes of developing animals suggest that these changes are more complicated than alteration in the ratios between the two androgens, and of particular interest was the detection of marked 5α-reductase activity in testes of young animals [66—68]. Detailed studies of Steinberger and Ficher [69] and Ficher and Steinberger [70] revealed a striking biphasic character of the changes in the steroid biosynthetic pathways in testes of developing rat. At birth the testes avidly convert progesterone to testosterone. As the development progresses, the rate of progesterone metabolism is not altered, but the major end products are 5α-reduced androgens, so that at twenty days of age about 80 % of progesterone is converted to androstanediol and androsterone, and only traces of testosterone are formed. As the development progresses, the 5α-reductase activity diminishes and formation of testosterone increases (Fig. 4). This *in vitro* demonstrated biphasic character of the capacity of developing testes to form testosterone has been confirmed by measurement of testosterone concentration in testicular tissue and plasma of developing rats [71]. The physiologic implications of this phenomenon are not clear. That the induction of 5α-reductase activity is, most likely, not under gonadotropin control has recently been demonstrated [72]. Whether the 5α-reduced androgens are important in feedback mechanisms dealing with initiation of pituitary gonadotropin activity can only be conjectured at this time.

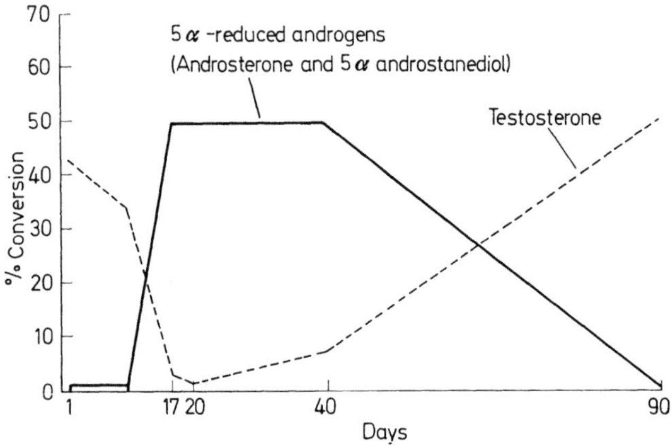

Fig. 4. *In vitro* metabolism of progesterone by incubates of testicular tissue from rats 1 to 90 days of age. Note the mirror image shape of curves, reflecting testosterone and 5α-reduced androgen formation

References

1. Aristotle: a) Historia Animalium (History of Animals), Creswell, R., translat., Book IX, Chap. 37, p. 277. London: Henry G. Bohn 1862. — b) Generation of Animals, Peek, A. L. translat., Book I, sec. II, p. 13. Page, T. E., Capps, E., Rouse, W. H. D., Prost, L. A. and Warmington, E. H., eds. Loeb Classical Library, London: Heinemann 1943.
2. Berthold, A. A.: Arch. Anat. Physiol. (Leipzig) **16**, 42 (1849).
3. Ruzicka, L.: Helv. chim. Acta **19**, E 89 (1936).
4. Srere, P. A., Chaikoff, I. L., Treitman, S. S., Burstein, L. S.: J. biol. Chem. **182**, 629 (1950).
5. Brady, R. O.: J. biol. Chem. **193**, 145 (1951).
6. Savard, K., Dorfman, R. I., Poutasse, E.: J. clin. Endocr. **12**, 935 (1952).
7. — — Baggett, B., Engel, L. L.: J. clin. Endocr. **16**, 1629 (1956).
8. — Besch, P. K., Restivo, S., Goldzieher, J. W.: Fed. Proc. **17**, 1199 (1958).
9. Slaunwhite, Jr., W. R., Sandberg, A. A., Jackson, J. E., Staubnitz, W. J.: J. clin. Endocr. **22**, 292 (1962).
10. Axelrod, L. R.: Biochim. biophys. Acta (Amst.) **97**, 551 (1965).
11. Lucas, W. M., Whitmore, F. W., West, C. D.: J. clin. Endocr. **17**, 465 (1957).

12. MORRIS, M. D., CHAIKOFF, I. L.: J. biol. Chem. **234**, 1095 (1959).
13. GERSON, T., SHORTLAND, F. B., DUNCKLEY, G. G.: Biochem. J. **92**, 385 (1964).
14. HALL, P. F.: Biochemistry **2**, 1232 (1963).
15. KRUM, A. A., MORRIS, M. D., BENNETT, L. L.: Endocrinology **74**, 543 (1964).
16. MAJOR, P. W., ARMSTRONG, D. T., GREEP, R. O.: Endocrinology **81**, 19 (1967).
17. SALOKANGAS, R. A., RILLING, H. C., SAMUELS, L. T.: Biochemistry **3**, 833 (1964).
18. TSAI, S.-C., YING, B. P., GAYLOR, J. L.: Arch. Biochem. **105**, 329 (1964).
19. SALOKANGAS, R. A., RILLING, H. C., SAMUELS, L. T.: Biochemistry **4**, 1606 (1965).
20. YING, B. P., CHANG, Y. J., GAYLOR, J. L.: Biochem. biophys. Acta (Amst.) **100**, 256 (1965).
21. TSAI, S.-C., GAYLOR, J. L.: J. biol. Chem. **241**, 4043 (1966).
22. NIGHTINGALE, M. S., TSAI, S.-C., GAYLOR, J. L.: J. biol. Chem. **242**, 341 (1967).
23. MASON, N. S., SAMUELS, L. T.: quoted in: Biogenesis of Androgens. Proc. 5th Intern. Cong. Biochem. Moscow, Vol. 7, 368, 1963.
24. GAYLOR, J. L., TSAI, S.-C.: Biochim. biophys. Acta (Amst.) **84**, 739 (1964).
25. — CHANG, Y. J., NIGHTINGALE, M. S., RECIO, E., YING, B. P.: Biochemistry **4**, 1144 (1965).
26. NEHER, K., WETTSTEIN, A.: Helv. chim. Acta **43**, 1628 (1960).
27. WEST, C. D., HOLLANDER, V. P., KRITCHEVSKY, T. H., DOBRINER, K.: J. clin. Endocr. **12**, 912 (1952).
28. HALL, P. F.: Endocrinology **74**, 201 (1964).
29. MENON, K. M. J., DORFMAN, R. I., FORCHIELLI, E.: Steroids Suppl. **2**, 165 (1965).
30. STAPLE, E., LYNN, JR., W. S., GURIN, S.: J. biol. Chem. **219**, 845 (1956).
31. HALL, P. F.: In: The Androgens of the Testis, p. 73. Ed. K. B. EIK-NES. New York: Marcel Dekker, Inc. 1970.
32. EIK-NES, K. B.: Endocrinology **71**, 101 (1962).
33. BARDIN, C. W., PETERSON, R. E.: Endocrinology **80**, 38 (1967).
34. FORCHIELLI, E., MENON, K. M. J., DORFMAN, R. I.: In: The Gonads, p. 519. Ed. K. W. MCKERNS. New York: Appleton-Century-Crofts 1969.
35. TAMAOKI, B., INANO, H., NAKANO, H.: In: The Gonads, p. 547. Ed. K. W. MCKERNS. New York: Appleton-Century-Crofts 1969.
36. EIK-NES, K. B.: In: The Androgens of the Testis, p. 1. Ed. K. B. EIK-NES. New York: Marcel Dekker, Inc. 1970.
37. HALL, P. F., EIK-NES, K.B.: Biochim. biophys. Acta (Amst.) **63**, 411 (1962).
38. — NISHIZAWA, E. E., EIK-NES, K. B.: Proc. Soc. exp. Biol. (N. Y.) **114**, 791 (1963).
39. HALKERSTON, I. D. K., EICHHORN, J., HECHTER, O.: J. biol. Chem. **236**, 374 (1961).

40. Koritz, S. B.: Biochim. biophys. Acta (Amst.) **56**, 63 (1962).
41. Hall, P. F.: Endocrinology **78**, 690 (1966).
42. — Young, D. G.: Endocrinology **82**, 559 (1968).
43. Samuels, L. T., Helmreich, M. L.: Endocrinology **58**, 435 (1956).
44. Schoen, E. J., Samuels, L. T.: Acta Endocr. **50**, 365 (1965).
45. Hagerman, D. D.: Biochem. J. **105**, 1119 (1967).
46. Shikita, M., Hall, P. F.: Biochim. biophys. Acta (Amst.) **136**, 484 (1967).
47. Dominguez, O. V., Samuels, L. T., Huseby, R. A.: Ciba Foundation Colloquia on Endocrinology **12**, 231 (1958).
48. Fevold, H. R., Eik-Nes, K. B.: Gen. comp. Endocr. **3**, 335 (1963).
49. Schoen, E. J.: Acta Endocr. **56**, 56 (1967).
50. Tamaoki, B.-I., Shikita, M.: In: Steroid Dynamics, p. 493. (Pincus, G., Nadano, T., Teit, J. F., Eds.) Academic Press. New York: 1966.
51. Carstensen, H. C. H.: Acta Soc. Med. upsalien. **66**, 129 (1961).
52. Steinberger, E., Ficher, M., Smith, K. D.: In: The Human Testis, p. 439. (Rosemberg, E., Paulsen, C. A., Eds.) New York: Plenum Publ. Corp. (1970).
53. Wattenberg, L. W.: J. Histochem. Cytochem. **6**, 225 (1958).
54. Albert, A.: In: Sex and Internal Secretions, 3rd Ed., Vol. 1, p. 305. Baltimore: Williams & Wilkins Co. 1961.
55. Baillie, A. H., Ferguson, M. M., Hart, D. M.: In: Developments in Steroid Histochemistry. New York: Academic Press 1966.
56. Christensen, A. K., Mason, N. R.: Endocrinology **76**, 646 (1965).
57. Steinberger, E., Steinberger, A., Vilar, O., Salamon, I. I., Sud, B. N.: Ciba Colloquium on Endocrinology of the Testis **16**, 56 (1967).
58. — — Ficher, M.: In: Recent Progress in Hormone Research. **26**, 547 New York: Academic Press (1970).
59. Christensen, A. K., Gillim, S. W.: In: The Gonads, p. 415. McKerns, K. W., Ed. New York: Appleton-Century-Crofts 1969.
60. Toren, D., Menon, K. M. J., Forchielli, E., Dorfman, R. I.: Steroids **3**, 38 (1964).
61. Shikita, M., Kakinzaki, H., Tamaoki, B.: Steroids **4**, 521 (1964).
62. — Tamaoki, B.: Endocrinology **76**, 563 (1965).
63. Lindner, H. R., Mann, T.: J. Endocr. **21**, 341 (1960).
64. Hashimoto, I., Suzuki, Y.: Endocrinology (Japan) **13**, 326 (1966).
65. Becker, W. G., Snipes, C. A.: Biochem. J. **107**, 35 (1968).
66. Snipes, C. A., Becker, W. G., Migeon, C. J.: Steroids **6**, 771 (1965).
67. Inano, H., Tamaoki, B.-I.: Endocrinology **79**, 579 (1966).
68. Nayfeh, S. N., Barefoot Jr., S. W., Baggett, B.: Endocrinology **78**, 1041 (1966).
69. Steinberger, E., Ficher, M.: Steroids **11**, 351 (1968).
70. Ficher, M., Steinberger, E.: Steroids **12**, 491 (1968).
71. Resko, J. A., Feder, H. H., Goy, R. W.: J. Endocr. **40**, 485 (1968).
72. Steinberger, E., Ficher, M.: Biol. Reprod. **1**, 119 (1969).

Biosynthesis, Secretion and Biological Action of the Female Sex Hormones

MORTIMER B. LIPSETT

Endocrinology Branch National Cancer Institute, National Institutes of Health Bethesda, MD

The ovary contains several compartments that are characterized by different patterns of biosynthesis. The epithelial cells of the follicle synthesize predominantly estradiol and estrone, luteinized granulosa cells progesterone, stroma and interstitium androstenedione and testosterone. The respective secretion rates of these hormones are, depending on the stage of the cycle: 30—200 μ/24 h for estradiol, 0.6–20 mg/24 h for progesterone, and 0.2—1 mg/24 h for androstenedione. The mentioned steroids are not only secreted but they are also produced from precursors in the liver and other tissues. Hence the major part of the plasma estrone in men and postmenopausal women originates from peripheral conversion of androstenedione secreted from the adrenal cortex. In men estradiol is additionally produced from testosterone secreted from the Lydig cells.

Of paramount biochemical interests is the phenomenon of luteinization. A sequal of this process is a substantial reduction of the biosynthetic capacity of the cell. Only progesterone can still be produced. Moreover, the life time of the luteinized cell is confined to the narrow limits of 11–15 days.

In all probability the sex hormones exert their lasting effects already at the level of transscription. Experiments on the Fallopian tubes of chicken have shown that estrogens induce the production of new m-RNA and increase the activity of m-RNA. Progesterone alters the DNA-dependent RNA synthesis. These results and also investigations showing the presence of steroid hormones in cell nuclei of sensitive tissues support the hypothesis that steroid hormones have an influence on transscription.

The Saga of Mammalian Sperm from Ejaculation to Syngamy

J. Michael Bedford

Department of Anatomy and International Institute for Human Reproduction, College of Physicians and Surgeons, Columbia University 630 West 168th Street, New York, N.Y. 10032

With 19 Figures

Introduction

Although the rather dramatic sense conveyed by the word "saga" may seem out of place in the present context, this term is perhaps not entirely inappropriate. For, it is well known that although many millions of sperm are introduced into the female at ejaculation, most perish en route to the site of fertilization and only very few ever reach the surface of the egg. Thus, after their deposition into the vagina of women, cows, ewes, and female rabbits, for instance, "successful" spermatozoa must journey past the selective barriers of the uterine cervix and utero-tubal junction. They are exposed in the uterus to phagocytes and, before they can fertilize the ovum, they must first traverse the local barriers around the egg formed by the granulosa cell mass and zona pellucida. Moreover, before these last events can be accomplished, the sperm must also complete a final phase of physiological maturation (capacitation) under the aegis of uterine and tubal secretions, which confers upon it finally the ability to undergo the acrosome reaction and to reach and penetrate the zona pellucida.

This present discussion reviews some of the facets of this "saga" which are of current interest and which are under investigation in the author's laboratory.

Passage of Sperm Into and Through the Cervix

Spermatozoa deposited into the vagina at ejaculation are presented with several physical and physiological obstacles which must be overcome before fertilization can be accomplished. In some species, sperm are projected directly into the uterine lumen at ejaculation (myomorph rodents, pig, horse, dog), and their passage as far as the utero-tubal junction is assured. By contrast, the sperm of man and several other species (rabbit, cow and sheep have all been studied extensively) are deposited solely into the vagina, and the cervix is the first "barrier" which faces the fertilizing sperm. This account is devoted mainly to the latter group.

The first question concerns the rate at which functionally significant numbers of sperm leave the vaginal sperm pool to enter the relative security of the cervical canal. The vaginal milieu in the human female is inimical to spermatozoa, particularly so in midcycle when the vaginal pH reaches its nadir around the time of ovulation. This is not necessarily the case in all mammals, and sperm recovered from the isolated rabbit vagina (separated from the uterus by cervical ligation) remained fertile when recovered 15 hours after coitus (BEDFORD, 1967b). For this reason, moreover, the protective role of seminal plasma for sperm in the vagina is likely to be more important in some species than in others; this is not of primary importance in the rabbit, in which vaginally inseminated epididymal sperm are as fertile as those in the ejaculate (DOTT and CHANG, 1966). Undoubtedly, some spermatozoa invade the cervix very soon after their deposition in the anterior vagina. In examination of mucus samples taken from the human cervix, SOBRERO and MACLEOD (1962) confirmed the observation of SIMS in 1868 that penetration of some sperm into the cervical canal is almost immediate, since 39 of 47 women studied had active sperm in the endocervix only 90 seconds to 3 minutes after ejaculation. Maximum numbers of sperm are established in the cervix of the ewe 15 minutes after mating, and some gain the Fallopian tubes as early as 8 minutes after coitus (MATTNER and BRADEN, 1963), though it is not known whether these vanguard sperm would ensure a normal fertilization rate upon arrival of the egg. In the rabbit, we have tried to determine the functional importance of the vaginal sperm pool by selectively destroying all sperm in the anterior vagina with 1 % lauryl sulphate, at various times after coitus. Control experiments showed that the inseminated detergent killed all vaginal sperm immediately; but the fluid was excluded from the cervical canal, as noted previously by WALTON (1930). Such destruction of all vaginal sperm 2 minutes after a single mating subsequently allowed only a 25 % fertilization rate, with essentially no accessory sperm about eggs. This was increased to a normal fertilization rate of 93 %, with a mean of 6.6 perivitelline sperm/egg when killing of the vaginal sperm was delayed until 5 minutes after a single mating. The rate of early passage of sperm into the security of the cervical canal was much enhanced by imposition of a second successive coital stimulus, since destruction of all vaginal sperm only 2 minutes after *double* mating allowed a 90 % fertilization rate and a mean of 5.5 perivitelline sperm/egg. The fact that comparable results were achieved when the second intromission was accomplished by a vasectomized male points to the importance here of the second coital stimulus rather than the greater sperm numbers provided by double mating (BEDFORD, 1970c). Thus, it may be inferred from these results that, in the rabbit at least, sperm numbers sufficient to ensure normal fertilization of all ova often reach the endocervix almost immediately after insemination, and that the coital stimulus can affect the ingress of sperm into the cervix. This finding, allied with the observation of the rapid rate of sperm passage into the human cervix (SOBRERO and MCLEOD, 1962) emphasizes the doubtful value of the post-coital douche as a means of preventing conception in women.

The question "How do sperms get into the uterus ?" was posed several years ago by HARTMANN (1957) and still has not been answered satisfactorily. It is well known that biochemical and physical changes occur in cervical mucus with change in the endocrine status of the female — these have recently been reviewed by SCHUMACHER (1969) and by VICKERY and BENNETT (1968), and will not be considered further. The estrous mucus secretion forms a highly suitable milieu for maintenance of the metabolic activity, and motile sperm have been found in the cervical secretion 70 hours and 120 hours after insemination in the ewe and woman, respectively. The chief components which render estrous mucus suitable as a "culture medium" for sperm have not yet been certainly identified, though hexosamines and carbohydrates which exist in a free or polysaccharide form may prove to contribute to sperm longevity in the cervix. During the ovulatory or estrous phase of the cycle, mucus is secreted by the cervix in greater amounts, becomes less viscous, and acquires anisotropic properties which ensure that sperm entering the mucus move in a uni-directional manner along lines of strain, and thus differently from the random fashion seen, for instance, in seminal plasma (TAMPION and GIBBONS, 1962; ODEBLAD, 1968).

Progesterone domination depresses the efficiency of sperm transport from the vagina to the tubal ampulla. The progestational phase brings both change in composition and in "molecular orientation" of the mucus which is then less easily penetrated by sperm. This seems to be important in the rabbit, since fertilization usually fails to occur and sperm can rarely be recovered from the uterus or oviducts of pseudopregnant animals inseminated vaginally (AUSTIN, 1949), yet 60 % of tubal eggs were fertilized when sperm were inseminated directly into the uterus of pseudo-pregnant rabbits (MURPHREE, WARWICK, CASIDA, and MCSHAN, 1947). According to FERIN (1948) and others, human sperm may penetrate into the cervical mucus to some degree at all stages of the cycle. However, in the progestational phase, sperm frequently fail to reach the body of the uterus. A similar failure of sperm transport in women has been advanced to partly explain the contraceptive effect of compounds such as chlormadinone acetate at 0.5 mg levels, which do not necessarily inhibit ovulation – under this regime few sperm can be recovered from the uterus or Fallopian tubes (MARTINEZ-MANAUTOU et al., 1967; ZANARTU et al., 1968).

Cervical mucus is thus a highly variable secretion whose composition may mirror minor changes in the endocrine state of the female. This is well illustrated in the study by MATTNER and BRADEN (1970) of the effect of the time of insemination on distribution of spermatozoa in the genital tract of ewes. Ewes were inseminated either in early or in late estrus; insemination during late estrus resulted in fewer spermatozoa in the cervix, uterus and Fallopian tubes, and in a lower fertilization rate, *notwithstanding the fact that these animals still showed a state of behavioral estrus at the time of insemination.* The lower fertility after insemination late in estrus was apparently due to a decreased penetrability of the cervical mucus at this time. Thus, subtle variation in the quality of mucus throughout the estrus or ovulatory period, can affect subsequent migration of sperm to the Fallopian tubes. This may have a bearing on sub-optimal sperm transport and fertilization in ewes and cows synchronized with synthetic progestins (ROBINSON, 1967, 1968; ZIMBLEMAN,

1968). Furthermore, this could explain the very poor conception rate in cattle inseminated late in or after behavioral estrus, at which time ovulation still would not be expected to occur for another 10 hours.

The cervix acts as one selection point in prohibiting the passage of large numbers of sperm to the tubal ampulla. It is likely that swimming activity of the sperm and intrinsic movement of the cervix together contribute to the passage of sperm through the cervical canal; their relative importance probably varies between species, and in the different phases of sperm transport following coitus. In the cow and sheep, there is a rapid phase of sperm transport which must depend mainly on contractile activity of the tract (VanDemark and Hays, 1954; Mattner and Braden, 1963); moreover, vaginally inseminated immotile particles can sometimes reach the oviducts (Amersbach, 1930; Egli and Newton, 1961 – man; Mattner and Braden, 1963 – sheep). In some species such rapid transport may be enhanced by the action of oxytocin, the release of which at coitus has been inferred from increased uterine activity (VanDemark and Hays, 1952) and milk ejection (Hammond, 1936; Hays and VanDemark, 1953; Harris and Pickles, 1953). Oxytocin has been detected in the peripheral blood of women after orgasm (Fox and Knaggs, 1969) and following stimulation of the ovine vulva (Roberts and Shaer, 1968).

It is questionable whether oxytocin plays a significant role at this point in the normal rabbit, since there is no evidence for its release following coitus in this species, and active motility seems to be a requisite for passage of rabbit sperm from the vagina to the uterus (Noyes, Adams, and Walton, 1958). Indeed, the weight of present evidence points to a role for the motility of the sperm cell in successful colonization of the cervix in other species also. There appears to exist a degree of selectivity within the human cervix which tends to eliminate a high proportion of the abnormal forms of spermatozoa found so commonly in the human ejaculate. This selectivity may well be based on a difference in the motility of these abnormal sperm, since, in animals, the ratio of motile to immotile sperm has been found to be higher in the cervix than in other parts of the female tract. The influence of motility on the distribution of sperm in the ovine cervix has been clarified recently by Mattner and Braden (1969). If living sperm were inseminated vaginally, these soon distributed themselves throughout the mucus and also particularly between the villi or within the cervical glands of the mucosa, persisting there for long periods. By contrast, dead sperm advanced only a short distance into the central mucus core, and their number then decreased markedly during the first four hours after insemination. Thus, it appears that if sperm are not motile,

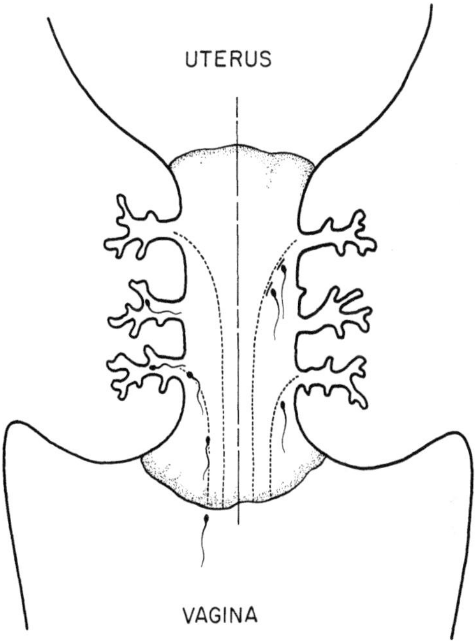

Fig. 1. Generalized diagram of the cervix to illustrate how sperm are led along the lines of strain in the mucus toward the mucosa and into the crypts. It is difficult for sperm to break across the oriented lines of strain

they do not reach the mucosa or the cervical crypts, and are rapidly eliminated from the cervix; this loss could occur by phagocytosis, as well as by loss into vagina. The mechanism of sperm storage in the cervix can be explained if it is realized that the lines of orientation in the mucus lead for the most part to their point of secretory origin, the mucosa and crypts of the cervical epithelium (MATTNER, 1966). Thus the strain lines tend to guide a majority of motile sperm to the mucosa and into the villous crypts (Fig. 1). Presumably only slow release of sperm from the mucosa is able to occur as the sperm eventually cross the many strain lines in the mucus to pass in a cranial direction toward the internal os of the cervix. This appears to be the mechanism by which motile sperm are able to be retained and later released into the uterus over a period of many hours.

It must be realized that our comprehension of the relationship between sperm and the cervix is by no means complete. For instance: Do enzymes play any part in allowing sperm to pass through the cervical secretion? Is there other than a simple frictional interaction between the sperm surface and the oriented molecules of the cervical secretion? FJALLBRANT (1968) found a definite correlation between high levels of sperm antibodies in men's blood, reduced penetration of cervical mucus by spermatozoa, and infertility, and that exposure of sperm to low concentrations of antibody reduced their motility in mucus, while higher concentrations also minimized the extent of sperm penetration (FJALLBRANT, 1969). Further evidence that antibody may also act other than through motility or sperm agglutination is provided by the apparent failure of rabbit sperm to pass the cervix after treatment with non-agglutinating univalent antibody, which does not depress motility *in vitro* (METZ and ANIKA, 1970).

There are indications of species specificity in the sperm/cervix relationship. When mixed insemination is made into the rabbit vagina, rabbit sperm are usually much more successful in traversing the cervix than are dog, cat, or human sperm, though many of these foreign sperm remain motile in the vagina for some hours (BEDFORD, unpublished). Guinea pig and rat sperm are less competent than rabbit sperm to pass through the rabbit cervix (CHANG and BEDFORD, 1961), and goat sperm are transported in lower proportion than sheep sperm to the Fallopian tube of the ewe (HANCOCK and McGOVERN, 1968).

Finally, there remains the question of whether seminal plasma accompanies sperm through the cervix to the uterine lumen. Whole semen definitely passes the cervix at coitus in myomorph rodents, the dog, pig and horse, but this probably does not occur to any degree in species in which insemination is only vaginal. Seminal fluid transport seems unlikely to occur in women, since there are many records of the specific discomfort which follows introduction of seminal plasma into the human uterus. Nonetheless, in view of the good evidence for initial rapid transport of small numbers of sperm to the Fallopian tube in the ewe and cow, the question of whether very small amounts of seminal fluid ever pass the cervix remains to be definitely resolved. Animal experiments have been performed using radiopaque and radioactive markers. Radiopaque materials fail to pass into the uterus of the cow when inseminated into the vagina (Rowson, 1955), but in the rabbit the evidence is somewhat conflicting. Although there are claims for the passage of seminal fluid in the rabbit, it is still difficult to reconcile such observations with evidence that rabbit sperm apparently do not appear in the lower uterus immediately after mating (Braden, 1953), that immotile rabbit sperm fail to reach the uterus (Noyes, Adams, and Walton, 1958), and that there may exist some degree of species specific selection at the cervix (see above).

In conclusion, it seems doubtful that any seminal plasma volume of significance for sperm metabolism normally reaches the uterus after coitus in woman, the sheep, the cow and the rabbit.

Sperm Transport Through the Uterus and Oviduct

This section treats mainly those species in which the sperm/ uterine relationship resembles that in man.

For convenience, sperm passage through the uterus and tube may be considered as occurring in two phases, i.e., a rapid phase, succeeded by another which is slower and prolonged. In the first, small numbers of sperm may be transported rapidly from the vagina to the tubal ampulla in some species (cow – in 2–4 minutes, VanDemark and Moeller, 1951; VanDemark and Hays, 1954; ewe – in 6–8 minutes, Starke, 1949; Mattner and Braden, 1963), and inert particles likewise (ewe – < 15 minutes, Mattner and Braden, 1963). Experimental studies have not yet been performed to any degree in women, but simple observations indicate that motile sperm (Rubenstein et al., 1951) or particles (Egli and Newton, 1961) can pass the length of the genital tract within 30 minutes. The minimum time may be considerably shorter than this.

There is some disparity between the calculated swimming rate of sperm, the distance from vagina to the tubal ampulla, and the observed transportation time between these points. For this reason, and taking into account the transport of immotile particles, the early rapid phase of transport can probably be considered to depend mainly on contractile activity of the myometrium and myosalpinx, and not on the swimming activity of sperm. These contractions may apparently be increased during "courtship" or through oxytocin released from the female neurohypophysis by the coital stimulus. In this context, it is important to be aware of species differences. For, although the estrous uterus of certain animals responds to oxytocin in this way, there are several reports which testify to the unresponsiveness of the non-pregnant human myometrium *in vivo* to a wide dose-range of oxytocin (see: KUMAR, 1967). Hence, while oxytocin can be released following orgasm in women (FOX and KNAGGS, 1969), seminal prostaglandins may have importance in our own species for sperm transport in the immediate post-coital phase (VON EULER and ELIASSON, 1967). Indeed, the sensitivity of the human uterus to prostaglandins does reach a peak between the late proliferative and early secretory phases of the menstrual cycle (BYGDEMAN and ELIASSON, 1963; BYGDEMAN, 1964), and of several species tested, only the semen of man, the sheep and goat contain physiologically significant amounts of prostaglandins (ELIASSON, 1959). However, the functional importance of these interesting active components in human semen remains to be clarified.

In the rabbit, in which the time between insemination and ovulation is fixed at about 10 hours, there is unlikely to have been selection pressure for a rapid phase of transport, and, in fact, sperm passage through the rabbit female tract is relatively slow.

The phase of slow transport, in which sperm pass up through the uterus and tubes over a period of some hours, demands the establishment of an adequate cervical sperm reserve. This in turn depends on the innate motility of sperm (MATTNER and BRADEN, 1969). Release of sperm from the cervical pool occurs sequentially over a prolonged period. Their passage thereafter as far as the utero-tubal junction is probably determined to a varying degree both by the activity of the myometrium and by the motility of the sperm.

Depressive psychosomatic episodes and other forms of stress are held to be responsible for certain forms of reduced fertility, and this has been believed to be due, in part at least, to inhibition of sperm transport in the disturbed female. Undoubtedly the phase of rapid sperm transport is depressed, as judged by the numbers of sperm reaching the ampulla, in ewes showing signs of discomfort or fright (MATTNER, 1963b; THIBAULT and WINTENBERGER-TORRES, 1967); it is likely that this depression is mediated through the release of adrenaline which in turn reduces the contractile response of the myometrium to oxytocin (HAYS and VANDEMARK, 1953), and perhaps through adrenergic receptors in particular in the tubal isthmus. There may also conceivably be a reduced response to prostaglandins, though this latter aspect has not been studied. On the other hand, it is very significant that stress does not markedly affect the entry and build-up of sperm numbers within the cervix after mating with normally fertile males. Since the cervical reserve serves as the source of sperm for the slower prolonged phase of transport, it is not surprising that stress at coitus has little influence on sperm numbers in the tubal ampulla several hours after insemination (MATTNER and BRADEN, 1963). This makes it difficult, in the case of ruminants at least, to understand how lowered fertility in stressed females can be due only to inhibition of transport to the site of fertilization. Nonetheless, there is a definite suggestion that in some individual ewes stress can act to reduce the efficiency of transport when smaller numbers of sperm are available, i.e., after mating to exhausted rams or following insemination of diluted ejaculates (MATTNER and BRADEN, 1963; THIBAULT and WINTENBERGER-TORRES, 1967).

In man, ruminants and the rabbit, the utero-tubal junction and the isthmus of the oviduct constitute a second barrier, which acts with the cervix to regulate the steady passage of relatively small numbers of sperm to the site of fertilization. In the pig, in which the bulk of the semen projected into the uterus at coitus disappears within the first 6 hours (MANN, POLGE, and ROWSON, 1956), this junctional zone probably functions as a point of sperm storage (RIGBY, 1966); and a comment by CHANG (1959) suggests that this may also hold for the rat. The functional characteristics of the utero-tubal junction probably depend on a complex valve-like configuration of the lumen throughhout the intramural region (HAFEZ and BLACK, 1969), but present evidence allows no clear statement of the relative importance of the intrinsic characteristics of the spermatozoon, i.e., motility, surface character, in passage through the junction and along the oviduct. Sperm motility is not essential (see: Hafez and Black, 1969), but homologous sperm of good motility definitely compete more effectively against foreign or dead sperm inseminated into the uterine lumen (Leonard and Perlman, 1949; Braden, unpublished – quoted by Mattner, 1963b; Bedford, unpublished). The specific failure of sperm to pass into the oviducts of mice mated with males carrying the sterility t-allele (Braden and Gluecksohn-Waelsch, 1958; Olds, 1970) also suggests that sperm characteristics have a significant role at this juncture. It seems, therefore, that the quality of the sperm has some part to play in their passage through the utero-tubal junction during the hours following copulation.

Sperm passage along the tube is not a simple function which is easily analyzed, and at present there is little real comprehension of the exact way in which this is accomplished, or of the relative importance of various factors which determine its efficiency. A major contribution by tubal cilia seems unlikely since their direction in the mammalian oviduct is essentially ab-ovarian, though, as proposed by Parker (1931), the arrangement of some of the folds of the tubal epithelium might act to modify these currents. The pattern of muscular contraction in the tube has not been identified clearly in any species, but it is reasonable to think that the activity of the circular muscle, giving rise to segmental movements, is probably a major factor in transportation of sperm through the oviduct (Black and Asdell, 1958). An example of the complexity of this phase is shown in the fact that sperm inseminated into the oviduct of the sheep behave differently in the motile and non-motile state, respectively; the former pass out of the fimbria and also down into the uterus, whereas dead sperm pass out solely through the fimbrial ostium (Dauzier, 1955; Mattner, 1963c). Moreover, sheep sperm can pass from the uterus up into the tubal ampulla ligated at the ovarian end, but accumulated fluid in the ligated tube apparently does not then pass in the opposite direction (Edgar and Asdell, 1960; Mattner, 1963b). Recent data on the fate of tubal secretions do not clarify the situation; for, in the normal estrous female, a major part (80 %) is lost via the fimbrial opening, but the remainder evidently drains into the uterus (Bellve and MacDonald, 1968).

In animals in which the fimbria does not form a permanent bursa around the ovary, sperm do not necessarily linger for many hours at the site of fertilization in the ampulla, and may pass on into the peritoneal cavity. Hence, it is sometimes possible to recover sperm in fluid aspirated from the cul-de-sac in women (HORNE and THIBAULT, 1962; AHLGREN, 1969).

Unfortunately, little is known of the effects on sperm transport of stimulation or inhibition of the α and β adrenergic receptors which have been identified in significant numbers in the isthmus of the rabbit and human Fallopian tube. However, tubal sperm transport probably can be influenced in a sensitive way by change in steroid status of the female, since sperm are transported more effectively along the tube in some rodents (YANAGIMACHI and CHANG, 1963a; BRADEN and AUSTIN, 1954) and perhaps even in rabbits (TURNBULL, 1966), in the hours following ovulation. There is, nonetheless, a dearth of information on the relative efficiency of sperm transport in different parts of the female tract under different endocrine conditions; furthermore, one must separate the effects on transport in the cervix and in the uterine and tubal regions, respectively. For instance, the failure of sperm to gain the Fallopian tubes in ewes feeding on oestrogenic pastures seems likely to lie largely at the cervical level (LIGHTFOOT, CROKER, and NEIL, 1967).

The results of DAUZIER and WINTENBERGER (1952b) reveal a relative depression of transport in an-estrous or di-estrous ewes, but, on occasion, sperm apparently can pass through the tract under widely differing estrogen and progesterone levels. Moreover, some sperm have been recovered on occasion from the uterus and tubes of women at all stages of the cycle, though sperm transport in women is clearly brought about most efficiently around the time of ovulation. In the rabbit, a deficiency of estrogen minimizes the efficiency of sperm transport, but this effect is by no means absolute (NOYES, ADAMS, and WALTON, 1959). Much of the adverse effect of progesterone on sperm transport is exerted at the cervix, and, in the progesterone-dominated rabbit, a proportion of sperm inseminated into the uterus will ascend to the oviduct and there fertilize eggs; however, the fertilization rate in these circumstances is only 40 to 60% (BOYARSKY et al., 1947; MURPHREE et al., 1947; CHANG, 1967), compared with 80–90% in control estrous animals inseminated similarly. Since capacitation is unlikely to be affected in the tube (see below), this suggests that sperm transport through the utero-tubal junction and perhaps tube is also less efficient in progesterone-dominated conditions.

There is no indication in man and other mammals studied that the IUD seriously interferes with sperm transport to the tubal ampulla (see: Corfman and Segal, 1968). The one significant exception is the sheep in which sperm transport and fertilization are usually inhibited in the presence of an IUD (Hawk, 1965). These effects in the sheep are often associated with signs of an inflammatory reaction in response to the device.

To conclude, in the animals in which ejaculation occurs into the vagina only, sperm transport depends first upon efficient penetration of sperm into the cervix. Thence, with the exception of the rabbit, in the immediate post-coital period some sperm may be transported passively to the tubal site of fertilization in a very few minutes, though the minimal time has not been determined accurately in man. A subsequent, slower phase, which involves both the activity of the tract and sperm motility, continues thereafter for a prolonged period. This latter phase is seen as a gradual release and forward passage of sperm from the cervical reservoir past the "selective barrier" of the utero-tubal junction to the tubal ampulla, and in many cases into the peritoneal cavity.

In most feral animals, and in efficient farming operations, the period of receptivity is normally detected by the male early in estrus; and insemination is most likely to occur some hours before ovulation. Hence, in most cases, the phase of slower sperm transport would seem of greatest importance for fertility, except in those instances when coitus happens to take place at or soon after the time of ovulation. It is not yet really clear how sperm negotiate the utero-tubal junction and tubal isthmus, nor is it certain as to how steroidal and other hormonal factors interact to modify the transport of sperm in the uterus and oviduct.

Functional Life and Fate of Sperm in the Female

The question of the longevity of sperm in the female has an obvious bearing on the organization of breeding programs in animals and on the "rhythm" method of contraception in man. In most mammals, the period of receptivity of the female is closely related to ovulation. One might expect to find, therefore, that the fertile life of sperm would at least span the interval between the normal time of insemination and termination of the fertile life of the egg. In the female tract, dog sperm sometimes survive for up to 11 days (DOAK, HALL, and DALE, 1967), and horse (DAY, 1942; BURKHARDT, 1949) and human sperm (PERLOFF and STEINBERGER, 1964; Southam, personal communication) for 6 days. The prolonged life of sperm might be considered as a correlate of the relatively long period of receptive estrus before ovulation in the mare and bitch. Indeed, CHANG (1965) has pointed out that "there would be no horse, as a species, if the fertilizing life of horse sperm in the female tract was shorter than six days, since estrus in mares can be longer than seven days." In the ferret and rabbit, both induced ovulators, the fertile life of the sperm is retained for approximately twice the period which elapses between insemination and deterioration of the unfertilized egg (100 hours or more in the ferret – CHANG, 1965; about 30 hours in the rabbit – HAMMOND and ASDELL, 1926; NOYES and THIBAULT, 1962). The fertile life of rodent sperm is shorter in an absolute and in a relative sense, but nevertheless persists for a period which largely spans the period of receptivity plus the fertile life of the ovum (rat – 14 hours, MERTON, 1939; YOCHEM, 1929; SODERWALL and BLANDAU, 1941; mouse – 12 hours, MCGAUGHEY, MARSTON, and CHANG, 1968). In such rodents functional numbers of sperm often do not begin to pass into the ampulla of the oviduct until about the time of, or slightly after, ovulation (LEONARD, 1950; BRADEN and AUSTIN, 1954; YANAGIMACHI and CHANG, 1963a). The significance of this timing is not clear, but it may relate to capacitation, since fluid from the ovarian follicle is able to capacitate rodent spermatozoa (BARROS and AUSTIN, 1967; YANAGIMACHI, 1969a; IWAMATSU and CHANG, 1969). Finally, as an extreme case, one may note the situation in the bat, in which sperm survive for five to seven months in the female (WIMSATT, 1944). The assumption that the sperm are quiescent for the very great part of this period in the bat uterus raises an interesting question as to the mechanism of their reactivation.

In studies involving fertility, measurement of the fertilizing life of sperm is not always meaningful. For, while ageing of the sperm in the female rabbit has little effect on the post-implantation survival rate (TESH, 1969), eggs fertilized by sperm aged in the male tract later show a relatively high rate of embryonic death (YOUNG, 1929; TESH and GLOVER, 1969), as do rabbit eggs fertilized by sperm taken from the corpus epididymidis (ORGEBIN-CRIST, 1969). One should perhaps also make a distinction between "motile life" and "fertile life," since the motility of sperm often persists for some hours after their fertilizing ability has declined (see below).

It has not yet been possible to determine accurately the limits of the fertile life of sperm in the reproductive tract of the human female. From various types of evidence, CARY (1936), COHEN and STEIN (1950), MOENCH (1939) and BELONOSCHKIN (1959) suggest an upper limit of about 48 hours as being a realistic figure, though human sperm often remain motile for much longer periods. On the other hand, statistical analysis of the duration of the intermenstruum, frequency of intercourse, and rate of conception per menstrual cycle led TIETZE (1960) to a figure of 12 to 24 hours; FARRIS (1950) also arrived at a similar figure derived from experience with A.I. donors. POTTER (1961) has offered reasons for considering the Farris/Tietze estimate as somewhat low and, taking all lines of evidence into account, proposed a time of rather less than 48 hours; how near to 24 hours this should be remains to be determined experimentally. In the sheep, DAUZIER and WINTENBERGER (1952a) found motile sperm in the cervix some 70 to 80 hours after mating, but fertilization was never obtained later than about 30 hours. Motile sperm have been observed in the rabbit uterus about 50 hours after their insemination (NOYES and THIBAULT, 1962), but these also do not remain fertile for more than about 30 hours (HAMMOND and ASDELL, 1926; NOYES and THIBAULT, 1962; CHANG and PINCUS, 1964). In the guinea pig, likewise, there is a difference between the maximum fertile life (22 hours) and motile life of sperm (41 hours, YOCHEM, 1929). By contrast, the fertilizing ability and motility of mouse and rat sperm, respectively, occupy a more or less equivalent, relatively short period of about 12 hours (MCGAUGHEY, MARSTON, and CHANG, 1968) and 14 to 17 hours (YOCHEM, 1929; SODERWALL and BLANDAU, 1941).

NOYES and THIBAULT (1962) suggested that the earlier loss of fertilizing ability in otherwise motile sperm might reflect the deterioration of the capacitated state. The specific reason for the apparent early failure of fertilizing ability is not known, but the inference of its association with capacitation has some support. For instance, capacitation was delayed somewhat in sperm incarcerated in a ligated rabbit uterus, but the fertile life of these sperm seemed to be prolonged beyond that in the control uterus (SOUPART and

ORGEBIN-CRIST, 1966). Rabbit sperm can begin but cannot complete their capacitation in the rat uterus, and in this foreign situation retain their fertilizing ability for the whole of their motile life; this, sometimes 50 to 55 hours, is significently longer than the fertile life of sperm in the rabbit uterus (BEDFORD, 1967c). This postulated relationship between capacitation and subsequent *loss* of fertilizing ability is not yet clear in other species. For instance, ferret sperm can first become capacitated in about 3 to 4 hours in the Fallopian tube (CHANG and YANAGIMACHI, 1963), but normally must await the egg for 30 hours (HAMMOND and WALTON, 1934) and remain fertile in some cases for more than 100 hours (CHANG, 1965). Pig sperm appear capacitated in the tube within 2 hours of their vaginal insemination, as judged by the time of penetration of oviducal eggs (HUNTER and DZIUK, 1967), and yet some sperm may remain fertile in the sow 30 hours or more after insemination. To resolve this enigma, we need to know more in different species about the capacitation potential of the uterus and Fallopian tube, respectively. It is possible to speculate that in some species spermatozoa become completely capacitated only when they pass into the oviduct (e.g., hamster, HUNTER, 1969), and that, overall, the oviduct has a dominant role in capacitation (BEDFORD, 1969a). In this respect it may be noted as mentioned above that the relatively short-lived sperm of the myomorph rodents tend for the most part to be first admitted to the tubal ampulla around the time of ovulation. In the pig, the utero-tubal junction appears to serve as a reservoir from which sperm are then able to move up into the oviduct over a prolonged period (RIGBY, 1966).

There is no clear view, as yet, of the specificity of the sperm/female relationship. Rabbit sperm will swim actively for periods of up to 48 hours in the estrous rat uterus, but die rather quickly (within about 12 hours) in the uterus of the estrous cat, guinea pig and ferret (BEDFORD and SHALKOVSKY, 1967). Ram and human sperm survive well in the rabbit uterus, whereas bull and boar sperm cannot maintain good motility in this situation (COGGINS and BAKER, 1968). Thus, there are selective effects on the sperm of different species, in an environment in which presumably neither the temperature nor the the osmolality of the fluids differ significantly between species. The question of the specificity of capacitation is considered later in this discussion.

An interesting point which has not been resolved concerns the "dilution effect" and sperm in the female tract (see: NELSON, 1967). It has long been known that reduction in sperm cell concentration has an adverse effect on the viability of sperm, approximately proportional to the dilution rate, and NEVO and MOHAN (1969) have pointed out that there is a high mortality rate among sperm which migrate into a sperm-free physiological medium *in vitro*. Clearly, sperm are subjected to a similar dilution effect in the cervix, uterus, and oviduct. Controlled experimental observations suggest that cervical mucus annuls the adverse effects of dilution by virtue of its structural or physical properties, though it is possible that a low oxygen tension in the cervical secretion may also act to mitigate the effects of dilution in this site (MATTNER, 1969b). Since passage through cervical mucus does not confer on sperm any lasting protection against dilution (MATTNER, 1969b), there must be other, as yet unrecognized, factors in the uterus and perhaps in the oviduct which allow motile sperm to retain their functional integrity for many hours in this "diluted" situation.

Spermatozoa are disposed of in the female by loss from the uterus and tubes into the vagina or peritoneal cavity, and by phagocytosis; there is no real evidence for extensive dissolution of sperm within the female tract. In rats, mice and hamsters, the uterus is invaded by leucocytes within a few hours of insemination, and, especially in the mouse, active phagocytosis soon begins (AUSTIN, 1957; YANAGIMACHI and CHANG, 1963b). Between 14 and 20 hours after mating, the cervix relaxes and the uterine content (sperm, debris, and phagocytes) is expelled into the vagina (BLANDAU and ODOR, 1949; AUSTIN, 1957; YANAGIMACHI and CHANG, 1963b). On occasion, some spermatozoa undoubtedly penetrate the tubal mucosa in certain species and have been found lying within the uterine glands in representatives of six different orders of mammals, though not in man (AUSTIN, 1960b).

In man, cow, sheep, and rabbit there is no coordinated expulsion of spermatozoa from the female tract. A continuous loss probably occurs as sperm move randomly back into the vagina, or, in small numbers, on into the peritoneal cavity (see: AHLGREN, 1969). Dead sperm are carried back passively through the cervix in uterine secretions. Active phagocytosis of sperm within the uterus (MENGE, TYLER, and CASIDA, 1962; BEDFORD, 1965; MOYER, KUNITAKE, and

NAKAMURA, 1965; HAYNES, 1967; HOWE, 1967), and in cervical mucus (MATTNER, 1969a), undoubtedly contributes to sperm disposal, though it is difficult to be sure of its relative importance. There is good evidence that the activity of leucocytes in the female tract is modified at different stages of the cycle (KILLINGBECK, HAYNES, and LAMMING, 1963) and that sperm are ingested less efficiently in the progesterone-dominated phase (MENGE, TYLER, and CASIDA, 1962; HAYNES, 1967). A differential response of leucocytes to intact and damaged sperm in the estrous rabbit uterus and in the pseudo-pregnant rabbit uterus and other sites (intact sperm were ingested only in the estrous uterus) was attributed to the development of some difference in the surface properties of the sperm head in the estrous uterus, perhaps associated with capacitation (BEDFORD, 1965). As far as the author is aware, nothing of consequence is known about the qualitative or quantitative aspects of sperm phagocytosis in man and the monkey.

Capacitation of Spermatozoa

In 1951, it was shown experimentally by CHANG and AUSTIN that fertile rabbit and rat sperm must first reside in the female tract for several hours before they finally acquire the capacity to penetrate the zona pellucida and fertilize the ovum. Recognition of the fact that this functional change in sperm, termed "capacitation" (AUSTIN, 1952), occurs as a consequence of exposure to the milieu of the uterus and oviduct has allowed more rational study of fertilization *in vivo*, as well as the development, of techniques for the fertilization of mammalian ova *in vitro*.

The process of capacitation has obvious implications which might enter into the study of hitherto unexplained infertility, and in the approach to contraception. The discussion which follows attempts to synthesize the more reliable data and useful speculations relating to capacitation and to explain its significance for fertilization.

a) The Need for Capacitation in Different Mammals

The need for capacitation prior to fertilization has been determined definitely for a few species only, and it remains uncertain as to whether this is a ubiquitous phenomenon among the Eutheria.

So far, however, in no mammalian species have the sperm been shown to be able to fertilize immediately, without some prior period of incubation *in vivo* or *in vitro*.

Rabbit sperm which pass from the uterus into the oviduct can become fully capacitated in about 5 to 6 hours (Chang, 1955; Adams and Chang, 1962). In the Fallopian tube alone, a period of 10 to 11 hours is required for functional capacitation (Adams and Chang, 1962), and estimates of the time at which sperm first become competent to penetrate eggs in the uterus suggest that capacitation of rabbit sperm in the uterus alone is completed in about 10 hours (Bedford, 1969a).

In the rat, studies of the time relations of fertilization and pronucleus formation *in vivo* indicate that capacitation is achieved in 2 to 3 hours in sperm passing from the uterus to the oviduct (Austin, 1951; Austin and Braden, 1954; Noyes, 1953). Analysis of the time relations of fertilization in the mouse indicates that a very short time is required for capacitation. This is probably of the order of 1 hour *in vivo* (Austin and Braden, 1954); capacitation of mouse sperm can be achieved *in vitro* in about 2 hours in a medium of Tyrode's solution containing 50% of heated bovine follicular fluid (Iwamatsu and Chang, 1969). In considering the hamster, it is apparent from observations of fertilization *in vivo* (Strauss, 1956; Chang and Sheaffer, 1957) and from convincing experimental studies of fertilization *in vitro* (Yanagimachi and Chang, 1964; Yanagimachi, 1966, 1969a, b; Barros and Austin, 1967; Barros, 1968) that functional capacitation of golden hamster sperm requires $2^1/_2$ to 3 hours, either *in vivo* or *in vitro*.

Ferret spermatozoa sometimes begin to penetrate eggs after only $3^1/_2$ hours when inseminated into the Fallopian tube (Chang and Yanagimachi, 1963) and since sperm were found earlier on unfertilized eggs, this time can probably be considered as minimal for capacitation in the oviduct alone. The contribution of the ferret uterus to capacitation remains yet to be settled. A similar situation exists with respect to the sheep, in which sperm have been shown to develop the capacity for penetration of tubal ova only $1^1/_2$ hours after their tubal insemination (Mattner, 1963a), but not before this time. Likewise, the role of the sheep uterus in capacitation during the normal ascent of the sperm after mating has not been determined. As mentioned earlier, pig sperm develop the capacity to penetrate tubal eggs within 2 hours of vaginal insemination (Hunter and Dziuk, 1968).

The question of capacitation in man and in other primates has not been closely studied. In the rhesus monkey, a pronucleate egg was found among a total of three penetrated eggs collected only 6 hours after vaginal insemination (Marston and Kelly, 1968) – an indication that capacitation can be completed in 2 to 3 hours after vaginal insemination. The situation in man has been obscured by many claims of fertilization *in vitro*, in which the results were never well substantiated, and in which little attention was given to the possibility of the occurrence of artificial activation. The most convincing data on man are found in the recent report of Edwards, Bavister, and Steptoe (1969), whose claims for *in vitro* penetration of human oocytes seem well founded in light of their further demonstration, in some cases, of the presence of the fertilizing sperm tail in the ooplasm (Bavister, Edwards, and Steptoe, 1969). In these studies, oocytes were seminated with ejaculate spermatozoa washed for various lengths of time in physiological media, with and without human follicular fluids. If capacitation does occur in these conditions, this suggests a similar capacitation requirement *in vitro* as for hamster and mouse spermatozoa.

To conclude, the need for capacitation has been well defined in only one or two mammals, but in several other species there is a good indication that a comparable interval occurs before sperm may begin to penetrate the zona pellucida. The time required for capacitation varies, being only 1 to 2 hours in some species. In no case, however, have ejaculated or epididymal spermatozoa of any eutherian mammal been observed to begin penetration of the zona immediately after their introduction into the vicinity of fertile eggs, such as occurs, for instance, in the sea urchin.

b) Environmental Requirement for Capacitation

That the details of capacitation should differ somewhat in the few species studied is not surprising in view of the rather variable nature of the sperm/female relationship in different animals. For instance, there is variation in the site of deposition of sperm (vagina or uterus), and the timing of insemination in relation to ovulation. These difficulties are compounded by the fact that, with the rabbit particularly, investigators have not always used standard assays.

Rabbit. Rabbit sperm can be completely capacitated in the uterus or the Fallopian tube (CHANG, 1951a, 1955, 1959; AUSTIN, 1951). Since sperm taken from the uterus 6 hours after insemination fertilized tubal ova (CHANG, 1955), and sperm inseminated into the rabbit uterus were able to pass into the oviduct and there fertilize ova about 6 hours after insemination (ADAMS and CHANG, 1962), the idea has grown that capacitation is completed in the rabbit uterus in about 6 hours. Recent experiments now indicate that a longer period is required for complete capacitation in the rabbit uterus alone, for eggs placed in the uterus at various times after vaginal insemination were not penetrated by sperm until about 10 to 11 hours after such insemination; after this time sperm penetration occurred normally and promptly (BEDFORD, 1969a). Capacitation in the rabbit oviduct alone requires about 10 to 11 hours (ADAMS and CHANG, 1962). The fact that capacitation in the uterus or tube alone demands about 10 hours, and yet is achieved in 5 to 6 hours in normal passage of sperm from uterus to tube, suggests that such efficient capacitation in the latter case may depend upon a synergism of two different activities, one of which is more potent in the uterus and the other more so in the tube. The likelihood that capacitation of rabbit sperm can be analyzed in this way, and is not a sudden all-or-none phenomenon, is supported by evidence that partial but incomplete capacitation of rabbit sperm can occur in a variety of homologous and heterologous situations (see: BEDFORD, 1967b). Complete or functional capacitation of rabbit sperm *in vitro* does not appear to have been achieved as yet.

Rodents. Although initial studies in the rat were partly responsible for establishing the need for capacitation (AUSTIN, 1951; NOYES, 1953), relatively little has been done since to clarify the role of the various regions of the tract in this respect. There is considerable doubt as to whether capacitation can be completed in the hamster or rat uterus. The first *in vitro* fertilization experiments of YANAGIMACHI and CHANG (1964) seemed to show that hamster uterine sperm hold some advantage over those taken directly from the epididymis, but this has since been shown to be slight (HUNTER, 1969). Furthermore, BARROS and AUSTIN (1967) found that hamster uterine sperm could not begin to penetrate eggs immediately *in vitro*, and HUNTER (1968) was never able to fertilize hamster eggs in the uterus of inseminated females. Likewise, rat eggs were never fertilized when placed in the uterus 7 to 8 hours post coitus even though the eggs recovered later had many motile sperm adhering to the zona (BEDFORD, unpublished). The uterus may have been harmful to the egg, but this nevertheless allows the suspicion that sperm cannot easily complete their capacitation in the uterus of the rat and hamster. Mouse spermatozoa taken from the uterus 1 to 2 hours after coitus have been used successfully for *in vitro* fertilization (WHITTINGHAM, 1968), though the percentage fertilization was rather low. It is possible, in this instance, that the mouse uterine spermatozoa continued their capacitation *in vitro* before fertilization, as occurs in the hamster. This question could be answered by critical comparison of the time after semination at which eggs first begin to be penetrated *in vitro* by uterine sperm and by sperm capacitated *in vitro* by follicular fluids (IWAMATSU and CHANG, 1969).

There is still a question as to the specific value of the fluids in the oviduct and follicle for capacitation of hamster sperm. For, BAVISTER (1968) managed to achieve a high rate of fertilization *in vitro* with epididymal sperm in a defined medium in which the fluid/cumulus mass ratio was 50 : 1 or greater. In these experiments, emphasis was laid on the physical nature of the environment created, rather than on its biologically active organic components.

Other Species. The sheep (MATTNER, 1963a) and the ferret (CHANG and YANAGIMACHI, 1963) are the other species in which estimates have been given of the time required for capacitation in any one region of the tract. These studies show that perhaps only $1^1/_2$ hours in the sheep and 3 to 4 hours in the ferret are needed for capacitation in the oviduct. It is intriguing, in light of this brief tubal capacitation period, that some sheep sperm can remain fertile for 30 to 40 hours (DAUZIER and WINTENBERGER, 1952a), and ferret sperm up to 126 hours (CHANG, 1965) in the female tract. For, the advent of the capacitated state appears to exert some limitation on the fertile life of rabbit sperm (see pp. 138–139).

To find a coherent pattern in these results, more information is needed about the capacitation potential of the uterus and oviduct in different species. Inevitably, the use of fertilization as an assay measures only the minimal time required for capacitation. As suggested earlier, it may be that the sperm of some species cannot easily complete their capacitation in the uterus alone, and that this occurs efficiently only when sperm pass into the Fallopian tube. In this vein, one wonders whether there is any functional correlation between the proven efficacy for capacitation of the mixture of tubal and follicular fluid (BARROS and AUSTIN, 1967) and the fact that most sperm entering the ampulla of the rodent oviduct do so at the time of or after ovulation.

Species-specificity of Capacitation. Although the female tract of certain species is hostile to some foreign sperm, for the most part there is no exacting species-specificity for the capacitation of spermatozoa. As judged by the occurrence of cross-fertilization, reciprocal capacitation occurs between related species, i.e., sheep and goat, rabbit and hare, cottontail rabbit and domestic rabbit, and mink and ferret (CHANG and HANCOCK, 1967). Capacitation of hamster sperm has been achieved with follicular fluids from the mouse and rat, though not as efficiently as with the homologous fluids (BARROS and AUSTIN, 1967; BARROS, 1968; YANAGIMACHI, 1969a). By contrast with rodents, the rabbit seems to have developed a relatively closer dependence upon the homologous female for complete capacitation (BEDFORD, 1967b), although partial capacitation of rabbit sperm can take place in a foreign uterus (BEDFORD and SHALKOVSKY, 1967; HAMNER and SOJKA, 1967).

The Influence of Steroids on Capacitation. As might be expected, the capacitation potential of the reproductive tract can be modified by the endocrine state of the female. Experimental data have been obtained so far only in the rabbit, in which the sperm have a relatively close dependence on the reproductive tract for functional capacitation. It appears that the capacitation potential of the uterus can be modified to a greater degree than that of the Fallopian tube. Progesterone almost completely depresses the capacitation activity of the rabbit uterus (CHANG, 1958; SOUPART, 1967; BEDFORD, 1967b; HAMNER, JONES, and SOJKA, 1968). Capacitation is achieved in the estrous uterus in about 10 hours; in the ovariectomized rabbit, sperm are only partially capacitated at this time, and almost 20 hours is required before sperm can begin to penetrate ova in this environment (BEDFORD, 1970a). The capacitation potential of the estrogendeficient uterus can be restored by administration of exogenous estrogen (SOUPART, 1967).

In contrast to the uterus, the rabbit oviduct apparently remains competent to capacitate sperm regardless of the endocrine state of the female. The capacitation potential of the estrogen-deficient and/or progesterone-dominated tube is still maintained at a level adequate to cope with larger numbers of sperm than normally reach this site after natural mating (BEDFORD, 1970a). The rabbit oviduct thus possesses an inherent facility for capacitation which can be enhanced by estrogen, but which cannot be destroyed by progesterone, even in the absence of ovarian estrogen. Bearing in mind the fact that rabbit sperm appear to have a relatively close dependence upon the secretions of the female tract for completion of capacitation, these results make it unlikely that contraceptive control of capacitation *per se* can be achieved simply by modification of the steroid status of the female.

Decapacitation Effect of Seminal Plasma. Seminal plasma from the rabbit, bull and man has been shown to reversibly inhibit the fertilizing ability of capacitated rabbit sperm (CHANG, 1957), and a similar property has been found subsequently in stallion, boar and monkey seminal plasma, but not in that of the dog (DUKELOW, CHERNOFF and WILLIAMS, 1967), in which there is no seminal vesicle. The active decapacitation factor (DF) is of some interest as a substance that is not toxic and has rather specific effects at the stage of the sperm/egg interaction. As yet, the identity of the active factor has not been established.

The mode of action of DF is not entirely clear, but it does not affect sperm motility. There are reasons to think that DF may operate at the sperm surface, perhaps to modify the stability of or to block receptor sites on the plasma membrane over the acrosome region, thus preventing the acrosome reaction (see below) (BEDFORD, 1969b, 1970b). The partially purified active factor will inhibit certain sperm enzymes extracted presumably from the acrosome (ZANEVELD and WILLIAMS, 1970), but it is unlikely that, *in vivo*, this inhibition could occur before the acrosome reaction is initiated.

What are the Changes Which Occur in Sperm During Capacitarion? As yet, it has not been possible to identify any change in the sperm which can be shown unequivocally to be a necessary part or direct manifestation of the process of capacitation, *per se.* The question of structural alteration in the sperm has naturally received attention, and the idea has grown among some workers that such changes do take place in the acrosome during capacitation. In the writer's opinion, no morphological changes of note are involved in capacitation as such, but the capacitation process does appear to be instrumental in preparing sperm to be able to undergo the acrosome reaction at the appropriate time, in the vicinity of the egg (see below).

Insofar as the rabbit is concerned, light and electron microscope studies indicate that no structural modification is involved in capacitation (Fig. 2). Moreover, the fact that seminal plasma

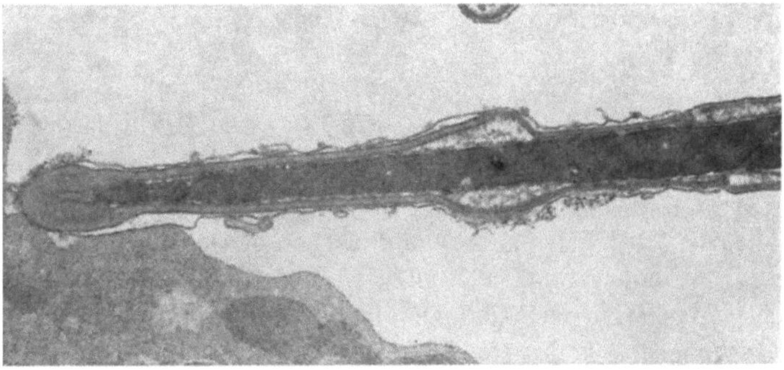

Fig. 2. Electron micrograph of anterior region of (capacitated ?) rabbit sperm head taken from the uterus 14 hours post-coitum. There is no change in the morphology of the head associated with capacitation. Note the intact acrosome and plasma membrane which is adhering to a leucocyte

apparently may reversibly inhibit the capacitated state militates against the idea of irreversible structural changes as a facet of capacitation. By contrast with the rabbit, studies of sperm in the female tract of various rodents have revealed changes in the acrosome cap which have been considered by some to be part of the capacitation process (Austin and Bishop, 1958; Yanagimachi, 1969b). It is important to realize, however, that these investigations were carried out in conditions manifestly able to bring about capacitation and also, subsequently, to trigger the acrosome reaction. Many invertebrate species undergo a similar type of acrosome reaction to that seen in mammals (Colwin and Colwin, 1967), but apparently have no requirement for capacitation. Since neither rabbit nor rodent spermatozoa can undergo the acrosome reaction without a prior period of capacitation, it seems preferable from a semantic and zoological standpoint therefore to consider capacitation as encompassing only the phase or process which prepares the sperm for, and takes place before, the acrosome reaction.

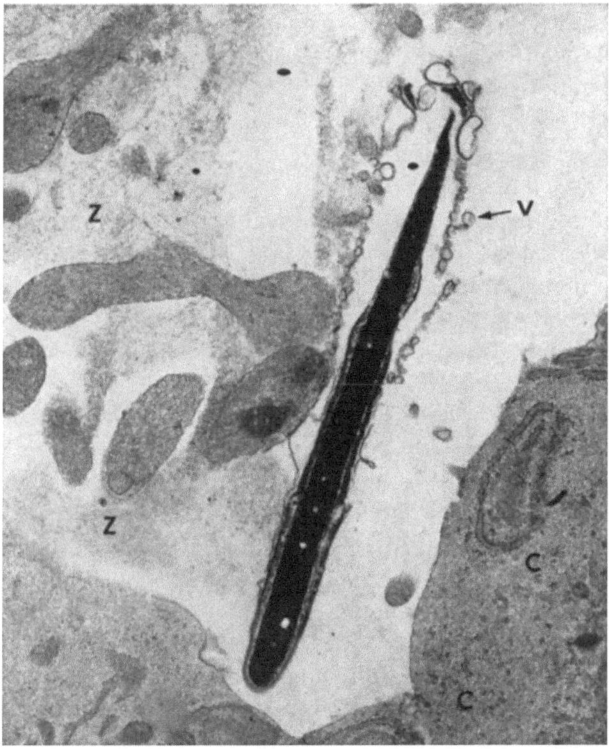

Fig. 3. Capacitated rabbit sperm undergoing the acrosome reaction at the surface of the zona pellucida (*Z*) which is seen here as diffuse material interspersed between the base of the corona cells (*C*). Note that the acrosome has now become transformed as a shroud of vesicles (*V*) around the front of the sperm head

There is no direct evidence that changes occur in the sperm plasma membrane during capacitation. However, one must consider the particular character of the membrane breakdown in the acrosome reaction (see below, Fig. 3), and the susceptibility of the capacitated state to the decapacitation factor which, if covalently linked to a large molecule in rabbit semen, must presumably act at the sperm cell surface. This type of evidence makes it reasonable to infer the occurrence of certain modifications in the surface of sperm during capacitation. More recently, YANAGIMACHI and NODA (1970) have shown that, after enzymic removal of the hamster zona, sperm stick to and fuse with the vitelline surface by the post-acrosomal region (see below) only when capacitated sperm are used. This implies that, in the hamster, capacitation may also involve change in the surface of the post-acrosomal region, though this may not be necessary in the rat (TOYODA and CHANG, 1968). It is doubtful that capacitation involves simply the removal of some surface coating akin to DF. For, evidence that the rabbit uterus and tube may act synergistically in this respect (BEDFORD, 1969a), and the demonstration that "partial capacitation" can occur in the rabbit (BEDFORD, 1967b), suggest the involvement of more than one such simple step before a state of functional capacitation is achieved.

Finally, in this context, it may be relevant to draw attention to an apparent difference in surface characteristics between mature sperm of anthropoids, including man, and those of other eutheria. In physiological media, the sperm of the latter (including the prosimae) display a marked tendency for head-head agglutination (Fig. 4), particularly so in the presence of normal serum. Furthermore, among the species having sperm of "tadpole" form (bull, ram, and rabbit), the mature sperm cells orientate quickly in an electrophoretic field, with the tail directed toward the anode (NEVO, MICHAELI, and SCHINDLER, 1961; BANGHAM, 1961), presumably as another reflection of regional differentiation of the sperm surface. By contrast, neither of these phenomena – head agglutination or tail-anode orientation – is shown by human and anthropoid monkey sperm (BEDFORD – in preparation). One wonders whether such manifest differences in the character of the sperm surface have any bearing on the question of capacitation in the monkey and in man.

Modifications in energy metabolism have been shown to occur in sperm as a result of exposure for some hours to secretions of the female tract (HAMNER and WILLIAMS, 1963; see IRITANI et al., 1969, for refs.), and it is possible that these may be allied to the capacitation process. There is no direct evidence yet that such metabolic changes are important in the economy of the fertilizing sperm, and these may of course merely reflect incidental changes in the sperm membranes allowing substrate and ions to pass more freely. Nonetheless, it seems germane at this point to draw attention

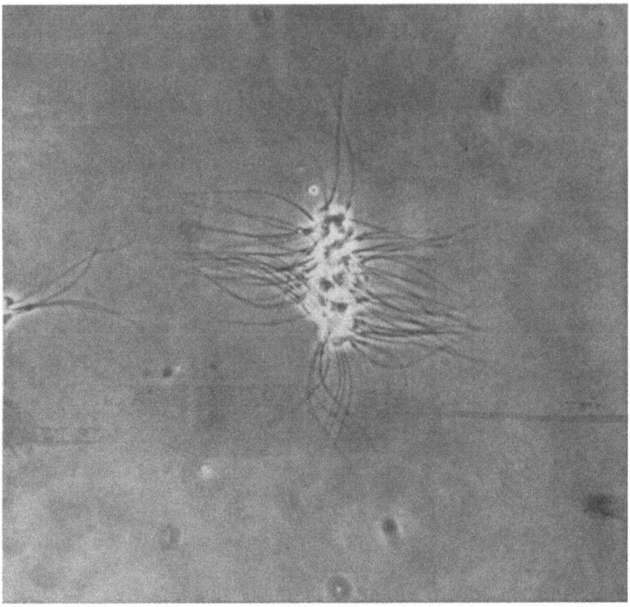

Fig. 4. Mature sperm from the vas deferens of the Tree shrew (Tupaia glis) in a medium of Ringer/serum. Note that the sperm have become agglutinated (non-specifically) solely by the head region

to observations of changes in the swimming pattern of hamster (YANAGIMACHI, 1969b) and mouse sperm (IWAMATSU and CHANG, 1969) coincident with the appearance, *in vitro*, of capacitated sperm. During capacitation, sperm may develop a type of movement giving optimum forward thrust or moment of vibration, but further careful studies in other species are needed before these interesting findings can be interpreted as such.

Fertilization

In recent years, the topic of mammalian fertilization has been discussed at length (Austin and Walton, 1959; Austin, 1968; Blandau, 1961; Piko, 1969). The present treatment is therefore confined to consideration of the most recent developments in our understanding of, as well as uncertainties or disagreement about, the way the sperm approaches and penetrates the egg.

a) Preliminaries to Sperm Penetration

Upon the arrival of the ovum in the tubal ampulla, the potential fertilizing spermatozoon is faced with a significant obstacle in the form of a granulosa cell mass – the cumulus oophorus (Fig. 5); in

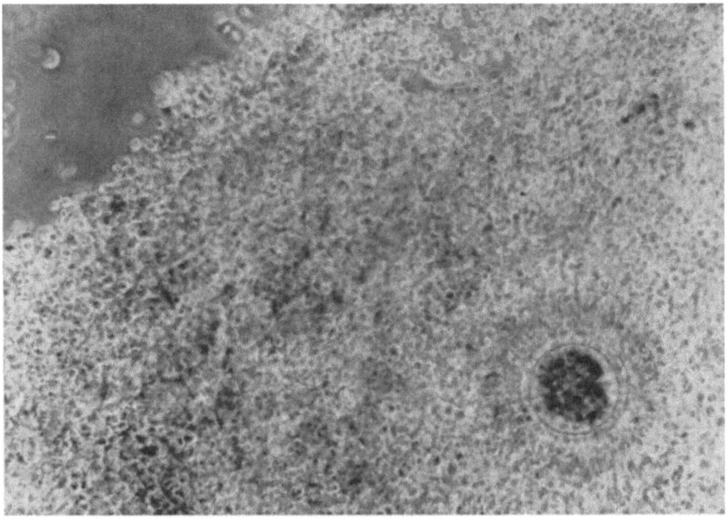

Fig. 5. Freshly ovulated rabbit egg surrounded by a halo of corona cells; outside this corona lies a mass of granulosa cells — the cumulus oophorus

many animals this must be traversed before sperm contact with the zona can be established. In the mouse, in particular, there is evidence that the matrix which binds the cells of the cumulus oophorus undergoes some form of "maturation" in the first 2 hours or more after ovulation (BRADEN and AUSTIN, 1954; BRADEN, 1959; ZAMBONI, 1970), since mouse sperm, capacitated or not, are unable to penetrate the cumulus mass before this time. Comparable changes may possibly occur in the immediate post-ovulatory period in the cumulus matrix of other species; if so, as judged by the time of sperm penetration following ovulation, they must take place within a relatively shorter period.

There are marked species differences in the rate at which these follicular cells are dispersed, but in the species which have been studied most frequently such dispersion does not occur for several hours in the absence of fertilization (see: DICKMANN, 1963). Some years ago it was commonly believed that hyaluronidase, emanating from a phalanx of tubal sperm, acts to disperse the granulosa cell barrier around the ovum, thereby allowing the fertilizing sperm access to the zona surface. Such denudation was thought to be a necessary preliminary to sperm penetration. This notion is no longer acceptable; first, because – at least in the laboratory species studied – sperm very often penetrate as far as the vitellus before any noticeable dissipation of the follicle cells has occurred (LEWIS and WRIGHT, 1935; AUSTIN, 1948a), and also because careful counts have shown that there are relatively few sperm in the ampulla of the oviduct at any one moment (AUSTIN, 1948b; HAMMOND and WALTON, 1934; BLANDAU and MONEY, 1944; CHANG, 1951b; STEFANINI, OURA, and ZAMBONI, 1970). In the electron microscope, sperm have been observed occasionally within corona cells (Fig. 6); this has been recorded in rabbits after tubal insemination (BEDFORD, 1968) and also after natural mating (unpublished). Sperm with or without intact acrosomes are picked up in this way; hence, it is more likely that the corona cell is the active element rather than the sperm.

The factors which influence the likelihood of successful sperm/egg contact in mammals are not entirely clear. Probably no more than one sperm is needed to ensure fertilization. There is some evidence, however, that whereas capacitated sperm attain the egg surface relatively easily, it is difficult for ejaculate or epididymal sperm to penetrate the matrix which occupies the interstices of the cumulus oophorus. This was first pointed out by AUSTIN (1960a), who observed that while non-capacitated sperm failed to penetrate the periphery of the cumulus *in vitro*, sperm were passing freely through the substance of the cumulus oophorus collected soon after ovulation from mated animals. Later, Bedford (1967a) examined this question in the rabbit by inseminating one tube with capacitated sperm and the other with greater numbers of epididymal

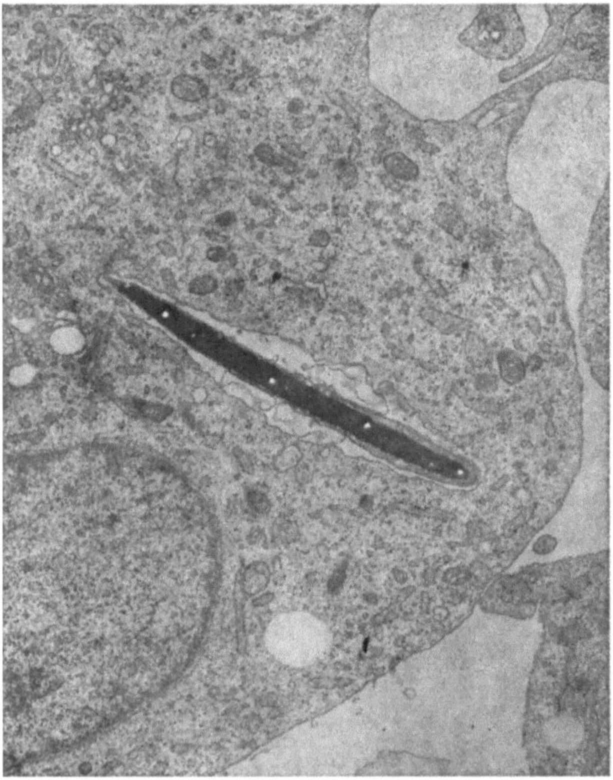

Fig. 6. Rabbit sperm head lying in a corona cell of an egg recovered 13 hours after natural mating

sperm at the time of ovulation. The eggs recovered some 6 hours after exposure to the capacitated samples were mostly fertilized and had significantly more sperm associated with the egg (31.6 $\pm$ 5.7 sperm/egg) than those exposed under similar conditions to mature epididymal sperm (1.9 ± 0.22 sperm/egg); the latter become highly competent (103.4 ± 24.2 sperm/egg), however, when first allowed time for capacitation. Since the enveloping follicular cells could have been shed from these eggs only a short while previously, the overt superiority of capacitated samples to establish sperm/egg contact would seem to lie mainly in their greater ability to penetrate the cumulus and corona radiata. In mammals, sperm/egg contact is probably established by random movement, and there is no strong evidence that mammalian sperm are capable of a chemotactic response. The idea that the cumulus oophorus might serve to guide sperm in the vicinity of the egg toward the zona (AUSTIN and BRADEN, 1954) receives support from a recent study of HARPER (1970). In his experiments, eggs with and without cumulus were placed, respectively, into the opposite tubes of rabbits about $9^1/_2$ to $9^3/_4$ hours *post-coitum*; 24 hours later, 67 percent of eggs in cumulus were penetrated compared with only 35 to 42 percent of those in which the cumulus had been removed before their tubal instillation. That this difference was eliminated when eggs were instilled later, 13 to 14 hours after mating, may reflect the availability of greater numbers of fully capacitated sperm at this time.

b) The Acrosome Reaction

The advantage conferred on capacitated spermatozoa during their approach through the cumulus oophorus may lie in their ready competence to undergo the acrosome reaction which in turn allows the release of acrosomal enzymes. It is now apparent that, besides hyaluronidase, the sperm head contains extractable amounts of other enzymes, which will disperse the cumulus and corona cells. Those identified so far are a lipo-glycoprotein (HARTREE and SRIVASTAVA, 1965), a trypsin-like enzyme (STAMBAUGH and BUCKLEY, 1968, 1969), and an enzyme which can be differentiated from the latter, apparently, by selective inhibitors (ZANEVELD and WILLIAMS, 1970). All of these will remove the corona radiata and/or the cumulus oophorus, depending upon the species being used, and they probably play a role in assisting the sperm to traverse the cumulus mass.

In mammals, the normal physiological mode of acrosomal breakdown – the acrosome reaction – apparently occurs by fusion at several points between the overlying plasma membrane and outer acrosome membrane, such that a section of the sperm head appears to be shrouded by a series of vesicles (Barros, Bedford, Franklin, and Austin, 1967) (Fig. 3). The process of fusion between plasma and outer acrosome membranes extends caudally as

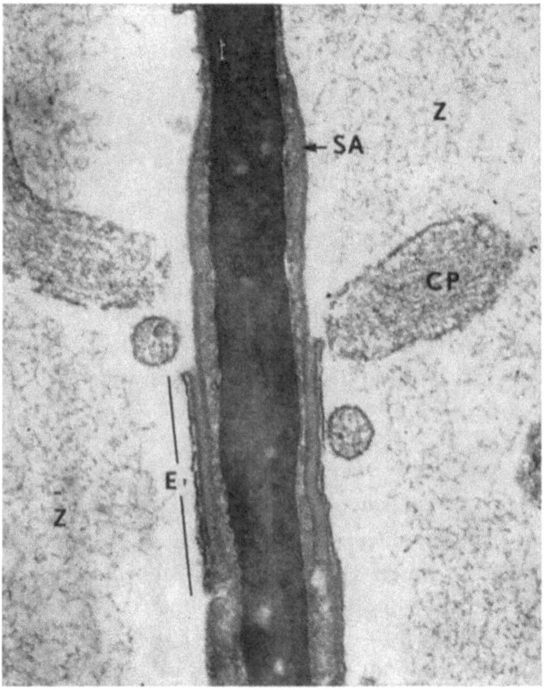

Fig. 7. Section of the mid region of a rabbit sperm head entering the zona pellucida (*Z*) between corona cell processes (*CP*). The equatorial segment (*E*) persists throughout zona passage and penetration of the vitellus. Note the accummulation of sub-acrosomal substance (*SA*) in the mid-region anterior to the equatorial segment, as well as in the post-acrosomal region

far as the anterior limit of the persistent "equatorial" region of the acrosome, which remains unchanged (Fig. 7). The vesiculated products of this membrane fusion are lost from the sperm head before penetration of the zona begins (Bedford, 1968). At this stage it is now the *inner* membrane of the acrosome which has come to bound the leading surface of the sperm head (Fig. 8). This type of membrane fusion and vesicle formation has been observed to take place *very occasionally* during sperm degeneration in a variety of situa-

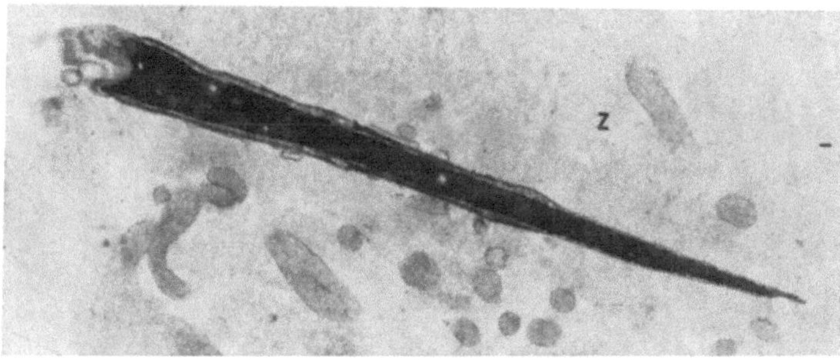

Fig. 8. Rabbit sperm head, devoid of any vesiculated remnant at the acrosome, beginning to enter the zona pellucida (*Z*)

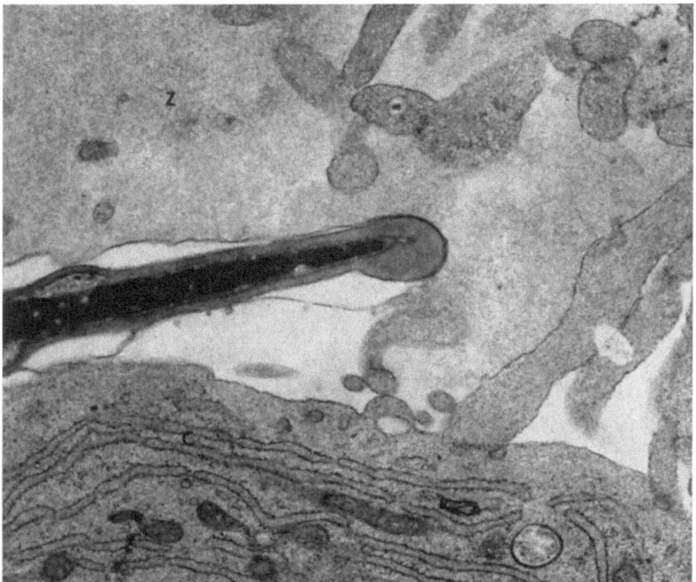

Fig. 9. Unreacted rabbit sperm at the zona surface. This picture of the intact sperm is seen when non-capacitated sperm interact with ova (cf. Fig. 3)

tions from the testis to the female tract; it must be made clear, however, that pathological breakdown of the acrosome almost involves haphazard disintegration of the plasma and outer acrosomal membranes as separate entities (Bedford, 1969b).

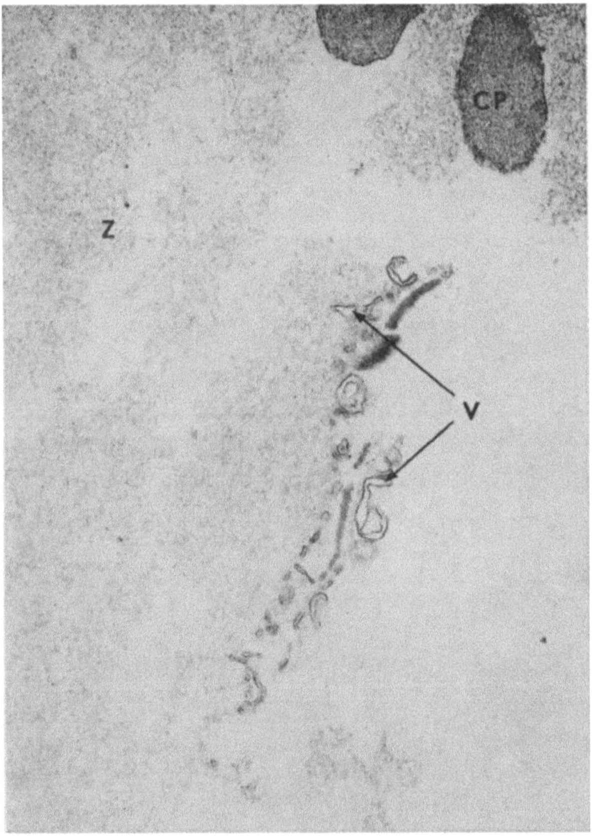

Fig. 10. Vesiculated remnants of the acrosome remaining at the surface of the zona pellucida

Capacitation appears to be an important prerequisite for accomplishment of the normal acrosome reaction at the surface of the rabbit egg. Non-capacitated sperm, aged *in vitro* for 10 to 12 hours and then inseminated at the time of ovulation into the Fallopian tube, can sometimes be found sticking to denuded areas of the zona after about 3 hours, but show no ultrastructural modification of the plasma or acrosome membranes (BEDFORD, 1969b) (Fig. 9). By contrast, following natural mating, or after tubal insemination 12 to 15 hours before ovulation, sperm observed in the vicinity of or attached to the tubal egg 2 to 3 hours after ovulation often show the classical vesiculation response (Fig. 3) or have already lost the outer acrosome complex and are now bounded rostrally only by the intact inner membrane of the acrosome. Most of the sperm at the zona surface can be seen undergoing the acrosome reaction or often have already lost the vesiculated elements, which sometimes adhere to and remain at the zona surface after the sperm has begun to penetrate its substance (Fig. 10).

c) Penetration of the Zona Pellucida

Notwithstanding several recent studies of this aspect of fertilization, many of the important questions still remain. A major question concerns the identity of the hypothetical sperm head zona lysin. Sperm enzymes which probably originate in the acrosome have been shown to be capable of dissolving the zona (STAMBAUGH and BUCKLEY, 1968, 1969; SRIVASTAVA, ADAMS, and HARTREE, 1965). Yet, although various commentators have suggested that these enzymes facilitate zona penetration, it is reasonably certain that the outer acrosome membrane and the acrosomal content are discharged *before* zona penetration begins (Fig. 8). Furthermore, no visible remnants of this material persist on the sperm surface as it begins to penetrate the substance of the zona pellucida, with the exception of that in the equatorial region (Fig. 1). It is possible that enzyme remains bound to the inner membrane of the acrosome in a form not readily visible in the electron microscope; this point might be studied profitably with histochemical methods at the ultrastructural level. Evidence that material lying between the nuclear and inner acrosomal membranes (Fig. 7) (the sub-acrosomal material) is enzymic (TEICHMANN and BERNSTEIN, 1969) makes this a possible candidate as the source of the zona lysin. However, this sub-acrosomal material is not visibly dissipated during sperm transit through the zona, and persists in the fertilizing sperm which has been drawn into the vitellus (see below).

The suggestion that the penetrating sperm within the zona pellucida develops an acrosomal filament comparable to that seen at the apex of reacted spermatozoa of some invertebrates has not been substantiated in the electron microscope. In the writer's opinion, the impression of such a filament in the phase contrast

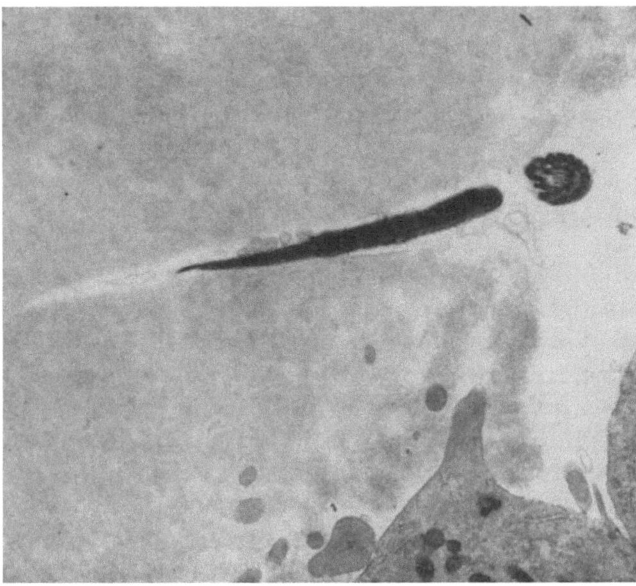

Fig. 11a

Fig. 11a and b. Rabbit sperm heads penetrating the zona of an ageing egg collected 20 hours after mating. Note the rupture of the zona substance in advance of the sperm head

microscope may be produced by the development of zonal fissures which project from the apex of sperm within the substance of the *ageing* zona (Fig. 11).

The functional significance of the "equatorial" region of the acrosome still remains as an enigma. Many observations show that there is no significant diminution in the bulk or electron density of the content of the "equatorial" segment during its transit across the zona pellucida (Fig. 7), and it appears to enter the vitellus unchanged.

Nothing new can be added here to explain the nature and species variability of the "zona reaction." This refractory response to the entry of subsequent spermatozoa, stimulated by the fertilizing

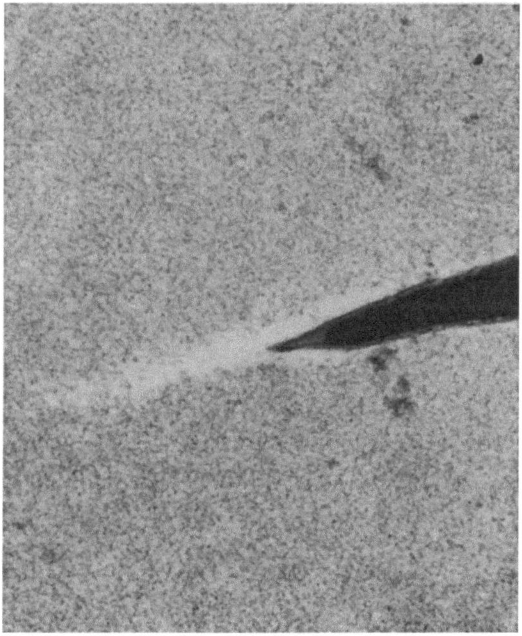

Fig. 11b

spermatozoon, is manifested more strongly in some species (e.g., the hamster) than in others in which there appears to be no significant alteration in the penetrability of the zona during the fertilizable life of the ovum. However, even in the rabbit, subtle differences have been observed in the rate at which trypsin will dissolve the zona, before and after fertilization (SMITHBERG, 1953; CHANG and HUNT, 1956). The time seems ripe for serious study of the chemical components and their physical relationships, i.e., the nature of the cross linkages or bonds, which ultimately determine the biological character of the zona substance.

Sperm Entry into the Vitellus

The dynamic aspects of the union of the sperm with the vitelline surface have been carefully studied with the phase contrast microscope (see: AUSTIN and WALTON, 1959). Soon after it reaches the peri-vitelline space, the fertilizing sperm makes contact with and stimulates the egg plasma membrane. Coincidentally with the establishment of sperm/egg contact, the sperm tail ceases to move, though why this should happen is not really clear. Only the head of the sperm establishes a bond with the egg surface at this early stage but, since the sperm head and tail are bounded by a continuous cell membrane, it is possible that the initial reaction of the

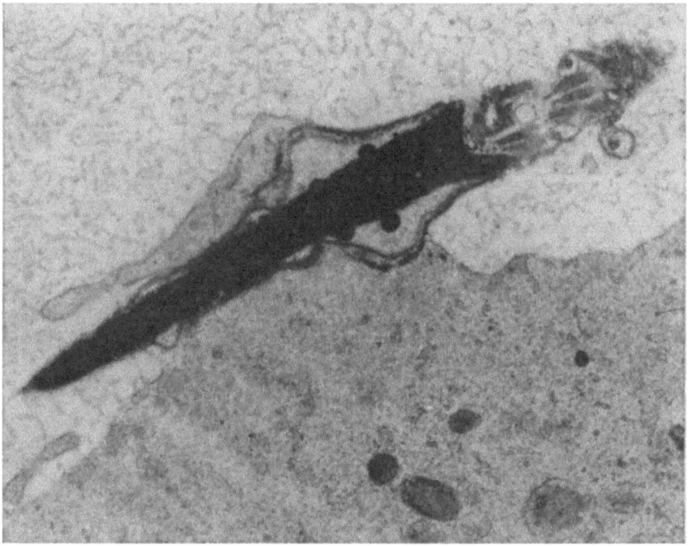

Fig. 12. Fertilizing rabbit sperm head in the initial stages of fusion with the vitelline surface. Note that the interaction occurs over the mid-posterior region of the sperm head

post-acrosomal region with the egg (see below) brings about a propagated depolarization of the remainder of the sperm surface.

Ultrastructural studies in the hamster (BARROS and FRANKLIN, 1968; YANAGIMACHI and NODA, 1970), the mouse (STEFANINI, OURA, and ZAMBONI, 1969), the rat (PIKO, 1964) and the rabbit (BEDFORD, 1970b) have shown that the sperm makes initial contact with the egg only over the post-acrosomal surface of the head (Fig. 12). This apparently marks a definite departure from the pattern of fertilization observed in several invertebrates (COLWIN and COLWIN, 1967), in which the apical region of the sperm head makes the initial con-

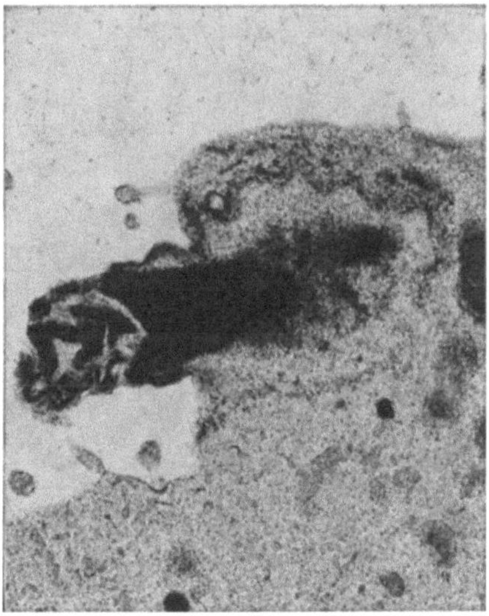

Fig. 13a. Posterior nucleus and neck region of rabbit sperm entering the egg. Note that the nucleus is beginning to decondense even at this very early phase

tact with the vitelline membrane. Studies by YANAGIMACHI and NODA (1970) using nude (zona-less) hamster ova suggest that capacitation in this species may also involve some change in the post-acrosomal surface of the sperm head, since only capacitated sperm were able to adhere successfully to the exposed vitelline surface. On the other hand, rat epididymal sperm will fuse with and penetrate the nude ovum in which the zona pellucida has been removed by chymotrypsin (TOYODA and CHANG, 1968).

The details of the process of sperm incorporation have not yet been clearly established, but recent studies in the hamster indicate that the first stage involves a series of point fusions of the sperm plasma membrane over the post-acrosomal region with bleb-like processes on the vitelline surface. The plasma membrane around the post-acrosomal region of the sperm-head disappears, and this region of the nucleus becomes enveloped by the ooplasm. In the

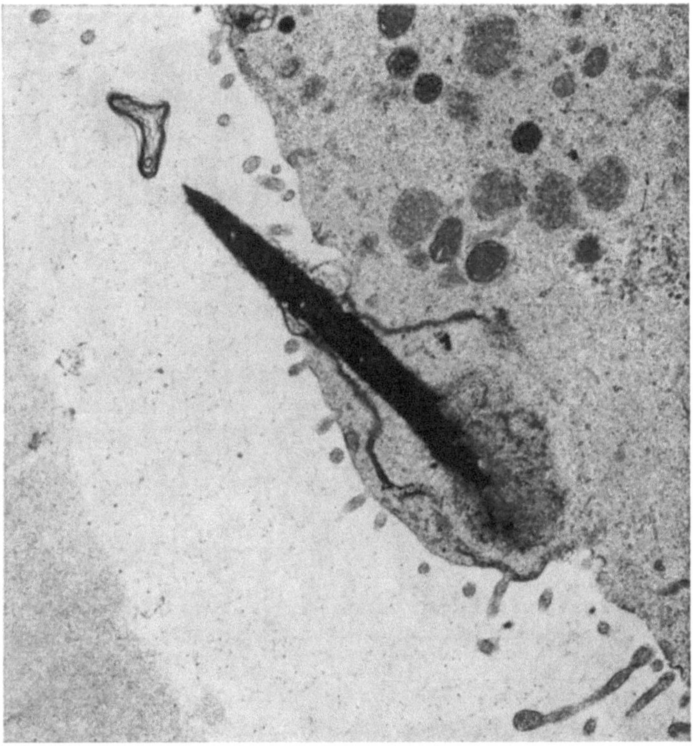

Fig. 13b. Another section through the sperm head in Fig. 13a above to show the mode of decondensation of the sperm nucleus. The anterior region of the nucleus still projects into the perivitelline space

rabbit, dispersion of the dense nuclear chromatin sometimes begins rapidly in the middle-posterior region of the head even before the anterior and caudal regions have entered the ooplasm (Fig. 13a, b) (Bedford, 1970b). Incorporation of the most caudal part of the head, the neck, and the sperm tail is definitely accompanied by loss of the sperm plasma membrane (SZOLLOSI and RIS, 1961), but it is not clear as yet whether this membrane becomes incorporated as a part of the egg surface. At all events, in the recently fertilized egg the posterior part of the sperm head and the tail organelles are

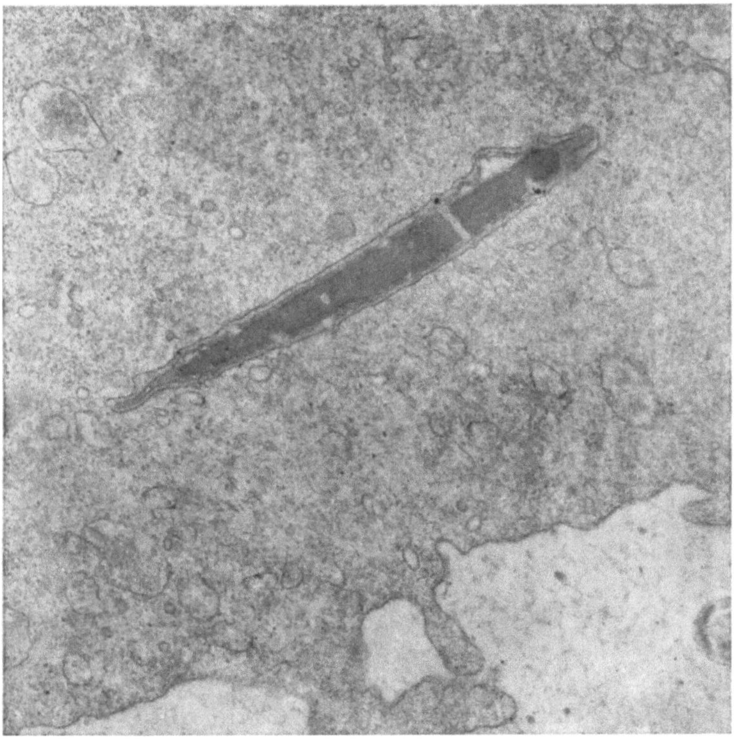

Fig. 14. Transverse section through anterior (acrosomal) region of rabbit sperm head inside the vitellus. Note that the nucleus is enveloped by two membranes. The inner membrane is the inner acrosomal membrane of the sperm; the outer is the invaginated section of the vitelline surface which enters the vitellus around the anterior half of the sperm head

always devoid of an encompassing membrane. By contrast, the sub-acrosomal region of the nucleus appears to be drawn into the egg within a "phagocytic" vacuole, since this part of the fertilizing sperm head within the rabbit and rat egg is definitely covered by two distinct membranes (Fig. 14, 15) (Piko, 1969; Bedford, 1970b). The inner membrane almost certainly represents the persistent inner membrane of the acrosome including that of the "equatorial" segment, while the outer membrane is probably derived from egg

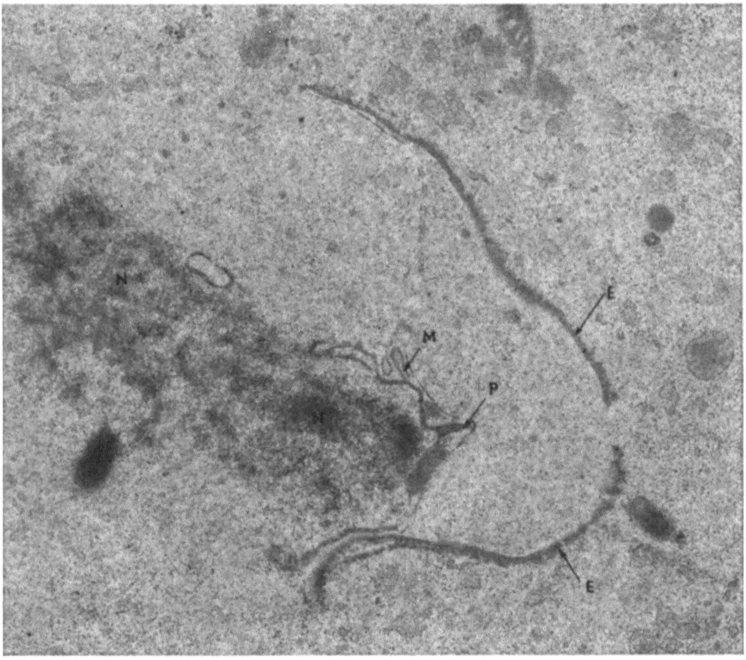

Fig. 15. Section taken from rabbit egg recovered 13 hours after ovulation. This shows the partially transformed fertilizing sperm head within the ooplasm. N. decondensed nucleus. M. membranes of sperm head. E. remains of equatorial segment. P. perforatorium

plasma membrane carried in around the sperm. Thus, in the writer's present estimate, incorporation of the sperm head into mammalian egg is a twofold process involving membrane fusion over the caudal half and ingestion of the anterior sub-acrosomal region within a membrane-limited vesicle. Why the mammalian sperm head should enter the egg in such a complex manner is not understood. The phagocytized membranes persist within the ooplasm throughhout the nuclear swelling or decondensation stage (Fig. 15) which requires approximately 1 hour, but their function and their fate thereafter have not been determined. One may also ask why the sperm first associates with the egg surface by its post-acrosomal region? This is not certain, but it is worthy of note that, after initial dispersion of the sperm plasma membrane covering this region, the egg is then immediately exposed to the active material

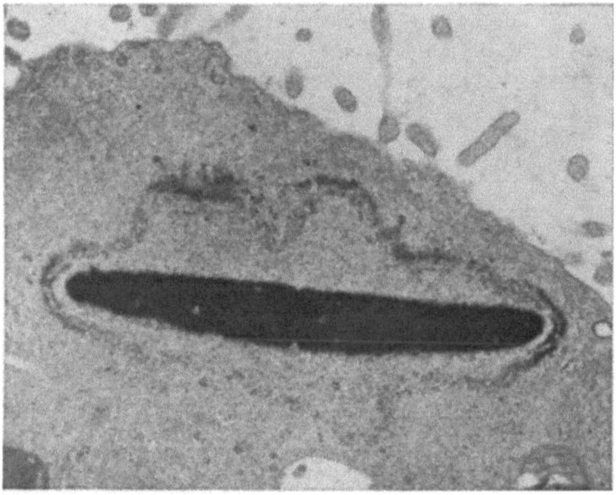

Fig. 16. Posterior region of fertilizing sperm head showing the diffuse halo of material situated originally in the post-acrosomal region (the post nuclear cap of the light microscopist)

which constitutes the "post-nuclear cap" of the light microscopist (Fig. 16). This enzymically active material lying between the nuclear membrane and plasma membrane extends rostrally beneath the acrosome cap (NICANDER and BANE, 1966; BEDFORD, 1967d; FAWCETT and PHILLIPS, 1969) and appears to be progressively revealed to the ooplasm as the ingested membranes are reflected away from the anterior half of the nucleus. Peri-nuclear material may play some part in activation of the egg, or in the "nucleolytic" process, but such speculation must await some element of solid evidence.

The final paragraphs of this discussion are concerned with the fate of the sperm within the egg. In many species the organelles of the sperm tail enter the vitellus with the head, but they probably do not contribute further. Rat sperm mitochondria may persist after the four-cell stage within the ooplasm (Szollosi, 1965), but it is doubtful that their DNA survives and replicates in the embryo

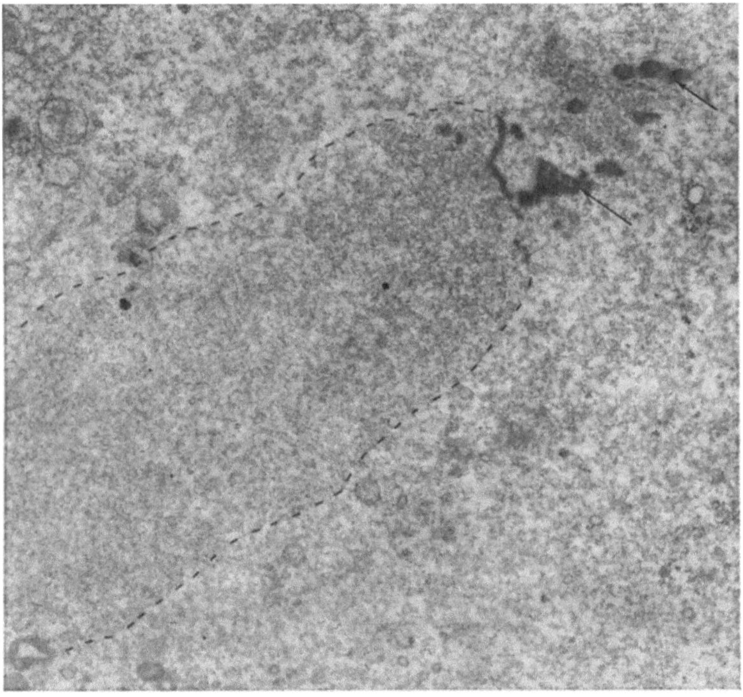

Fig. 17. Organelles of the neck region (arrowed) still in apposition to the decondensed nucleus of the fertilizing sperm. Note that the nucleus (outlined) hardly differs in texture at this stage from that of the ooplasm

(Merker and Neubert, 1969). The swelling phase of the fertilizing sperm nucleus marks a dramatic transformation of the dense chromatin which first becomes highly condensed in the late spermatid, and remains visibly unchanged until contact is made with the ooplasm. The initial erosion of the mid-region of the nucleus extends in both directions until, ultimately, the whole nucleus appears swollen. At this time the intact "perforatorium" can be seen within the egg (Fig. 15) and with the phase microscope may sometimes be observed in recently fertilized rodent ova. Viewed in the phase contrast microscope, the swollen sperm head becomes

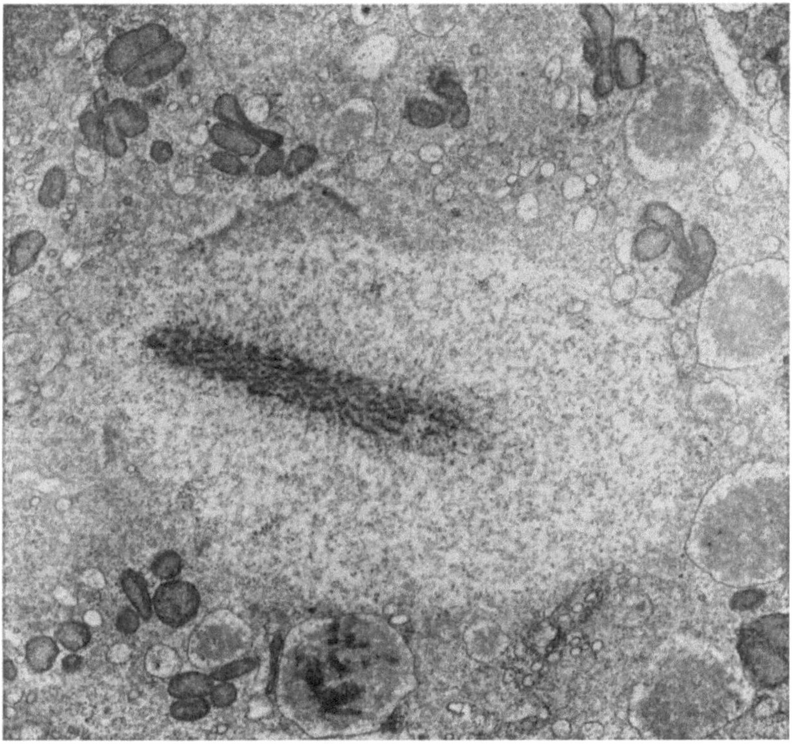

Fig. 18. Stage in decondensation of the nucleus of the fertilizing sperm (rabbit) within the ooplasm

virtually invisible for a short period before the eventual appearance of a definitive male pronucleus (Chang and Hunt, 1962). In the electron microscope, on the other hand, its identity can be followed sequentially, though at the final swelling stage there is sometimes relatively little visible distinction between the sperm nucleoplasm and the surrounding ooplasm (Fig. 17). The factors in the ovum which bring about decondensation of the nucleus (Fig. 18) have not been identified. However, the finding that the stability of the nucleus involves di-sulphide linkages and that uniform swelling and

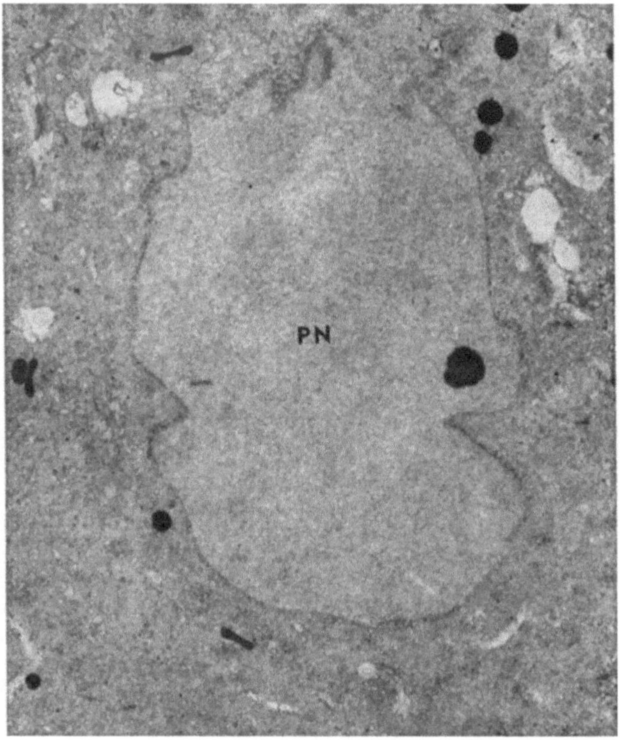

Fig. 19a

Fig. 19. Figures a and b show different magnifications of the male pronucleus (PN) and dense nucleoli (NL) which are formed 4—5 hours after entry of the fertilizing sperm

disintegration of the sperm nucleus can be achieved *in vitro* in media containing Cleland's reagent – di-thiothreitol (Calvin and Bedford – unpublished) may mean that the egg must bring about cleavage of di-sulphide bonds within the nuclear chromatin.

After sperm penetration in the sea urchin, breakdown of the nuclear envelope is incomplete and portions of the nuclear membrane remaining over the apical and caudal regions of the head contribute to the newly-formed limiting membrane of the male

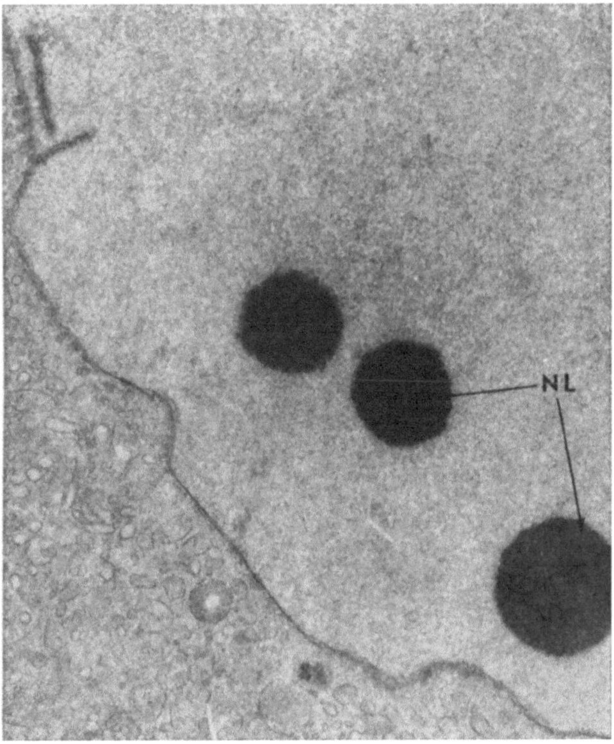

Fig. 19b

pronucleus (Longo and Anderson, 1968). In the rabbit, however, observations by the writer suggest that none of the sperm nuclear envelope remains after the swelling phase of chromatin decondensation, and that the limiting membrane of the presumptive male pronucleus arises *de novo*. Subsequent stages of maturation of the rabbit pronucleus in the succeeding 2 to 3 hours involve the elaboration of membrane, which in early stages may become highly invaginated. This is followed by swelling, with the appearance of rounded dense nucleoli (Fig. 19).

The subsequent events leading to syngamy have been described for mammals in the light microscope (see: Austin and Walton, 1959) but, with the exception of a study reported in abstract by Longo and Anderson (1969), there has been little investigation at the ultrastructural level. It has been noted in the phase microscope that during pronucleus formation the organelles of the neck and midpiece often remain in close proximity to the male pronucleus. Nonetheless, there is no obvious aster formation such as has been described in the sea urchin (Longo and Anderson, 1968), and the nature of the vectors which bring about its central migration is not certain. In the rabbit, after the male and female pronucleus come into apposition, at about 6 hours after sperm penetration, the adjacent margins apparently first extrude buds of nucleoplasm and later produce interdigitating surface extrusions. Mixing of the male and female genome first occurs when points of fusion and vesicles develop between the co-axial pronuclear membranes (Longo and Anderson, 1969) in a manner similar to that seen in the mammalian acrosome reaction (Barros et al., 1967). At the same time, nuclear material in this region condenses into clumps and, as seen in the electron microscope, these become focal points for micro-tubules which enter the disrupted nuclear membrane from the ooplasm. These are the presumptive spindle fibers of the first mitotic division (Longo and Anderson, 1969). The union of the male and female elements at this time marks the end of fertilization.

Acknowledgment

Unpublished work discussed in this review was supported by NIH Grant HD-03623.

References

Adams, C. E., Chang, M. C.: Capacitation of rabbit spermatozoa in the Fallopian tube and in the uterus. J. exp. Zool. **151**, 159 (1962).

Ahlgren, M.: Migration of spermatozoa to the Fallopian tubes and the abdominal cavity in women including some immunological aspects. Lund: Studentliteratur 1969.

Amersbach, R.: Sterilität und Frigidität. Münch. med. Wschr. **77**, 225 (1930).

Austin, C. R.: Function of hyaluronidase in fertilization. Nature (Lond.) **162**, 63 (1948a).

AUSTIN, C. R.: Number of sperm required for fertilization. Nature (Lond.) **162**, 534 (1948b).
— Fertilization and the transport of gametes in the pseudo-pregnant rabbit. J. Endocr. **6**, 293 (1949).
— Observations on the penetration of the sperm into the mammalian egg. Aust. J. Sci. Res. Ser. B. **4**, 581 (1951).
— The capacitation of the mammalian sperm. Nature (Lond.) **170**, 326 (1952).
— Fate of spermatozoa in the uterus of the mouse and rat. J. Endocr. **14**, 335 (1957).
— Capacitation and release of hyaluronidase. J. Reprod. Fertil. **1**, 310 (1960a).
— Fate of spermatozoa in the female genital tract. J. Reprod. Fertil. **1**, 151—156 (1960b).
— Ultrastructure of Fertilization. New York: Holt, Rinehart and Winston 1968.
— BISHOP, M. W. H.: Role of the rodent acrosome and perforatorium in fertilization. Proc. roy. Soc. B **148**, 234 (1958).
— BRADEN, A. W. H.: Time relations and their significance in the ovulation and penetration of eggs in rats and rabbits. Aust. J. Biol. Sci. **7**, 195 (1954).
— WALTON, A.: Fertilisation. Chap. 10. Marshall's Physiology of Reproduction. Vol. I., Part 2. Ed. by A. S. PARKES. London: Longman, Green and Co. 1959.
BANGHAM, A. D.: Electrophoretic characteristics of ram and rabbit spermatozoa. Proc. roy. Soc. B **155**, 292 (1961).
BARROS, C.: *In vitro* capacitation of golden hamster spermatozoa with Fallopian tube fluid of the mouse and rat. J. Reprod. Fertil. **17**, 203 (1968).
— AUSTIN, C. R.: *In vitro* fertilization and the sperm acrosome reaction in the hamster. J. exp. Zool. **166**, 317 (1967).
— BEDFORD, J. M., FRANKLIN, L. E., AUSTIN, C. R.: Membrane vesiculation as a feature of the mammalian acrosome reaction. J. Cell Biol. **34**, C1 (1967).
— FRANKLIN, L. E.: Behaviour of the gamete membranes during sperm entry into the mammalian egg. J. Cell Biol. **37**, C13 (1968).
BAVISTER, B. D.: Environmental factors important for *in vitro* fertilization in the hamster. J. Reprod. Fertil. (Abstr.) **6**, 245 (1968).
— EDWARDS, R. G., STEPTOE, P. C.: Identification of the mid-piece and tail of the spermatozoon during fertilization of human eggs *in vitro*. J. Reprod. Fertil. **20**, 159 (1969).
BEDFORD, J. M.: Effect of environment on phagocytosis of rabbit spermatozoa. J. Reprod. Fertil. **9**, 249 (1965).
— Importance of capacitation for establishing contact between eggs and sperm in the rabbit. J. Reprod. Fertil. **13**, 365 (1967a).
— Experimental requirement for capacitation and observations on ultrastructural changes in rabbit spermatozoa during fertilization. J. Reprod. Fertil. Suppl. **2**, 35 (1967b).
— Fertile life of rabbit spermatozoa in the rat uterus. Nature (Lond.) **213**, 1097 (1967c).

Bedford, J. M.: Observations on the fine structure of spermatozoa of the Bush Baby (*Galago Senegalensis*), the African Green Monkey (*Cercopithecus Aethiops*) and man. Amer. J. Anat. **121**, 443 (1967d).
— Ultrastructural changes in the sperm head during fertilization in the rabbit. Amer. J. Anat. **123**, 329 (1968).
— Limitations of the uterus in the development of the fertilizing ability (capacitation) of spermatozoa. J. Reprod. Fertil. Suppl. 8, 19 (1969a).
— Morphological aspects of capacitation. In: Advances in Biosciences. 4. Schering Symposium on Mechanisms Involved in Conception: Berlin, 1969, p. 36. London: Pergamon Press 1969b.
— The influence of oestrogen and progesterone on sperm capacitation in the female reproductive tract of the rabbit. J. Endocr. **46**, 191 (1970a).
— Sperm capacitation and fertilization in mammals. Biol. Reprod. Suppl. **2**, 128 (1970b).
— The limited importance for fertility of the vaginal sperm reserve in the rabbit. J. Reprod. Fertil. (In press). (1970c).
— Shalkovsky, S.: Species-specificity of sperm capacitation in the rabbit. J. Reprod. Fertil. **13**, 361 (1967).
Bellve, A. R., MacDonald, M. T.: Directional flow of Fallopian tube secretion in the Romney ewe. J. Reprod. Fertil. **15**, 357 (1968).
Black, D. L., Asdell, S. A.: Transport through the rabbit oviduct. Amer. J. Physiol. **192**, 63 (1958).
Blandau, R. J.: Biology of eggs and implantation. Chap. 14. Sex and internal secretions. Vol. II. Ed. by W. C. Young. Baltimore: Williams and Wilkins Co. (1961).
— Money, W. L.: Observations of the rate of transport of spermatozoa in the female genital tract of the rat. Anat. Rec. **90**, 255 (1944).
— Odor, D. L.: The total number of spermatozoa reaching various segments of the reproductive tract of the female albino rat at intervals after insemination. Anat. Rec. **103**, 93 (1949).
Boyarsky, L. H., Bayliss, H., Casida, L. E., Meyes, R. K.: Influence of progesterone upon the fertility of gonadotrophin-treated female rabbits. Endocrinology **41**, 312 (1947).
Braden, A. W. H.: Distribution of sperms in the genital tract of the female rabbit after coitus. Aust. J. Biol. Sci. **6**, 693 (1953).
— Sperm penetration and fertilization in the mouse. Symposio Genetica et Biologica Italia. Vol. IX. May, 1959. p. 1 (1959).
— Austin, C. R.: The number of sperms about the eggs in mammals and its significance for normal fertilization. Aust. J. Biol. Sci. **7**, 543 (1954).
— Gluecksohn-Waelsch, S.: Further studies of the effect of the T locus in the house mouse on male fertility. J. exp. Zool. **138**, 431 (1958).
Burkhardt, J.: Sperm survival in the genital tract of the mare. J. Agricult. Sci. **39**, 201 (1949).
Bygdeman, M.: The effect of different prostaglandins on human myometrium, *in vitro*. Acta physiol. scand. **63**, Suppl. 242, 5 (1964).
— Eliasson, R.: The effect of prostaglandin from human seminal fluid on the motility of the non-pregnant human uterus *in vitro*. Acta physiol. scand. **59**, 43 (1963).

CALVIN, H. I., BEDFORD, J. M.: Formation of disulfide bonds in the nucleus and accessory structures of mammalian spermatozoa during maturation in the epididymis. Biol. Reprod. (in press) (1970).
CARY, W. H.: Duration of sperm cell migration and uterine secretions. J. Amer. med. Ass. **106**, 2221 (1936).
CHANG, M. C.: Fertilizing capacity of spermatozoa deposited into the Fallopian tubes. Nature (Lond.) **168**, 697 (1951 a).
— Fertilization in relation to the number of spermatozoa in the Fallopian tubes of rabbits. Ann. Ostet. Ginec. **73**, 918 (1951 b).
— Development of fertilizing capacity of rabbit spermatozoa in the uterus. Nature (Lond.) **175**, 1036 (1955).
— A detrimental effect of rabbit seminal plasma on the fertilizing capacity of sperm. Nature (Lond.) **179**, 258 (1957).
— Capacitation of rabbit spermatozoa in the uterus with special reference to the reproductive phases of the female. Endocrinology **63**, 619 (1958).
— Aspects of mammalian fertilization and their vulnerability. Proceedings of the Sixth International Conference on Planned Parenthood: New Delhi, 1959, p. 129 (1959).
— Fertilizing life of ferret sperm in the female tract. J. exp. Zool. **158**, 87 (1965).
— Effect of progesterone and related compounds on fertilization, transportation and development of rabbit eggs. Endocrinology **81**, 1251 (1967).
— BEDFORD, J. M.: Effects of various hormones on transportation of gametes and fertilization in the rabbit. Proceedings of IV th International Congress of Animal Reproduction. The Hague. Vol. II, 367 (1961).
— HANCOCK, J. L.: Experimental hybridisation. In: Comparative aspects of reproductive failure. Ed. by K. BENIRSCHKE. Berlin-Heidelberg-New York: Springer 1967.
— HUNT, D. M.: Effect of proteolytic enzymes on the zona pellucida of fertilized and unfertilized mammalian eggs. Exp. Cell Res. **11**, 497 (1956).
— — Morphological changes of sperm in the ooplasm of mouse, rat, hamster and rabbit. Anat. Rec. **142**, 417 (1962).
— PINCUS, G.: Fertilizable life of rabbit sperm deposited into different parts of the female tract. 5th International Congress of Animal Reproduction. Vol. IV. Trento, p. 377 (1964).
— SHEAFFER, D.: Numbers of spermatozoa ejaculated at copulation, transported into the female and present in the male tract of the golden hamster. J. Hered. **48**, 107 (1957).
— YANAGIMACHI, R.: Fertilization of ferret ova by deposition of epididymal sperm into the ovarian capsule with special reference to the fertilizable life of ova and the capacitation of sperm. J. exp. Zool. **154**, 175 (1963).
COGGINS, E. G., BAKER, R. D.: Survival and transport of foreign spermatozoa in the genital tract of the rabbit. 6th International Congress of Animal Reproduction. Vol. I. Paris. July, 1968, p. 47 (1968).
COHEN, M. R., STEIN, I. F.: Sperm survival at the estimated ovulation time. Fertil. and Steril. **2**, 20 (1950).
COLWIN, L. H., COLWIN, A. H.: Membrane fusion in relation to sperm-egg association. Chap. 7. Fertilization. Vol. I. Eds. METZ, C. B., MONROY, A. London-New York: Academic Press 1967.

Corfmann, P. A., Segal, S. J.: Biological effects of intrauterine devices. Amer. J. Obstet. Gynec. **100**, 448 (1968).
Dauzier, L.: Recherches sur les facteurs de la remontee des spermatozoides dans les voies genitales femelles (trompes de Fallope): Etude chez la brebis. C. R. Soc. Biol. (Paris) **149**, 1941 (1955).
— Wintenberger, S.: Recherches sur la fecondation chez les Mammiferes: duree du pouvoir fecondant des spermatozoides de belier dans le tractus genital de la brebis et duree de la periode de fecondite de l'oeuf apres l'ovulation. C. R. Soc. Biol. (Paris) **146**, 661 (1952a).
— — Recherches sur la fecondation chez les Mammiferes: la remontee des spermatozoides dans le tractus genital de la brebis en dioestrus et anoestrus. C. R. Soc. Biol. (Paris) **146**, 663 (1952b).
Day, F. T.: Survival of spermatozoa in the genital tract of the mare. J. Agricult. Sci. **32**, 108 (1942).
Dickmann, Z.: Denudation of the rabbit egg: time sequence and mechanism. Amer. J. Anat. **113**, 303 (1963a).
Doak, R. L., Hall, A., Dale, H. E.: Longevity of spermatozoa in the reproductive tract of the bitch. J. Reprod. Fertil. **13**, 51 (1967).
Dott, H., Chang, M. C.: Fertilizing life of ejaculated and epididymal rabbit sperm in the female tract. Anat. Rec. **154**, 459 (1966).
Dukelow, W. R., Chernoff, H. N., Williams, W. L.: Properties of decapacitation factor and presence in various species. J. Reprod. Fertil. **14**, 393 (1967).
Edgar, D. G., Asdell, S. A.: Spermatozoa in the female genital tract. J. Endocr. **21**, 321 (1960).
Edwards, R. G., Bavister, B. D., Steptoe, P. C.: Early stages of *in vitro* fertilization of human oocytes matured *in vitro*. Nature (Lond.) **221**, 632 (1969).
Egli, G. E., Newton, M.: The transport of carbon particles in the human female reproductive tract. Fertil. and Steril. **12**, 151 (1961).
Eliasson, R.: Studies on prostaglandin (occurrence, formation and biological actions). Acta physiol. scand. **46**, Suppl. 158, 1 (1959).
Farris, E. J.: Human fertility and problems of the male. New York: Authors Press, White Plains 1950.
Fawcett, D. W., Phillips, D. M.: Observations on the release of spermatozoa and on changes in the head during passage through the epididymis. J. Reprod. Fertil. Suppl. **6**, 405 (1969).
Ferin, J.: Facteurs hormonaux et migration des spermatozoides dans l'uterus chez la femme. Ann. Endocr. **9**, 77 (1948).
Fjallbrant, B.: Sperm antibodies and sterility in men. Acta obstet. gynec. sacnd. **47**, Suppl. 4, 1 (1968).
— Cervical mucus penetration by human spermatozoa treated with antispermatozoal antibodies from rabbit and man. Acta obstet. gynec. scand. **47**, 1 (1969).
Fox, C. A., Knaggs, G. S.: Milk ejection activity (oxytocin) in peripheral venous blood in man during lactation and in association with coitus. J. Endocr. **45**, 145 (1969).

HAFEZ, E. S. E., BLACK, D. L. The mammalian utero-tubal junction. Chap. 4. The mammalian oviduct. Eds. HAFEZ, E. S. E. BLANDAU, R. J.. Chicago: University of Chicago Press 1969.

HAMMOND, J.: The physiology of milk and butter fat secretion. Veterinary Rec. **48**, 519 (1936).

— ASDELL, S. A.: The vitality of the spermatozoa in the male and female reproductive tract. J. exp. Biol. **4**, 155 (1926).

— WALTON, A.: Notes on ovulation and fertilization in the ferret. J. exp. Biol. **11**, 307 (1934).

HAMNER, C. E., JONES, J. P., SOJKA, N. J.: Influence of the hormonal state of the female on the fertilizing capacity of rabbit spermatozoa. Fertil. and Steril. **19**, 137 (1968).

— SOJKA, N. J.: Capacitation of rabbit spermatozoa; species and organ specificity. Proc. Soc. exp. Biol. (N. Y.). **124**, 689 (1967).

— WILLIAMS, W. L.: Effect of the female reproductive tract on sperm metabolism in the rabbit and fowl. J. Reprod. Fertil. **5**, 143 (1963).

HANCOCK, J. L., McGOVERN, P. T.: The transport of sheep and goat spermatozoa in the ewe. J. Reprod. Fertil. **15**, 283 (1968).

HARPER, M. J. K.: Factors influencing sperm penetration of rabbit eggs *in vivo*. J. exp. Zool. **173**, 47 (1970).

HARRIS, G. W., PICKLES, V. P.: Reflex stimulation of the neurohyppophysis (posterior pituitary gland) and the nature of posterior pituitary hormone(s). Nature (Lond.) **172**, 1049 (1953).

HARTMAN, C. G.: How do sperms get into the uterus? Fertil. and Steril. **8**, 403 (1957).

HARTREE, E. F., SRIVASTAVA, P. N.: Chemical composition of the acrosomes of ram spermatozoa. J. Reprod. Fertil. **9**, 47 (1965).

HAWK, H. H.: Inhibition of ovum fertilization in the ewe by intra-uterine plastic spirals. J. Reprod. Fertil. **10**, 267 (1965).

HAYNES, N. B.: The influence of the uterine environment on the phagocytosis of spermatozoa. In: Reproduction in the female mammal, p. 500. Eds. LAMMING, G. E., AMOROSO, E. C. London: Butterworths 1967.

HAYS, R. L., VAN DEMARK, N. L.: Effects of oxytocin and epinephrine on uterine motility in the bovine. Amer. J. Physiol. **172**, 557 (1953).

HORNE, H. W., THIBAULT, J. P.: Sperm migration through the human female reproductive tract. Fertil. and Steril. **13**, 135 (1962).

HOWE, G. R.: Leucocytic response to spermatozoa in ligated segments of the rabbit vagina, uterus and oviduct. J. Reprod. Fertil. **13**, 563 (1967).

HUNTER, R. H. F.: Attempted fertilization of hamster eggs following transplantation into the uterus. J. exp. Zool. **168**, 511 (1968).

— Capacitation in the golden hamster with special reference to the influence of the uterine environment. J. Reprod. Fertil. **20**, 223 (1969).

— DZIUK, P.: Sperm penetration of pig eggs in relation to timing of ovulation and insemination. J. Reprod. Fertil. **15**, 199 (1968).

IRITANI, A., GOMES, W. R., VAN DEMARK, N. L.: The effect of whole, dialysed and heated female genital tract fluids on respiration of rabbit and rat spermatozoa. Biol. Reprod. **1**, 77 (1969).

IWAMATSU, T., CHANG, M. C.: *In vitro* fertilization of mouse eggs in the presence of bovine follicular fluid. Nature (Lond.) **224**, 919 (1969).

KILLINGBECK, J., HAYNES, B. N., LAMMING, G. E.: Influence of isolated components of the uterine secretion on phagocytosis. Nature (Lond.) **199**, 255 (1963).

KUMAR, D.: Hormonal regulation of myometrial activity: clinical implications. Chap. 12. Cellular biology of the uterus. Ed. by WYNN, R. M., New York N. Y.: Appleton Century Crofts 1967.

LEONARD, S. L.: The reduction of uterine sperm and uterine fluid on fertilization of rat ova. Anat. Rec. **106**, 607 (1950).

— PERLMAN, P. L.: Conditions affecting the passage of spermatozoa through the utero-tubal junction of the rat. Anat. Rec. **104**, 89 (1949).

LEWIS, W. H., WRIGHT, E. S.: On the early development of the mouse egg. Carnegie Institute Contributions to Embryology **25**, 113 (1935).

LIGHTFOOT, R. J., CROKER, K. P., NEIL, H. G.: Failure of sperm transport in relation to ewe infertility following prolonged grazing on oestrogenic pastures. Aust. J. Agricult. Res. **18**, 755 (1967).

LONGO, F. J., ANDERSON, E.: The fine structure of pronuclear development and fusion in the sea urchin. Arbacia punctulata. J. Cell Biol. **39**, 339 (1968).

— — Fertilization of the rabbit egg: association of the maternal and paternal genomes. Anat. Rec. **163**, 314 (Abstract) (1969).

MANN, T., POLGE, C., ROWSON, L. E. A.: Participation of seminal plasma during passage of sperm in the female reproductive tract of the pig and horse. J. Endocr. **13**, 133 (1956).

MARSTON, J. H., KELLY, W. A.: Time relationships of spermatozoon penetration into the egg of the rhesus monkey. Nature (Lond.) **217**, 1073 (1968).

MARTINEZ-MANATAOU, J., JINER-VELAZQUES, J., AZUAS-RAMOS, M., LOZANO-BALDERAS, M., RUDEL, W. H.: Continuous administration of 500 μg of chlormadinone acetate as a method of regulating fertility without inhibiting ovulation. Proceedings of the 8th International Congress of the International Planned Parenthood Federation. Santiago, Chile, p. 241 (1967).

MATTNER, P. E.: Capacitation of ram spermatozoa and penetration of the ovine egg. Nature (Lond.) **199**, 772 (1963a).

— Spermatozoa in the genital tract of the ewe. II. Distribution after coitus. Aust. J. Biol. Sci. **16**, 688 (1963b).

— Spermatozoa in the genital tract of the ewe. III. Role of spermatozoon motility and of uterine contractions in transport of spermatozoa. Aust. J. Biol. Sci. **16**, 877 (1963c).

— Formation and retention of the spermatozoon reservoir in the cervix of the ruminant. Nature (Lond.) **212**, 1479 (1966).

— The survival of spermatozoa in bovine cervical mucus and mucus fractions. Reprod. Fertil. **20**, 193 (1969).

— BRADEN, A. W. H.: Spermatozoa in the genital tract of the ewe. I. Rapidity of transport. Aust. J. Biol. Sci. **16**, 473 (1963).

— — Comparison of the distribution of motile and immotile spermatozoa in the ovine cervix. Aust. J. Biol. Sci. **22**, 1069 (1969).

MATTNER, P. E., BRADEN, A. W. H.: Effect of time of insemination on the distribution of spermatozoa in the genital tract in ewes. Aust. J. Biol. Sci. (in press) (1970).

McGAUGHEY, R. W., MARSTON, J. H., CHANG, M. C.: Fertilizing life of mouse spermatozoa in the female tract. J. Reprod. Fertil. **16**, 147 (1968).

MENGE, A. C., TYLER, W. J., CASIDA, L. E.: Factors affecting the removal of spermatozoa from the rabbit uterus. J. Reprod. Fertil. **3**, 396 (1962).

MERKER, H. J., NEUBERT, D.: Das elektronenmikroskopische Bild der Rattenspermien nach der Befruchtung. Z. Zellforsch. Abt. Histochem. **95**, 594 (1969).

MERTON, H.: Studies on reproduction in the albino mouse. III. The duration of life of spermatozoa in the female reproductive tract. Proc. roy. Soc. Edinburgh **59**, 207 (1939).

METZ, C. B., ANIKA, J.: Failure of conception in rabbits inseminated with non-agglutinating, univalent antibody treated semen. Biol. Reprod. **2**, 284 (1970).

MOENCH, G. L.: The longevity of the human spermatozoa. Amer. J. Obstet. Gynec. **38**, 153 (1939).

MOYER, D. L., KUNITAKE, G. M., NAKAMURA, R. M.: Electron microscopic observations on phagocytosis of rabbit spermatozoa in the female genital tract. Experientia (Basel) **21**, 6 (1965).

MURPHREE, R. L., WARWICK, E. J., CASIDA, L. E., McSHAN, W. H.: Influence of reproductive stage upon the fertility of gonadotrophin-treated female rabbits. Endocrinology **41**, 308 (1947).

NELSON, L.: Sperm motility. Chap. 2. Fertilization. Vol. I. Eds. METZ, C. B., MONROY, A. London-New York: Academic Press 1967.

NEVO, A. C., MICHAELI, I., SCHINDLER, H.: Electrophoretic properties of bull and rabbit spermatozoa. Exp. Cell Res. **23**, 69 (1961).

— MOHAN, R.: Migration of motile spermatozoa into sperm free medium and the dilution effect. J. Reprod. Fertil. **18**, 379 (1969).

NICANDER, L., BANE, A.: Fine structure of the sperm head in some mammals with particular reference to the acrosome and the sub-acrosomal substance. Z. Zellforsch. Abt. Histochem. **72**, 495 (1966).

NOYES, R. W.: The fertilizing capacity of spermatozoa. West. J. Surg. **61**, 342 (1953).

— WALTON, A., ADAMS, C. E.: Capacitation of rabbit spermatozoa. J. Endocr. **7**, 374 (1958).

— ADAMS, C. E., WALTON, A.: Transport of spermatozoa into the uterus of the rabbit. Fertil. and Steril. **9**, 288 (1958).

— — — The passage of spermatozoa through the genital tract of female rabbits after ovariectomy and oestrogen treatment. J. Endocr. **18**, 165 (1959).

— THIBAULT, C.: Endocrine factors in the survival of spermatozoa in the female reproductive tract. Fertil. and Steril. **13**, 346 (1962).

ODEBLAD, E.: The functional structure of human cervical mucus. Acta obstet. gynec. scand. **47**. Suppl. 1, 57 (1968).

OLDS, P. J.: Effect of the T locus on sperm distribution in the house mouse. Biol. Reprod. **2**, 91 (1970).

Orgebin-Crist, M.: Studies on the function of the epididymis. Biol. Reprod. Suppl. **1**, 155 (1969).

Parker, G. H.: The passage of sperms and eggs through the oviducts in terrestrial vertebrates. Phil. Trans. B **219**, 381 (1931).

Perloff, R., Steinberger, E.: *In vivo* survival of sperm in cervical mucus. Amer. J. Obstet. Gynec. **88**, 439 (1964).

Piko, L.: Mechanism of sperm penetration in the rat and the Chinese hamster based on fine structural studies. 5th International Congress of Animal Reproduction. Vol. VII, Trento, p. 301 (1964).

— Gamete structure and sperm entry in mammals. Chap. 8. Fertilization. Vol. II. Eds. Metz, C. B., Monroy, A. London-New York: Academic Press 1969.

Potter, R. C.: Length of the fertile period. Milbank mem. Fd. Quart. **39**, 132 (1961).

Rigby, J. P.: The persistence of spermatozoa at the utero-tubal junction of the sow. J. Reprod. Fertil. **11**, 153 (1966).

Roberts, J. S., Share, L.: Oxytocin in plasma of pregnant, lactating and cycling ewes during vaginal stimulation. Endocrinology **83**, 272 (1968).

Robinson, T. J.: Control of the ovarian cycle in sheep. In: Reproduction in the female mammal, p. 373. Eds. Lamming, G. E., Amoroso, E. C.. London: Butterworths 1967.

— The synchronisation of the oestrous cycle and fertility. 6th International Congress of Animal Reproduction, Vol. II. p. 1347 (1968).

Rowson, L. E. A.: The movement of radio-opaque material in the bovine uterine tract. Brit. Vet. J. **111**, 334 (1955).

Rubenstein, B. B., Straus, H., Lazarus, M. L., Hawkins, H.: Sperm survival in women. Fertil. and Steril. **2**, 15 (1951).

Schumacher, G. F. B.: Biochemical and biophysical properties of cervical mucus in different hormonal states. In: Advances in biosciences, 4 — Schering symposium on mechanisms involved in conception: Berlin, 1969, p. 95. London: Pergamon Press 1969.

Smithberg, M.: The effect of different proteolytic enzymes on the zona pellucida of mouse ova. Anat. Rec. **117**, 554 (1953). (Abstr.).

Sobrero, A. J., MacLeod, J.: The immediate postcoital test. Fertil. and Steril. **13**, 184 (1962).

Soderwall, A. L., Blandau, R. J.: The duration of the fertilizing capacity of spermatozoa in genital tract of the rat. J. exp. Zool. **88**, 55 (1941).

Soupart, P.: Studies on the hormonal control of rabbit sperm capacitation. J. Reprod. Fertil. Suppl. **2**, 49 (1967).

— Orgebin-Crist, M. C.: Capacitation of rabbit spermatozoa delayed *in vivo* by double ligation of uterine horn. J. exp. Zool. **163**, 311 (1966).

Srivastava, P. N., Adams, C. E., Hartree, E. F.: Enzymic action of acrosomal preparations on the rabbit ovum, *in vitro*. J. Reprod. Fertil. **10**, 61 (1965).

Stambaugh, R., Buckley, J.: Zona pellucida dissolution enzymes of the rabbit sperm head. Science **161**, 585 (1968).

STAMBAUGH, R., BUCKLEY, J.: Identification and subcellular localization of the enzymes effecting penetration of the zona pellucida by rabbit spermatozoa. J. Reprod. Fertil. **19**, 423 (1969).

STARKE, N. C.: The sperm picture of rams of different breeds as an indication of their fertility. II. The rate of sperm travel in the genital tract of the ewe. Onderstepoort J. Vet. Sci. **22**, 415 (1949).

STEFANINI, M., OURA, C., ZAMBONI, L.: Ultrastructure of fertilization in the mouse. 2. Penetration of sperm into the ovum. J. Submicr. Cytol. **1**, 1 (1969).

— — — Ultrastructure of fertilization in the mouse. I. The content of the ampulla prior to sperm penetration. J. Reprod. Fertil. (in press) (1970).

STRAUSS, F.: The time and place of fertilization of the golden hamster egg. J. Embryol. exp. Morph. **4**, 42 (1956).

SZOLLOSI, D. G.: Cortical granules: a general feature of mammalian eggs. J. Reprod. Fertil. **4**, 223 (1962).

— The fate of sperm middle-piece mitochondria in the rat egg. J. exp. Zool. **159**, 367 (1965).

— RIS, H.: Observations on sperm penetration in the rat. J. Biophys. Biochem. Cytol. **10**, 275 (1961).

TAMPION, D., GIBBONS, R. A.: Orientation of spermatozoa in mucus of the cervix uteri. Nature (Lond.) **194**, 381 (1962).

— — Swimming rate of bull spermatozoa in various media and the effect of dilution. J. Reprod. Fertil. **5**, 259 (1963).

TEICHMANN, R. J., BERNSTEIN, M. H.: Regional differentiation in the head of human and rabbit spermatozoa. Anat. Rec. **163**, 343 (Abstr.) (1969).

TERNER, C.: Oxidation and biosynthetic utilization by human spermatozoa of a metabolite of the female reproductive tract. Nature (Lond.) **208**, 1115 (1966).

TESH, J. M.: Effects of the ageing of rabbit spermatozoa *in utero* on fertilization and pre-natal development. J. Reprod. Fertil. **10**, 299 (1969).

— GLOVER, T. D.: Ageing of rabbit spermatozoa in the male tract and its effect on fertility. J. Reprod. Fertil. **20**, 287 (1969).

THIBAULT, C., WINTENBERGER-TORRES, S.: Oxytocin and sperm transport in the ewe. Int. J. Fertil. **12**, 410 (1967).

TIETZE, C.: Probability of pregnancy resulting from a single unprotected coitus. Fertil. and Steril. **11**, 485 (1960).

TOYODA, Y., CHANG, M. C.: Sperm penetration of rat eggs *in vitro* after dissolution of zona pellucida by chymotrypsin. Nature (Lond.) **220**, 889 (1968).

TURNBULL, K. E.: Transport of spermatozoa in the rabbit doe before and after ovulation. Aust. J. Biol. Sci. **19**, 1095 (1966).

VANDEMARK, N. L., HAYS, R. L.: Uterine motility responses to mating. Amer. J. Physiol. **170**, 518 (1952).

— — Rapid sperm transport in the cow. Fertil. and Steril. **5**, 131 (1954).

— MOELLER, A. N.: Speed of spermatozoan transport in the reproductive tract of the estrous cow. Amer. J. Physiol. **165**, 674 (1951).

VICKERY, B. H., BENNETT, J. P.: The cervix and its secretion in mammals. Physiol. Rev. **48**, 135 (1968).

Von Euler, U. S., Eliasson, R.: Prostaglandins. London-New York: Academic Press 1967.

Walton, A.: On the function of the rabbit cervix during coitus. J. Obstet. Gynec. Brit. Empire **37**, 92 (1930).

Whittingham, D. G.: Fertilization of mouse eggs *in vitro*. Nature (Lond.) **220**, 592 (1968).

Wimsatt, W. A.: Further studies on the survival of spermatozoa in the female reproductive tract of the bat. Anat. Rec. **88**, 193 (1944).

Yanagimachi, R.: Time and process of sperm penetration into hamster ova *in vivo* and *in vitro*. J. Reprod. Fertil. **11**, 359 (1966).

— *In vitro* capacitation of hamster spermatozoa by follicular fluid. J. Reprod. Fertil. **18**, 275 (1969a).

— *In vitro* acrosome reaction and capacitation of golden hamster spermatozoa by bovine follicular fluid and its fractions. J. exp. Zool. **170**, 269 (1969b).

— Chang, M. C.: Sperm ascent through the oviduct of the hamster and rabbit in relation to the time of ovulation. J. Reprod. Fertil. **6**, 413 (1963a).

— — Infiltration of leucocytes into the uterine lumen of the golden hamster during the oestrous cycle and following mating. J. Reprod. Fertil. **5**, 389 (1963b).

— — *In vitro* fertilization of golden hamster ova. J. exp. Zool. **186**, 361 (1964).

— Noda, Y. D.: Behavior of gamete plasma membranes in fertilization in hamsters. 2nd Annual Meeting of the Society for Study of Reproduction. Davis, Calif. Sept., 1969. (Abstr.) (1970).

Yochem, D. E.: Spermatozoon life in the female reproductive tract of guinea pig and rat. Biol. Bull. **56**, 274 (1927).

Young, W. C.: A study of the function of the epididymis. II. The importance of an ageing process in sperm for the length of the period during which fertilizing capacity is retained by sperm isolated in the epididymis of the guinea pig. J. Morphol. Physiol. **48**, 475 (1929).

Zamboni, L.: Ultrastructure of mammalian oocytes and ova. Biol. Reprod. Suppl. **2**, 44 (1970).

Zanartu, J., Pupkin, M., Rosenberg, D., Guerrero, R., Rodriguez-Bravo, R., Garcia-Huidobro, M., Puga, J. A.: Effect of oral continuous progestogen therapy in microdosage on human ovary and sperm transport. Brit. med. J. **1968**, 266.

Zaneveld, L. J. D., Williams, W. L.: A sperm enzyme involved in the penetration of the corona radiata and its inhibition by decapacitation factor. Biol. Reprod. **2**, 363 (1970).

Zimbelman, R. G.: Oral progestogens in cattle for control of the estrous cycle. 6th International Congress of Animal Reproduction, Vol. II. Paris, p. 1385 (1968).

A Contribution to the Biochemistry and Biology of Seminal Plasma

G. Ruhenstroth-Bauer
Max-Planck-Institut für Biochemie, München

With 3 Figures

When mammalian spermatozoa migrate from the caput to the cauda epididymis, besides other changes, the electrical charge of their membranes is increased (Bedford, 1963). Concomitantly these spermatozoa mature gradually on their way so that caudal spermatozoa become capable of fertilization, in different species to a different extent (Salisbury and Van Demark, 1961; Bedford, 1966). Nevertheless, there are still functional differences between these and ejaculated spermatozoa. In rabbits, for example, fertilization by epididymal spermatozoa, as compared with ejaculated ones, is delayed, and embryo mortality in this case is clearly higher (Orgebin-Crist, 1968). It was assumed that these functional differences might depend on an action of seminal plasma on the spermatozoa during and after ejaculation. It is in this context, that we should like to make the following contribution.

In the very beginning we observed that the charge on the membranes of ejaculated spermatozoa was on average about 10 to 15 % lower than that on spermatozoa obtained from the caput epididymis (Fuhrmann et al., 1962; Forrester et al., 1969). When washed ejaculated or epididymal spermatozoa were incubated at 37° with seminal plasma, the charge on their membranes was diminished as well. Hence, we tried to isolate the relevant factor from the seminal plasma (Forrester et al., 1969).

Caput spermatozoa, treated in a standardized manner, were used for assaying. Centrifuged bull seminal plasma was used as starting material for preparing the active principle. By stepwise precipitation with ammonium sulfate, the fraction between

80–100% saturation proved to be active. Surprisingly the reduction of the electrical charge on the membrane of the spermatozoa by this material is greater than calculated from the whole seminal plasma. Moreover, it could be shown that the electrical charge of erythrocytes was reduced as well; furthermore erythrocytes treated with this fraction became reversibly agglutinated. This agglutination was accompanied by cell deformation, which could also be observed after the addition of polylysine (Katchalski et al., 1959). After washing with an excess of physiological saline, the original shape of the erythrocytes was also restored.

From this observation it was assumed that the unknown factor acted as a basic protein, reducing the electrical charge on the cell membrane by adsorption on the cell surface. Purification was tried by means of a CM-Sephadex-C 25 column (Fig. 1). Only peak B

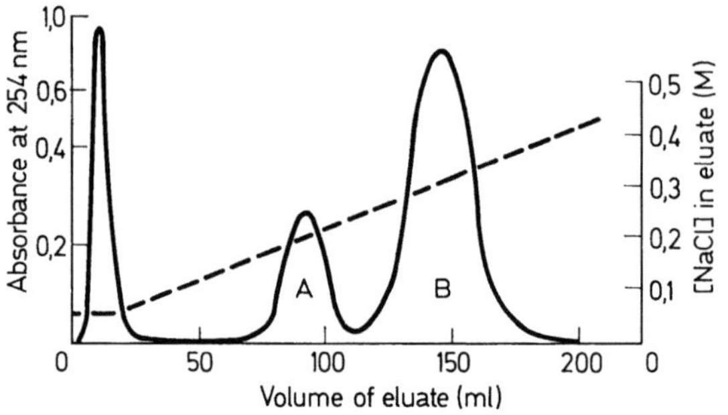

Fig. 1. Elution of material precipitated from bull seminal plasma between 75% and 100% saturation with ammonium sulphate after application to a CM-Sephadex column in 0.067 M phosphate buffer pH 7.2 and development with a linear NaCl gradient

turned out to be active: 50 μg/ml under standard conditions caused erythrocyte agglutination, whereas 25 μg/ml were ineffective. The yield corresponded to an original content in the seminal plasma of about 0.9 mg/ml.

In agarose-gel electrophoresis substance B migrated at pH 8.2 as a single and strongly positively charged peak with a mobility of 0.65 relative to human serum albumin. Electrophoresis on polyacrylamide gel reveals more neighbouring peaks in an amount of not more than 5 % of the main peak. From the eluted volume on a Sephadex G 100-column the molecular weight could be determined to be about 46,000, assuming the substance to be a spherical protein.

The amino acid composition can be seen from Table. As expected, the substance possesses a high concentration of basic amino acids, in particular lysine; almost $^1/_5$ of all the remaining

Table. *Amino acid composition of two samples of the seminal plasma agglutinin*

Amino acid	Sample I moles %	Sample II moles %
Aspartic acid	8.4	9.5
Glutamic acid	7.0	8.4
Glycine	5.5	5.4
Alanine	8.0	6.9
Valine	8.0	8.3
Leucine	1.8	7.2
Isoleucine	2.6	2.3
Phenylalanine	2.6	2.6
Tyrosine	2.6	2.7
Serine	12.4	11.4
Threonine	7.0	7.6
Proline	4.0	4.3
Methionine	4.5	3.9
$^1/_2$ Cystine	6.6	7.1
Arginine	3.6	3.4
Histidine	3.6	3.4
Lysine	11.8	11.3

amino acids are basic ones. On the other hand, about 16.5 mole-% aspartic acid and glutamic acid residues were found, mostly in amide form. Tryptophan could not be detected, and there was no neutral sugar or hexosamine.

By Edman-degradation lysine and alanine were detected as end groups in comparable quantities. So the protein under examination seems to be composed of two chains. The minimal optical rotatory dispersion (Fig. 2) is at 228–230 nm, excluding a greater proportion of an α-helix structure. The same followed from the Cotton effect, where only $^1/_5$ of its values would indicate an α-helix. The high content (Table) of lysine, valine, and serine residues is in accord with these results, as well as the observation that polylysine under similar conditions is β-configurated.

As to the mechanism of the reaction, the basic protein is reversibly adsorbed on the spermatozoan surface at 37°, yet on the contrary not at all at 4°. The negative charge of spermatozoa depends

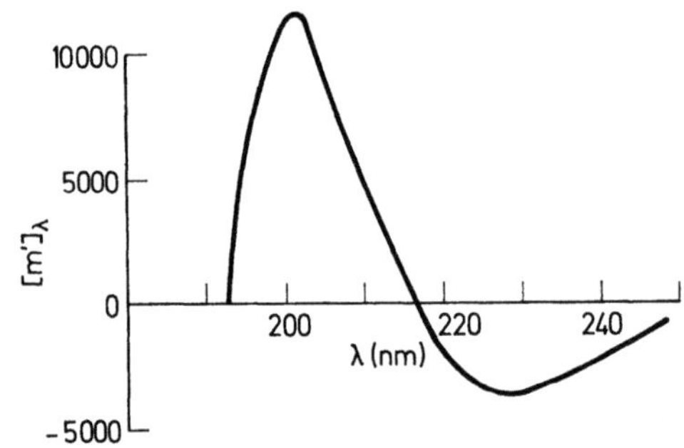

Fig. 2. Optical rotatory dispersion of the protein between 190 and 250 nm

on the presence, among others, of neuraminic acids (Fuhrmann et al., 1963). Therefore, it seemed necessary to state whether the membrane receptors for the basic protein might contain neuraminic acids. Nine different cell types, the electrophoretic mobility of which was reduced by the basic protein, were treated with neuraminidase. In all cases, the electrophoretic mobility of the enzyme-treated spermatozoa after addition of the basic protein remained unchanged. Accordingly the membrane receptors of the basic proteins of the nine cell types presumably contain neuraminic acids.

A small note should be added on the observation that selective x-irradiation of the testicle of rabbits after about 5 days leads to a gradual elevation in the electrophoretic mobility of normally ejaculated spermatozoa of these animals. This effect lasts for a 3- to 5-week period and then gradually reverts to the normal value (Fig. 3) (Ruhenstroth-Bauer and Hertel, 1965). These spermatozoa are not influenced in their electrophoretic mobility by additional normal seminal plasma. Apparently, during the second to the sixth week of spermiogenesis, the electrical surface pattern of the spermatozoan membrane develops. After x-irradiation this changed in a way that basic protein can no longer be bound. Such spermatozoa, nevertheless, are still capable of fertilization.

Rabbits may be immunized against the basic protein of bull seminal plasma. In gel electrophoresis the antiserum yields one peak active against the basic protein, whereas at least two confluent peaks against the whole bull seminal plasma. This could be due to a

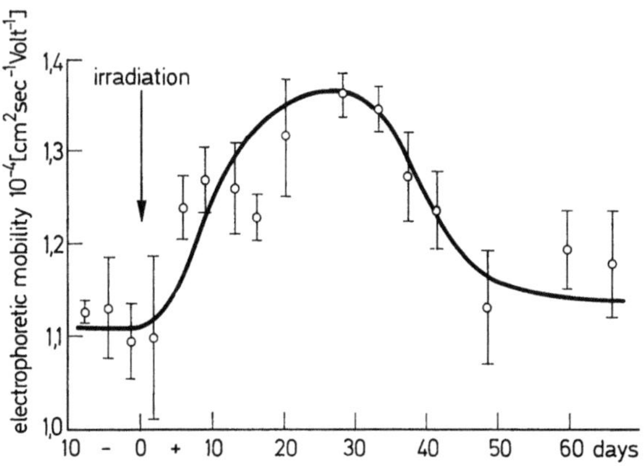

Fig. 3. Electrophoretic mobility of rabbit spermatozoa after a selective X-ray irradiation of testis with 400 R on day O

slight contamination of the immunogens or to a cross reaction from additional proteins in the seminal plasma.

By means of this antiserum it was possible to make antigen plates with which the basic protein in the seminal plasma could be quantitatively determined. With 20 bull specimens the concentration varied between 150–350 mg-%. There was no change in the concentration of the protein between a first and an immediately following second ejaculation. Storage of the seminal plasma at 4° during a 10-day period reduced the concentration of the protein continuously to about 10 %.

To clarify whether the protein was of importance for the fertilizing ability of the ejaculate, several samples of bull seminal plasma were halved. The concentration of the protein was increased until it was doubled and comparison made with the controls regarding the fertilization rate. In more than 400 experiments, the result in both groups was the same within experimental error. So, at the moment there is no proof of a relation between the fertilizing rate of bull semen and its content of basic protein.

References

BEDFORD, J. M.: Nature (Lond.) **200**, 1178 (1963).
— J. exp. Zool. **163**, 319 (1966).
FORRESTER, J. A., ARNOLD, R., RUHENSTROTH-BAUER, G.: Europ. J. Biochem. **11**, 341 (1969).
FUHRMANN, G. F., GRANZER, E., BEY, E., RUHENSTROTH-BAUER, G.: Z. Naturforsch. **18**b, 236 (1963).
KATCHALSKI, A., DANNON, D., NEVO, A., DEVRIES, A.: Biochem. biophys. Acta (Amst.) **33**, 120 (1959).
ORGEBIN-CRIST, M. C.: J. Reprod. Fertil. **16**, 29 (1968).
RUHENSTROTH-BAUER, G., HERTEL, E.: Strahlentherapie **127**, 108 (1965).
SALISBURY, G. W., VON DEMARK, N. L.: Physiology of reproduction and artificial insemination of cattle. San Francisco-London: W. H. Freeman Comp. 1961.

Ovum Transport

H. Koester
Universitäts-Frauenklinik, D-6300 Gießen

With 29 Figures

The ovum is fertilized in the oviduct, and during the first days of pregnancy the oviduct is the physiological home of the early embryo. Also, remarkably enough, the oviduct is capable of transporting the male and female germ cells in opposite directions at practically the same time. It is only in the last few years that we have come to recognise clearly the complex function of this organ in instituting and regulating the earliest phases of pregnancy and, as a result of this knowledge, we are now in a better position to explain the causes of certain cases of sterility and infertility in women, and also to understand some peripheral mechanisms of action of hormonal contraceptives.

There are many interesting aspects of the interaction between the mother's body and the ovum or early embryo, but I shall confine myself to mentioning a few points about the mechanism responsible for transporting the ovum through the oviduct. There are still very many different explanations for this transport mechanism.

First, let me give you a brief outline of the morphological structure of this organ. The oviduct or uterine tube consists of two parts (Fig. 1), the lateral, wider ampulla and the medial, narrower isthmus. At the ovarian end, the ampulla ends in the fimbriated infundibulum, and medially the isthmus opens into the uterus via the pars isthmica interstitialis. In the human uterine tube, which is about 12 cm long, the transition between ampulla and isthmus is between the inner and middle third (Schröder, 1930); in the uterine tube of the rabbit, which is about the same length (Parker, 1931), and in most laboratory animals (Alden, 1942) it is in the middle of the organ. The wall of the uterine tube consists of three

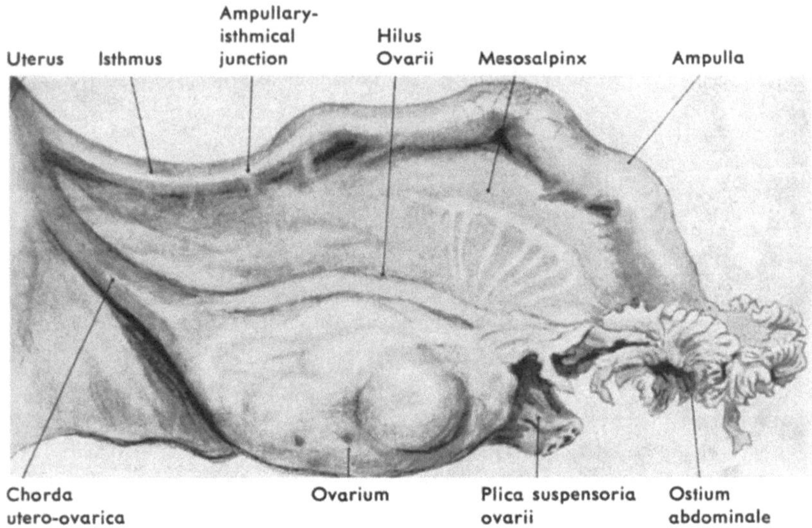

Fig. 1. Human oviduct (H. Martius, 1950)

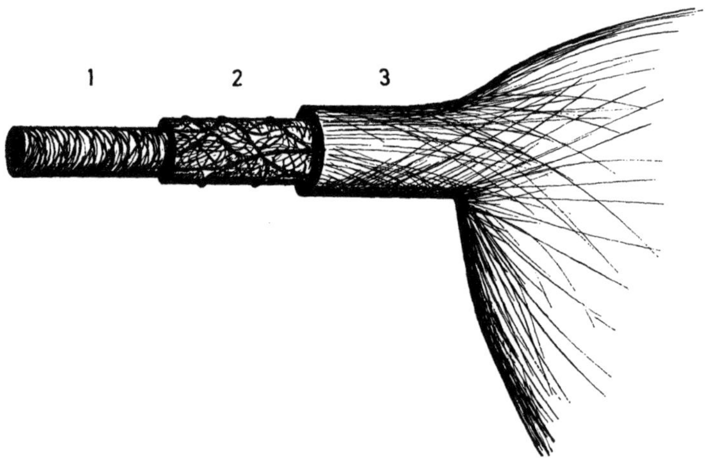

Fig. 2. Muscular layers of the oviduct. *1* circular layer, *2* vascular layer, *3* longitudinal layer (H. Erb, 1969)

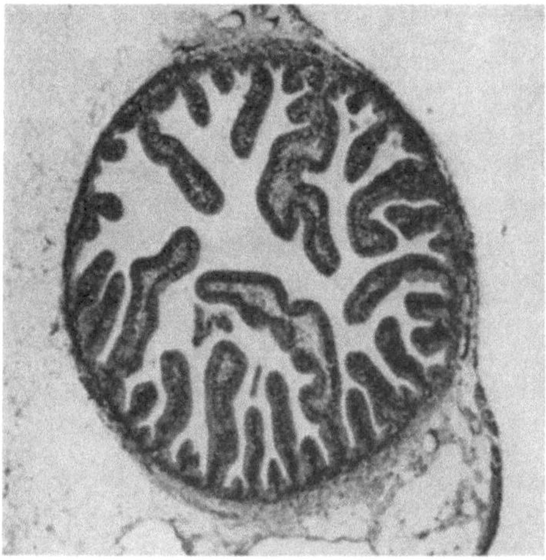

Fig. 3. Cross section of the ampulla of the rabbit oviduct (G. S. GREENWALD, 1961)

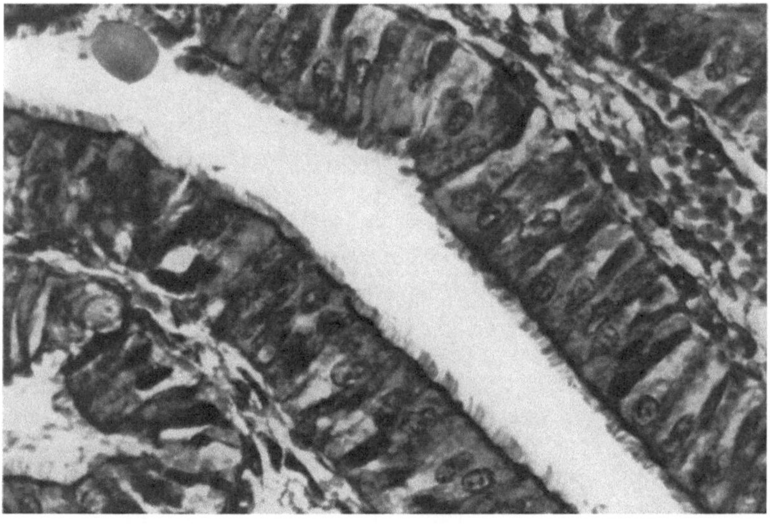

Fig. 4. Epithelium of the ampullary part of the rabbit oviduct (Masson trichome staining)

layers which merge into one another. The subperitoneal longitudinal muscle layer runs in the direction of the tube (Fig. 2). In the central layer, which contains blood vessels, muscle fibres and blood vessels run around the tube. The inner, autochthonous muscle layer is spiral in form (HORSTMANN, 1952). The three layers of muscle form a coherent system which is closely linked both from the anatomical and functional point of view (KIPFER, 1950). In the ampulla, the muscular wall of the tube is delicate and thin, but in the isthmus it is very much more powerful. In some species of animal the tube is rather tortuous.

The mucosa of the tube is thrown up into many folds and villi which almost fill the lumen of the tube (Fig. 3). However, the size of these folds may vary at different points in the same tube and also between the right and left tube (KIPFER, 1950).

The mucosa is lined with a simple epithelium which undergoes cyclic changes (Fig. 4). This epithelium consists of secretory cells, which produce the tubal secretion, and ciliated cells, the bunches of cilia of which all beat in the same direction. The cilia conform to the standard structural plan of kinocilia (STEGNER, 1962), whereby nine pairs of fibrils are always arranged concentrically around a central pair of fibrils (FAWCETT and PORTER, 1954).

The function of the oviduct in ovum transport begins when it picks up the ovum which has been released from the ovary at ovulation. In some species of animal, e.g., mouse, rat and hamster (HUMPHREY, 1969), the fimbriated end of the tube is connected to the ovary by a specially shaped peritoneal fold, the bursa ovarica, and this close contact between the two organs facilitates transfer of the ovum (Fig. 5). In man, rabbit, guinea pig, sheep and other animals there is no permanent close contact between the two organs. Thus, in these species the oviduct is not directly, connected to the ovary and for a long time it was not understood how the ovum reached the tube. After earlier, strange and rather exotic theories, WESTMAN (1926, 1929, 1937) made observations by laparotomy on rabbits and by laparoscopy on rhesus monkeys and later also in man, and was able to demonstrate the complicated interaction between oviduct and ovary. These observations revealed that the fimbriated end of the tube which, at the time of ovulation, is wide open and engorged with blood, is brought into close contact with the surface of the ovary by muscular activity of the mesotubarium,

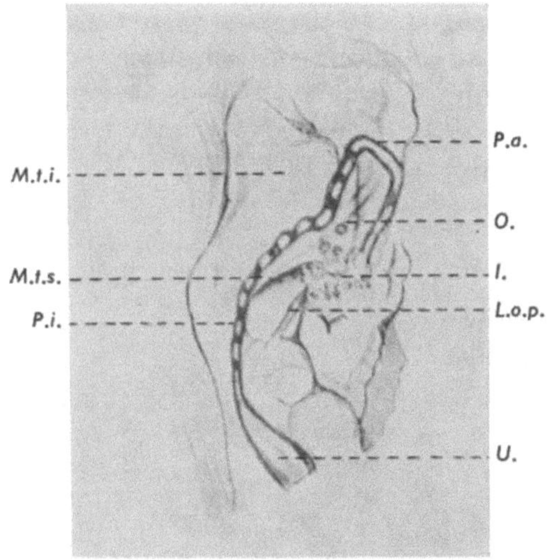

Fig. 5. Rabbit oviduct during estrus (A. Westman, in: F. v. Mikulicz-Radecki, 1931). *O.* Ovarium, *I.* Infundibulum, *U.* *Uterus*, *P.a.* Pars ampullaris, *P.i.* Pars isthmica, *M.t.i.* Mesotubarium inferius, *M.t.s.* Mesotubarium superius, *L.o.p.* Ligamentum ovarii propium

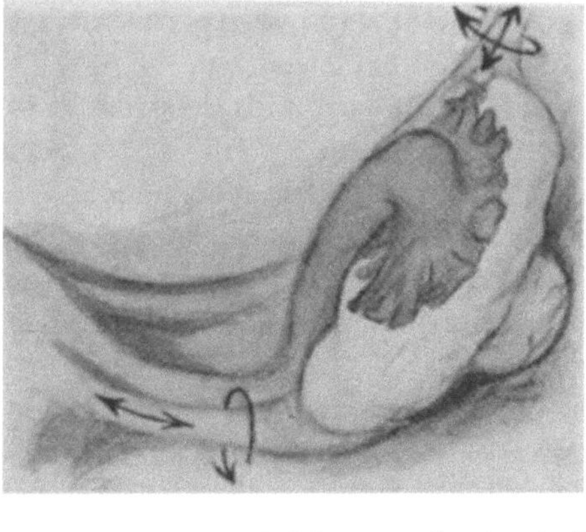

Fig. 6. Ovum pickup in man (H. Martius, Stuttgart: Thieme 1950)

particularly of the M. atrahens tubae which runs in the fimbria ovarica (Stange, 1952). At the same time, the ovary is moved slowly to and fro and about its longitudinal axis by contractions of the ligamentum ovarii proprium (Fig. 6) so that the fimbriated infundibulum is constantly feeling all over the surface of the ovary (Dyroff, 1932; Haselhorst, 1936; Caffier, 1936).

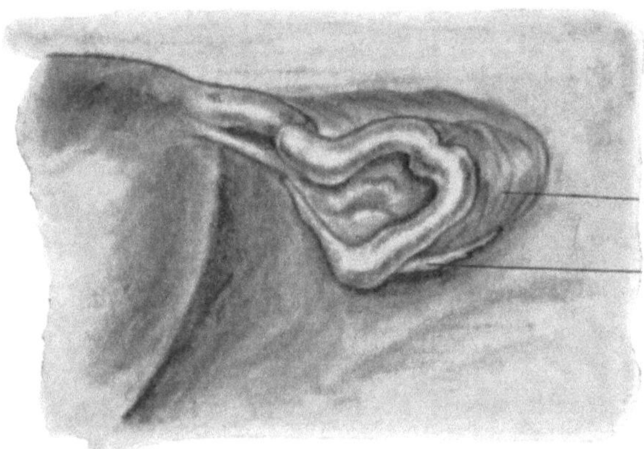

Fig. 7. Bursa ovarica position of the oviduct (F. v. Mikulicz-Radecki, 1931). *M.* Mesosalpinx, *F.* Fimbriae

Mikulicz-Radecki (1937) believed that the egg was taken up in this way. His results from animal experiments and X-ray studies show that, at the time of ovulation, the end of the tube takes up a special position (Fig. 7). It surrounds the ovary in such a way that, along with the mesotubarium inferius and superius, it forms a pocket into which the ovum slides and from where it can be taken up by the fimbriated infundibulum. Since this arrangement corresponds to the anatomically permanent bursa ovarica in other species of animal, Mikulicz-Radecki (1929) called it the "bursa ovarica position".

Elert (1947) examined women by laparoscopy and he found that the two mechanisms occur side by side. He observed that the exploration of the ovarian surface by the fimbriated infundibulum lasts for only a few minutes at a time and is probably repeated frequently. During the intermenstruum, the tubes sink temporarily into the pouch of DOUGLAS in the "bursa ovarica position" and remain there for a time. This evidently represents an additional safety measure (ELERT, 1947).

DUBREUIL (1944) and KNEER (1949) suggested that the ovum was drawn into the fimbriated infundibulum by a stream of liquid passing into the tube. However, in 1958 CLEWE and MASTROIANNI found that in rabbits the ovum consistently passes into the fimbriated infundibulum even if the tube is tied off directly behind the infundibulum. HARPER (1961) used rats to film the mechanism by which the ovum is taken up. He showed that once the ovum has been slowly forced out of a small aperture in the follicle (WESTMAN, 1936), a process which takes five to seven seconds (WALTON and HAMMOND, 1928; McCLEAN and ROWLANDS, 1942), the cilia on the fimbriae are sufficiently powerful to waft the ovum and cumulus oophorus, which is still attached, rapidly and steadily into the fimbriated infundibulum. The same happens with starch grains (ANDERES, 1941) and, in the mouse, with minute carbon particles (WIMSATT and WALDO, 1945). The ovum was still taken up even when, as an experiment in rabbits, the ovary was placed two to three centimetres away from the fimbriated end (WESTMAN, 1926; HARTMANN, 1939). Ova were not taken up if they had been injected intraperitoneally and hence were too far from the fimbriated infundibulum (HAFEZ, 1961).

On its migration through the twelve centimetres or so of tube, the ovum is very much under the control of its environment, since, unlike sperms, it is non-motile. However, surprisingly enough, there is still not complete agreement as to whether the ovum is moved by contractions of the tubal muscles or by the action of the epithelial cilia. Many investigators believe it is most likely to be the tubal muscles, although this does not explain one or two of the features of ovum transport. In the meantime, the nature of the muscular contractions has been examined in man (SECKINGER and SNYDER, 1924, 1926; KOK, 1925; v. MIKULICZ-RADECKI, 1926; WIMPFHEIMER and FERESTEN, 1939; KNEER, 1949; HAFEZ, 1959) and in many species of animal, including rabbits and pigs (SECKINGER, 1923; WISLOCKI and GUTTMACHER, 1924), cows, sheep, horses (KOK, 1925), mice (HUMPHREY, 1969), rats (ALDEN, 1942) and monkeys (SECKINGER and CORNER, 1923; WESTMAN, 1929). There were no great differences of tubal motility, and the type of movement appears to be the same in all species. At various points in the tube contraction-rings of longitudinal muscle form rapidly, and these occasionally run for a short distance in the direction of the uterus and then disappear, either rapidly or slowly (Black and Asdell, 1958). Antiperistaltic movements are observed only occasionally. Contractions in the ampulla are relatively weak, but they are more powerful and more regular in the isthmus. Contractions in the isthmus occur at intervals of seven to fifteen seconds and last for two to four seconds (Hafez, 1959). Since the inner circular fibres contract first, the tube stretches at the beginning of the contraction; it then shortens, owing to contraction of the outer fibres, before finally it relaxes again. When the contraction has subsided, the previously contracted segment becomes very engorged with blood, and this may cause that section of tube to become double the normal thickness (Hafez, 1959). Thus the whole tube appears to be very active, although, because of the many types of contraction which are possible, the individual movements often appear to be rather irregular and they are very difficult to analyse.

At the time of ovulation, under the influence of estrogens there is a sharp increase in the intensity of contractions, but by the second, third and especially the fourth, post-ovulatory day they again become weaker and more irregular as the progesterone level increases (Alden, 1942; Hafez, 1959; Black and Asdell, 1958, 1959 etc.). They increase again at the end of pregnancy. Thus, the tube goes through its first period of major activity during migration of the ovum.

This temporal association is certainly the reason for the functional connection assumed by many authors (Erb, 1969; Humphrey, 1969 etc.). After Parker (1931) had produced definite evidence that the cilia are capable of transporting the ovum, Harper (1961) produced what is, in my opinion, conclusive evidence of the effectiveness of ciliary action. He was able to show that the stream produced in the tube by the cilia is capable of moving the ovum even after the motility of the muscles has been eliminated experimentally. Wintenberger (1955) excised segments of ampulla and isthmus from sheep and sewed them back in the same or in the reverse direction. After healing, 82% of ova were retained by the reversed segments in which, of course, the cilia were also working in the reverse direction, whereas segments which had been replaced in the normal direction retained only 32% of all ova.

Thus, not only do the contractions of tubal muscles not appear to have the same effect as intestinal peristalsis in pushing forward the contents, but they actually seem to cause a slight delay of transport. From the nature of the muscular contractions described, it would appear that they produce thorough mixing of the tubal contents, promote fertilization (AUSTIN and BRADEN, 1952; BRADEN and AUSTIN, 1954) and help to denude the ova (SWYER, 1947; BURKS and DAVIS, 1964; MASTROIANNI and EHTESHAMZADEH, 1964). It is, however, the cilia which appear to be responsible both for uptake of the ovum by the fimbriated infundibulum and also for transport of the ovum.

Using a method developed by DALHAM for measuring ciliary activity, BORELL, NILSON, and WESTMAN (1957) were able to produce precise data on ciliary activity during passage of the ovum. They found that in the rabbit, the cilia beat towards the uterus at a rate of 1500 beats per minute during oestrus. After ovulation, as the level of progesterone rose rapidly, the frequency increased steadily and became 20% faster. However, this rise did not occur until 48 hours after ovulation, i.e. long after the ova had reached the isthmus, and when the intensity of the muscular contractions had already begun to decline. This increased ciliary activity was maintained until after implantation of the embryo.

Table 1. *Length of time taken for transport of ovum through the oviduct in some mammals* (*after* J. P. BENNET, 1969)

Animal	Time (hours)
Dog	168
Cat	148
Horse	98
Rat	96
Rhesus monkey	96
Cow	90
Guinea pig	82
Sheep	72
Mouse	72
Rabbit	60
Pig	50

The effect of strong tubal activity on the contents of the tube is a constant source of interest to research workers, mainly because, in spite of powerful ciliary movements and tubal contractions, the ovum needs a surprisingly long time to complete its journey. In most mammals, the ovum spends most of its pre-implantation phase of about three to four days in the tube (ANDERSEN, 1927;

Bennett, 1969 and other authors), in which time the fertilized ova may develop to the morula stage or, in some species, even the blastula stage (Table 1).

After ovulation, the ciliary activity rapidly and steadily wafts the ova through the fimbriated infundibulum and past the first loop of the tube, which is usually twisted. The forward movement of the ovum caused by the cilia then stops, and the ovum is constantly pushed backwards and forwards between the rings of contraction. Occasionally, the ovum will get past one of these contraction-rings and then it will again begin the rather violent "pendulum" movement in the next segment of tube.

The ovum is propelled at very different speeds in different sections of tube. It migrates through the ampullary part of the tube in a few hours (Gregory, 1930; Chang, 1951; Björk, 1959 and other authors) or, according to Harper (1961), in a few minutes. It is arrested for a time at the ampullary-isthmical junction. Delaying of the ova at this point has also been observed in the pig, in the cat (Andersen, 1927) and in the guinea pig (Squier, 1932). Wintenberger (1953) was able to demonstrate that, in sheep, ova reached the junction of ampulla and isthmus after two-and-a-half hours but entered the isthmus only after 49 hours, and then passed through this part of the tube very quickly. Black and Asdell (1958) found that although droplets of oil and Indian ink were distributed rapidly in the ampulla by pendular contractions of the tube, they entered the isthmus only slowly. According to Greenwald (1961), rabbit ova take only two hours to pass through the first half of the tube, but they can be detected in the isthmus only after 70 hours, and they pass through the last part of the tube very rapidly (Fig. 8). Harper, Bennett, Boursnell, and Rowson (1960) and Harper (1964) observed small radioactively-labelled glass beads the size of ova at the ampullary-isthmical junction eight hours after insertion, but they passed slowly into the isthmus only after 40 hours (Fig. 9). The ova enter the isthmus only after a fairly long delay at this point in the tube, and they pass through the last part of the tube so rapidly that they can hardly ever be detected there (Greenwald, 1961).

It was suggested long ago that the utero-tubal junction, i.e. the place where the tube opens into the uterus, has a regulatory function (Andersen, 1927; Lee, 1928; Parker, 1931; Alden, 1942;

SIEGLER, 1944; LEONARD and PERLMAN, 1949). BLACK and ASDELL (1958) were able to demonstrate just how far this is true with a very simple and impressive experiment. Using the rabbit as the experimental animal, they ligated the tube at the fimbriated end immediately after ovulation. After only 24 hours they found that

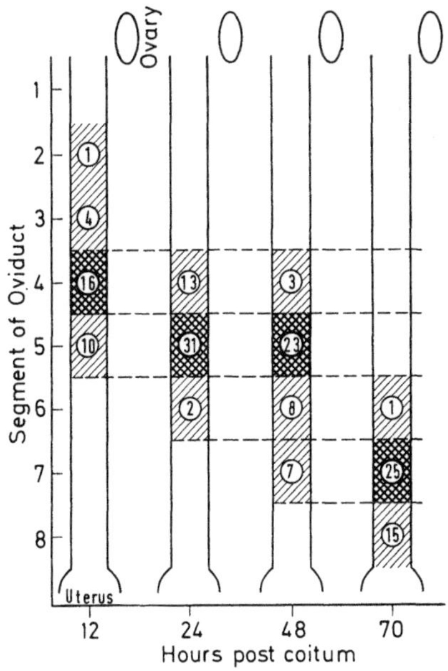

Fig. 8. Distribution of rabbit ova during the normal 3-day passage through the oviduct. Numbers indicate the number of ova within each segment of the oviduct (G. S. GREENWALD, 1961)

the less muscular part of the oviduct was extremely distended, indicating that no fluid was able to pass out of the oviduct through the utero-tubal junction (Fig. 10). Although the tube was grossly dilated, the fluid pressure developed within it was evidently insufficient to overcome the resistance. ANDERSEN (1927) stated this pressure to be 18 mm Hg. Only after 60 to 72 hours (BLACK and ASDELL, 1959), i.e. the usual time required for the ova to enter the uterus, did the resistance to fluid pressure fall sharply, evidently as a result of opening of the utero-tubal junction. If a short plastic

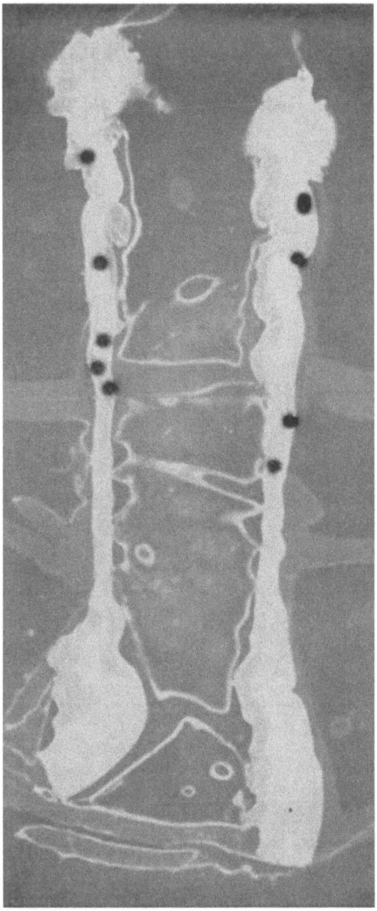

Fig. 9. Distribution of radioactive glass beeds within rabbit oviduct eight hours after ovulation (HARPER et al., 1960)

tube was placed in the terminal segment, the secretion did not accumulate. This experiment illustrated the control function of the utero-tubal junction and it also demonstrated a fact which was not previously known, that during migration of the ovum tubal secretion does not flow towards the uterus but in the opposite direction, back through the infundibulum into the peritoneal cavity. Moreover, this shows that ova move against the stream whilst sperms move with the stream. Since then the same mechanism has been demonstrated in sheep (EDGAR and ASDELL, 1960) and in the cow (BLACK and DAVIS, 1962). RAUSCHER (1968) has recently reported similar findings, obtained during surgery, in human beings. In

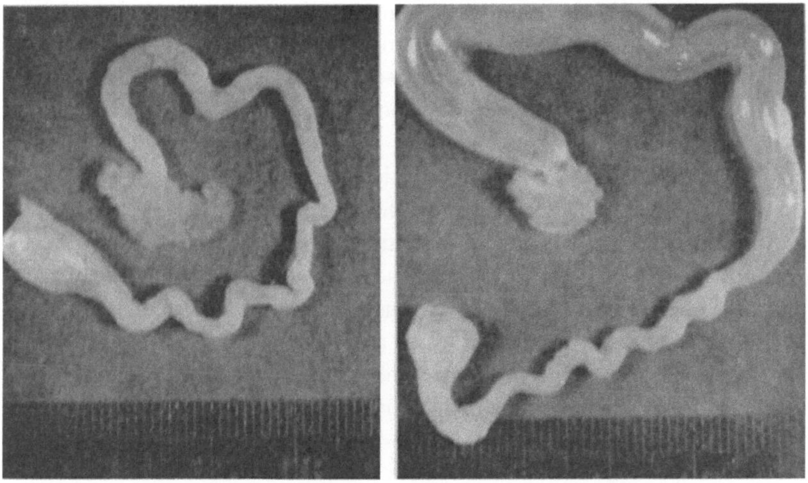

Fig. 10. Accumulation of fluid within rabbit oviduct after ligation of fimbrial end. Left: without ligature. Right: with ligature. (T. H. CLEWE and L. MASTROIANNI, JR., 1958)

rabbits, the utero-tubal junction opens only on the third and fourth day and on the 30th and 31st day of pregnancy, i.e. when the embryos pass out of the tube into the uterus and at the time of parturition (HAFEZ, 1963). Thus, it would seem that during these days the stream of fluid is at least partially reversed and the fertilized ova travel the remaining short distance into the uterus with the stream, whereas on their way through the tube they move against it.

ANDERSEN (1928) examined the morphology of the utero-tubal junction in 25 different mammals from ten different orders, and she found that the restrictive effect varied. Working with rabbits,

Black and Asdell (1959) concluded that estrogen-induced oedema was responsible for the occlusion, but they did not actually succeed in demonstrating it. Oestrogen-induced oedema has been shown in sheep in the subserous and muscular coats of the utero-tubal junction, and the lumen at this point was kinked so that the folds of the mucous membrane formed a valve-like occlusion (Edgar and Asdell, 1960).

However, studies in the cow (Black and Davis, 1962) indicate that the muscle tissue in this region also has an important part to play. In fact, if segments are removed from this point, the tube is no longer occluded, and this is also the case if the musculature of the tube is paralysed with nicotine.

Using serial cinephotography of large histological sections of specimens removed during operations on humans, Rocker (1964) was able to shed further light on the morphology of the utero-tubal junction and describe a probable functional mechanism. He showed that the musculature of the uterus continues into the utero-tubal junction, where the interlaced muscle fibres of the inner layers of the uterus become parallel and are arranged like a sphincter muscle. The lumen of the tube is completely straight as far as the vascular layer. Here the tube is sharply kinked and the lumen is only very small (100 γ). Thus, at the utero-tubal junction the tube has an outer layer of longitudinal muscle, an inner layer of sphincter, and inside this another layer of longitudinal muscle which is part of the uterine muscle. Thus, each contraction of uterine muscle can close the tube.

Thus, together with the muscular contractions, ciliary movement and (as I shall show later) tubal secretion, the utero-tubal junction is one of the factors which determine the movement and the timing of the movement of the ovum. Hafez (1963) and also Harper, Bennett, Boursnell, and Rowson (1960) and Harper (1964) take the view that the utero-tubal junction is controlled by changes in the oestrogen-progesterone ratio, and thus can easily be regulated. A constantly-changing relationship between these two hormones would appear to be necessary for this mechanism to function smoothly, but the exact quantitative scale of this relationship is still not known. Thus, after ovulation the ova are retained in the oviduct, as a result of closure of the utero-tubal junction by the effect of oestrogen produced in the ovary, until the uterus and endometrium have been properly prepared for implantation of the ovum by the increasing levels of progesterone. When, at the right moment, the oestrogen-progesterone ratio reaches a certain level, the path through the tube is opened up.

However, we still cannot explain the arrest of the ova in the middle of the tube, at the junction between the isthmus and the ampulla. Research workers were so surprised at the retention of the ova there, that GREENWALD (1963) suggested that there might be, at that point, an even more effective closure of the tube than the block shown by BLACK and ASDELL (1968) to occur at the utero-tubal junction. As a result of her experiments on sheep, WINTENBERGER-TORRÉS (1961) was also forced to assume that the occluding mechanism at the mid-point of the oviduct was more important.

In recent years several studies have set out to explain this observation. EDGAR and ASDELL (1961) suggested that there was occlusion of the tubal lumen owing to oedema of the subserosa and muscles both at the mid-point and also at the utero-tubal junction, but they were unable to demonstrate this type of oedema with any degree of certainty. HARPER (1964) postulated that this type of oedema occurs throughout the isthmus, and that peristalsis of the ampulla is weaker. BLACK and ASDELL (1958) considered that it was due to a decline in or failure of the activity of the sphincter fibres at this point, whereas WINTENBERGER-TORRÉS (1961) considered that powerful antiperistaltic waves in the isthmus are responsible for the 48-to-68-hour delay of the ova. GREENWALD (1961) made a vain attempt to demonstrate a muscular sphincter at this point by cutting serial sections; HUMPHREY (1969) also assumed that there was a muscular sphincter. Older explanations of this phenomenon were also referred to: that there is a special arrangement of the muscle fibres at this point (ALDEN, 1942) or that the amplitude and frequency of the muscular contractions in the ampulla and isthmus vary (HARTMANN, 1939).

GREENWALD (1963) studied the internal pressure in rabbit tubes and failed to find any change in the amplitude or frequency of the contractions in the ampulla during the three-day passage of the ovum. In the more muscular isthmus, under the effect of increasing levels of progesterone the power of contractions declined. In oophorectomized animals, estrogen administration restored the power of the contractions to the level seen during oestrus and the first day of migration of the ovum. This was achieved with a dose of 250 γ, and also with only 25 γ. Since earlier experiments had shown that whereas with the higher dose ova can be seen at the ampullary-isthmical junction for many days (the phenomenon described as "tube locking") but with the lower dose passage of the ovum is accelerated, yet both doses increase motility, GREENWALD (1963) believed that tube motility could be excluded as the cause of the block in the middle of the tube. As yet there has been no explanation of this "sphincter" mechanism at the ampullary-isthmical junction.

Therefore, we set out to discover whether, in addition to tubal motility and ciliary activity, tubal secretion is also an important factor in ovum transport. This appeared to us important, in that BLACK and ASDELL (1958) had discovered that the ovum migrates against the direction of fluid flow.

It was not until 1956 that BISHOP demonstrated that the tubal fluid is produced by a genuine secretory process of the tubal epithelium. BISHOP (1956) and later also MASTROIANNI et al. (1961) were able to show that the rate of secretion undergoes cyclic fluctuations. In sheep (BLACK, DUBY, and RIESSEN, 1963), monkeys (MASTROIANNI, SHAH, and ABDUL-KARIN, 1961) and other species of animal, it was possible to demonstrate that at the time of ovulation there was a sudden and sharp rise in the rate of secretion, and that as the corpus luteum developed the rate fell again. There was no increase of secretion in anovulatory cycles, and in castrated animals the amount of secretion fell by two-thirds. In rabbits in estrus, secretion is 0.92 to 2.31 ml/day (average 1.57 ml/day), and this falls by 50% by the third day after ovulation (MASTROIANNI, BEER, SHAH, and CLEWE, 1961). Thus, secretion in the oviduct is controlled hormonally, i.e. increased by estrogen and inhibited by progesterone.

We were interested in a report by FRIZ and MEY (1959), who believed they had demonstrated that the secretion of tubal fluid alters in each particular section of the oviduct as the position of the ovum alters, the secretion being most pronounced around the site of the ovum at any given time. In other words, the most powerful secretory activity travels like a wave with the ovum from the ovarian end to the uterine end of the oviduct. To investigate this theory we carried out experiments on rabbits.

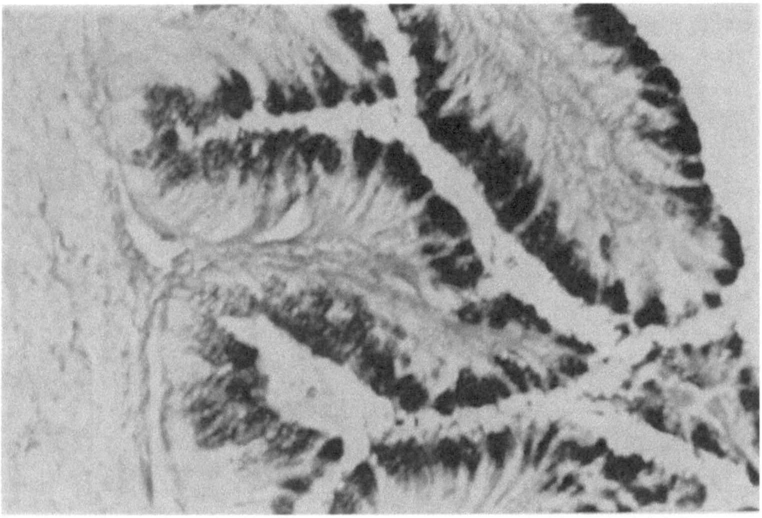

Fig. 11. Epithelium of the isthmus (Rabbit). Dark cells = secretory cells (Alcian blue staining)

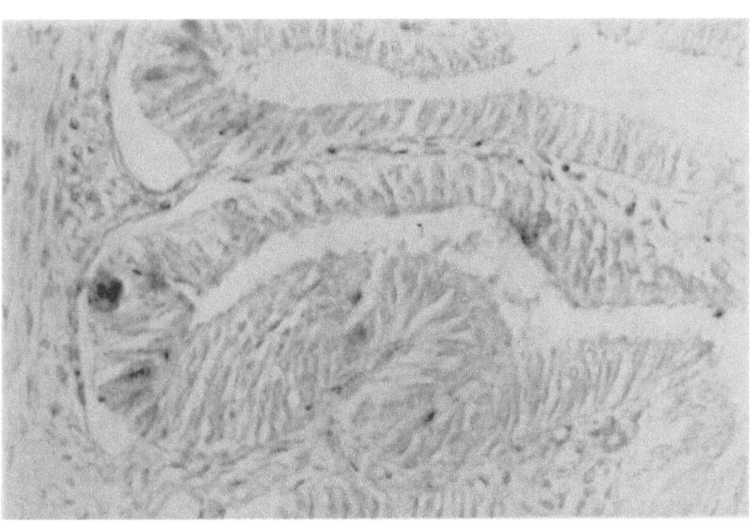

Fig. 12. Epithelium of the ampulla (Rabbit). Dark cells = secretory cells (Alcian blue staining)

In order to demonstrate a change in the type of secretion, the oviduct was examined 14 hours and 62 hours after ovulation, i.e. on the first and third days of the three-day passage of the ova through the oviduct. To determine the rate of secretion in various sections

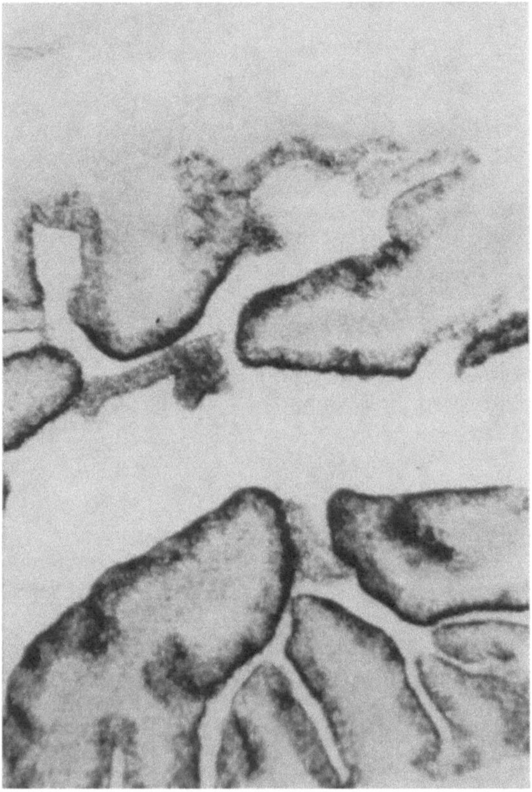

Fig. 13. Autoradiograph of the isthmus at first day after ovulation

of the oviduct, we employed parenteral administration of anorganic radioactively-labelled sulphur, which is practically selectively deposited in the mucopolysaccharides which represent a substantial part of the tubal secretion. A few hours after administering sulphur, an autoradiograph indicated activity in the secretory cells of the tubal epithelium, but not in the remaining tissues. Thus, the level of this radioactivity could be taken as a parameter of secretory activity. Seven hours after administering $Na_2{}^{35}SO_4$, the oviduct was divided into five equal segments and, after ignition, the radioactivity was assayed by a standardized automatic process.

In order to compare the activities in the different segments of tube, the radioactivity for each segment was related to the mean

Fig. 14. Autoradiograph of the ampulla at first day after ovulation

activity of the complete tube. The resulting figures for "relative segmental activity" were examined statistically. In this way we were able to compare a sufficiently large number of tubal segments with each other. The results showed us that tubal secretion did not move in a wavelike pattern from the ovarian to the uterine end of the tube. Secretory activity was always highest in the isthmus and lowest in the ampulla. This was because the epithelium of the isthmus (Fig. 11) comprises mainly secretory cells, with isolated ciliated cells, whereas the epithelium of the ampulla

(Fig. 12) is composed mainly of ciliated cells with few secretory cells.

An autoradiograph of the isthmus taken on the first day after ovulation (Fig. 13) shows intense darkening of the film over the whole epithelium. In an autoradiograph of the ampulla (Fig. 14)

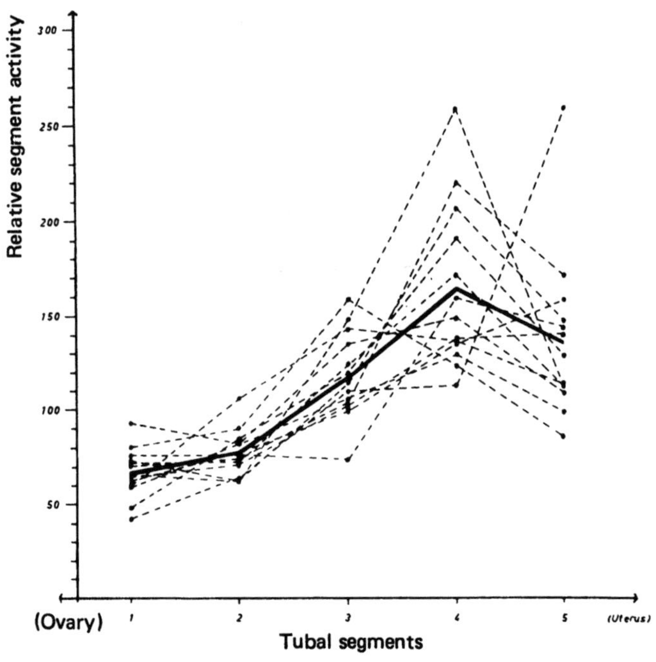

Fig. 15. Relative segmental activity at first day of normal pregnancy

there are only very few active secretory cells and graphs of the results (Fig. 15) reveal the same picture. The relative segmental activity of segments 4 and 5 of the isthmus is significantly higher than that of segments 1 and 2 of the ampulla.

On the third day after ovulation, i.e. on the last day on which the ova are passing through the tube, autoradiographs demonstrated that secretory activity is definitely drying up.

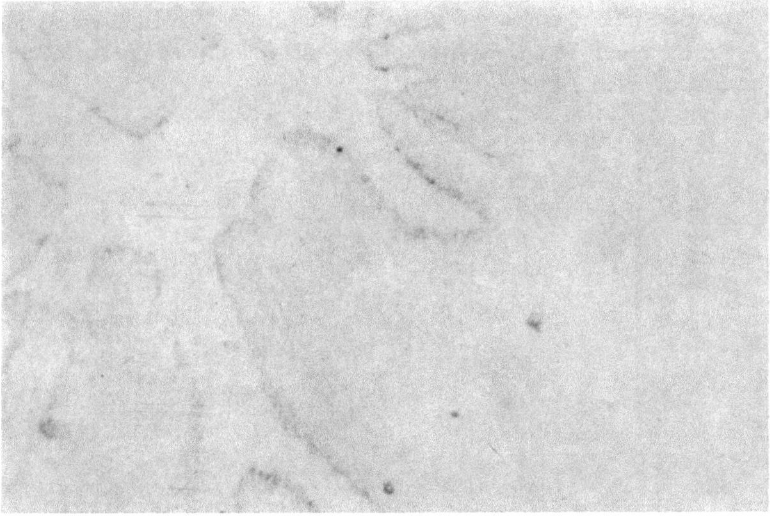

Fig. 16. Autoradiograph of the isthmus at third day after ovulation

Fig. 17. Autoradiograph of the ampulla at third day after ovulation

There is hardly any secretory activity in the isthmus (Fig. 16). Secretion has disappeared in the ampulla (Fig. 17). In the graphs (Fig. 18) the relative activities of ampullary and isthmic segments approach each other.

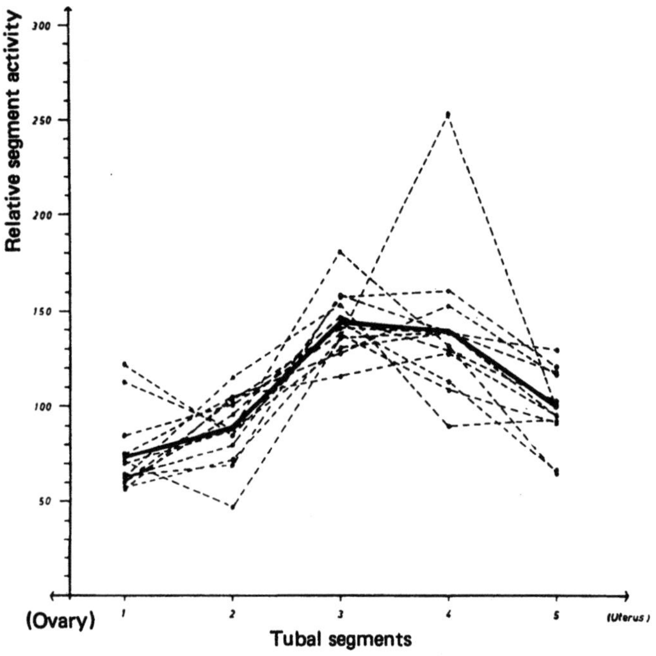

Fig. 18. Relative segmental activity at third day of normal pregnancy

Comparison of the summated curves reveals that there is a definite difference between the first and the third day (Fig. 19). There is a difference between the first and third day in that, at the end of ovum transport, the relative secretory activities of the two parts of the tube approach each other. The high density of secretory cells in the isthmus makes the fall of secretory output on the third day after ovulation, indicated by a fairly sharp fall in radioactivity over this section (as described by BISHOP and other authors), even more apparent. Thus, because there is a decline of total secretion,

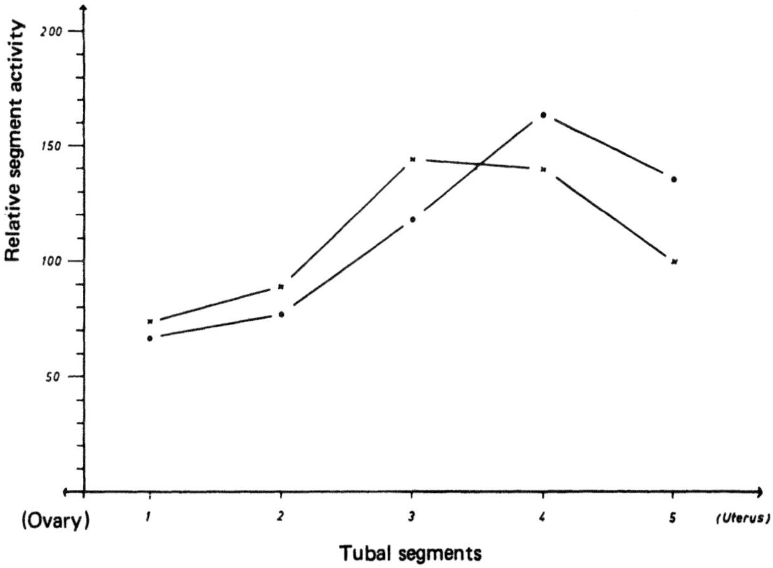

Fig. 19. Comparison of the relative segmental activities of first (○) and third (×) day of normal pregnancy

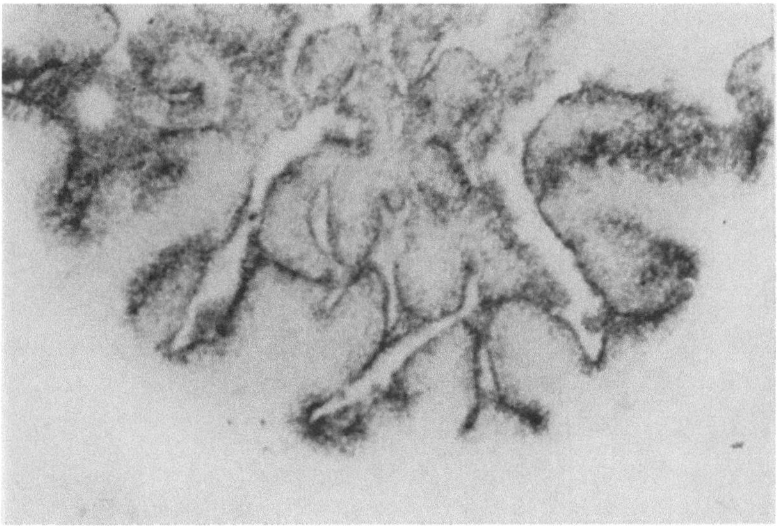

Fig. 20. Autoradiograph of the isthmus at third day after ovariectomy and application of estrogen

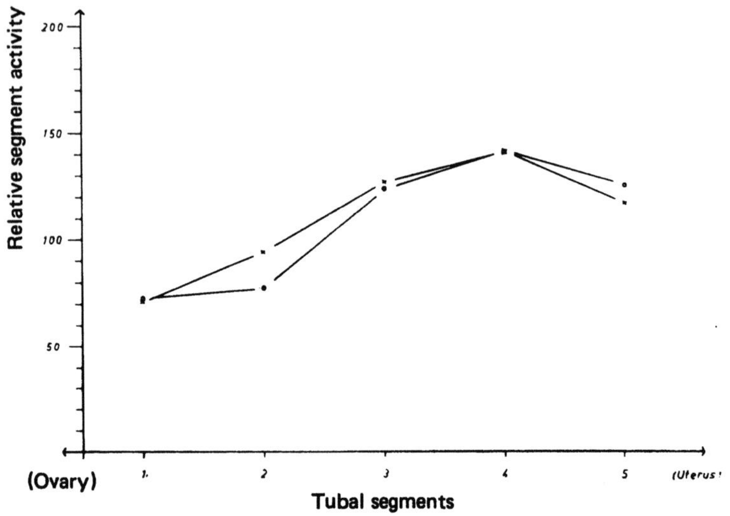

Fig. 21. Comparison of the relative segmental activities of first (○) and third (×) day after ovariectomy and application of estrogen

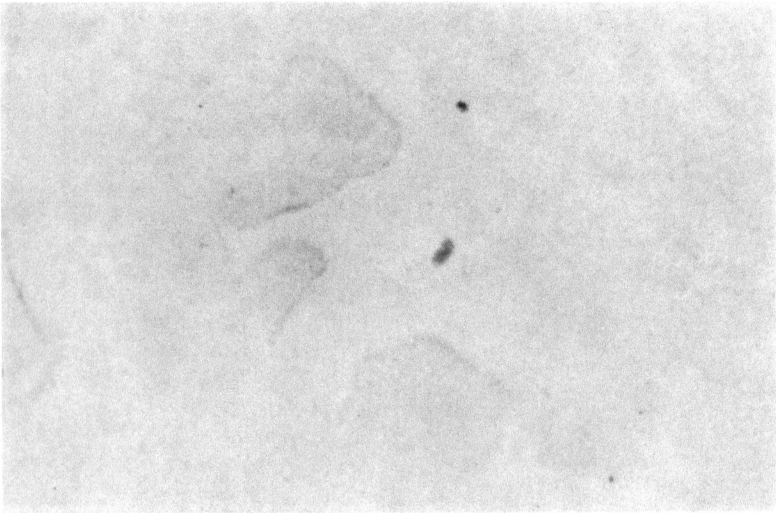

Fig. 22. Autoradiograph of the isthmus at third day after ovariectomy and application of progesterone

the relative activities of each segment approach each other. This finding is worth noting, and I shall refer to it again later.

The hormone-dependence of these changes can be demonstrated by castrating the animals after ovulation and then treating them with hormones. With daily administration of 100 γ 17-β-estradiol secretion is just as powerful even on the third day (Fig. 20).

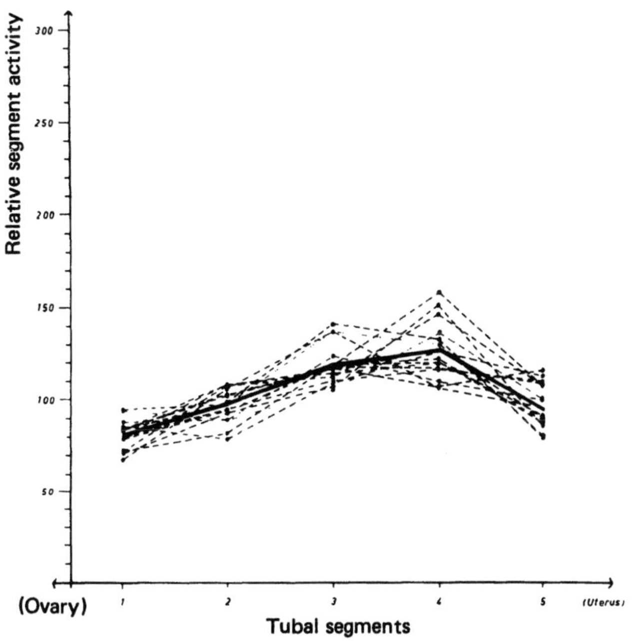

Fig. 23. Relative segmental activity at third day after ovariectomy and application of progesterone

Comparison of the summated curves (Fig. 21) also shows that the type of secretion on the first day is maintained by giving an estrogen.

If 10 mg progesterone (Proluton, Schering) is given daily after oophorectomy, an autoradiogram (Fig. 22) reveals hardly any secretion in the isthmus on the third day. This is manifest in the graphs (Fig. 23), in which the curves for ampulla and isthmus approach each other.

Hormone combinations, such as those used to inhibit ovulation, also elicit this effect (Fig. 24). In rabbits which, after oophorectomy, were treated with a combination of 5 γ ethinylestradiol and 50 γ norgestrel (Eugynon, Schering), there was a typical progesterone curve on the third day in spite of the estrogen component of the hormone mixture. Here again secretion dried up and the relative activities of the isthmus and ampulla changed and approached each other.

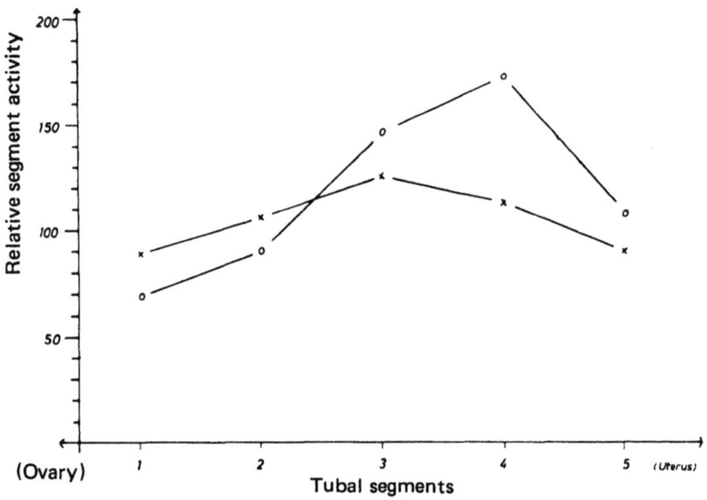

Fig. 24. Comparison of the relative segmental activities at third day after ovariectomy and application of ethinyl-estradiol (○) and a compound of ethinyl-estradiol with norgestrel (×)

Now, what is the significance of these observations in respect of the ova which are in the oviduct for these three days? The secretory picture described here may help to explain the as-yet-unexplained problem of the halting of the ovum at the ampullary-isthmical junction. It is possible to explain this if the secretory activity of each segment of the tube is related to its lumen.

The tubal lumen is irregular and narrowed by mucosal villi. Planimetric measurement of the actual tubal lumina of histological sections of the segments bordering on the ampullary-isthmical junction gave a figure of 0.95 mm^2 for the ampulla and 0.20 mm^2 for the isthmus. Thus, the cross-section of the ampulla is 4.7 times greater than that of the isthmus (Table 2).

Table 2. *Measurements of tubal lumen (rabbit)*

	Ampulla	Isthmus	Ratio
Diameter	1.72 mm	0.79 mm	2.2: 1
Cross-sectional area (without villi)	2.32 mm^2	0.49 mm^2	4.7: 1
Cross-sectional area (with villi)	0.95 mm^2	0.20 mm^2	4.7 : 1

Our investigations revealed that the most powerful secretory activity occurs in the isthmus whilst there is very much less activity in the ampulla. Most of the secretion is produced where the tube is narrowest. We know from BLACK and ASDELL (1958) that, in the three days that the ovum is in passage through the tube, the utero-tubal junction is closed. Thus, tubal secretion cannot pass into the uterus, but must flow from the isthmus into the ampulla and thence into the peritoneal cavity. Hence, there is a specific pattern of flow within the tube.

Because of the different diameters of the tubal lumen in the isthmus and ampulla there are different flow-rates for tubal secretion, and this is determined by the continuity-equation

$$\text{const.} = q_1 \cdot v_1 = q_2 \cdot v_2 \,.$$

That is to say, the rate of flow (v) is determined by the cross-section (q) of a tube, and so the same volume of fluid flows faster in a narrow tube than in a wide one.

If the tube was smooth, the rate of flow of secretion produced in the isthmus (isthmus segment 4) would, on the basis of the calculated cross-sections, be reduced to one-fifth after it entered the ampulla (Fig. 25).

In collaboration with the Department of Experimental Physics of the Giessen University (Professor A. SCHARMANN), we are now attempting to determine the viscosity of tubal secretion. The final values for the various days after ovulation are not available at the moment, but we can say that the results have indicated that the viscosity of tubal fluid is practically the same as that of water.

Although the absolute volume of tubal secretion produced on each of the days is known (Bishop, 1956; Mastroianni and Wallach, 1961), the rate cannot definitely calculated, since the tubal lumen is not smooth but is narrowed by mucosal villi and is constantly being changed by contractions. Even though the continuity-equation can only be regarded as an "approximation", it seems that we are justified in using it to explain the flow picture.

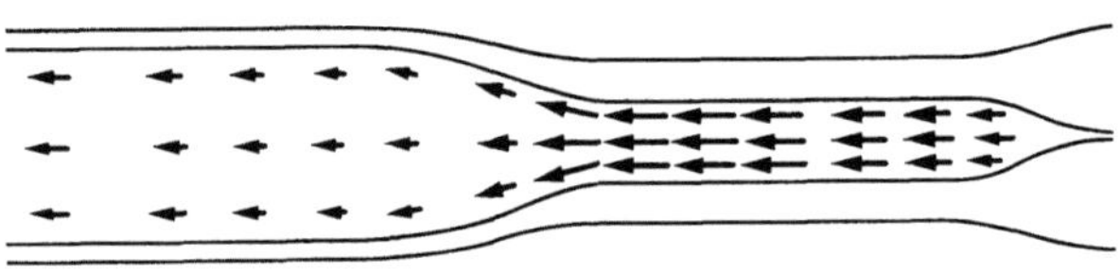

Fig. 25. Schema of tubal flow

Accordingly, tubal secretion flows faster in the narrower isthmus and slower in the wider ampulla, at least in those segments which border on the junction between isthmus and ampulla. If this fact is seen in the light of ciliary movements, which are also an important factor in ovum transport (Kneer and Cless, 1951; Borell et al., 1957; Harper, 1961), we have a possible explanation for the irregular rate of ovum migration in each of the segments, and, in particular for the stopping of the ova at the ampullary-isthmical junction.

The epithelium of the ampulla is made up mainly of ciliated cells, with scattered, isolated secretory cells. Here, the beats of the cilia rapidly move the egg cell forwards (Harper, 1961, 1964) against the flow of tubal secretion. However, the tubal secretion is moving so slowly in the ampulla that it is overcome by the power of the cilia. At the ampullary-isthmical junction the tubal lumen narrows and there is a sharp increase in the rate of flow. At the same time there is a sharp fall in the number of ciliated cells as compared with secretory cells. Thus, the power of the cilia driving the ovum forward is less effective against the counter-current of secretion at this point. All this occurs under the predominant influence of estrogens produced during estrus and on the first day after ovulation.

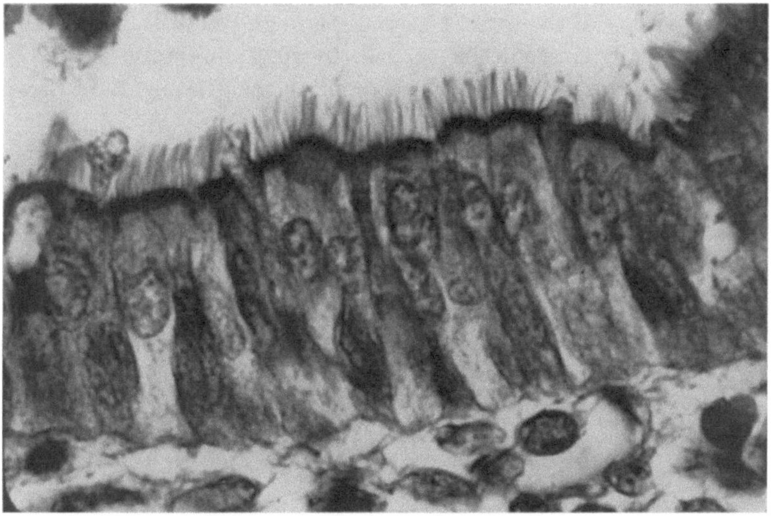

Fig. 26. Epithelium of the ampulla at first day after ovulation (Masson trichome staining)

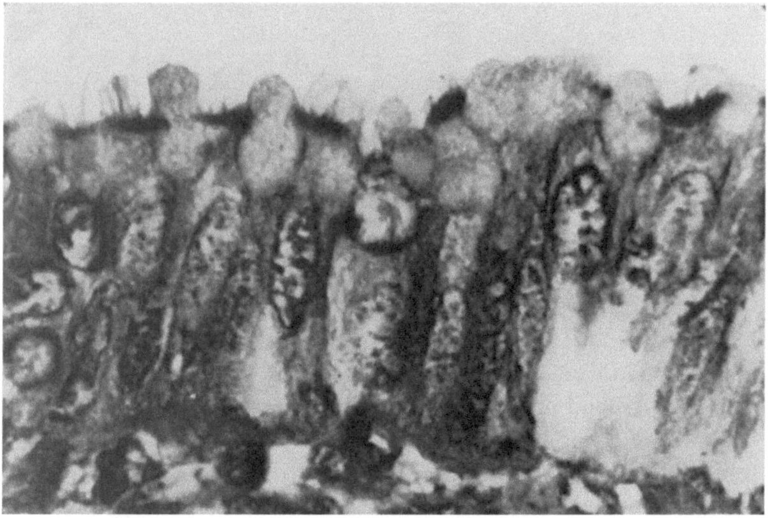

Fig. 27. Epithelium of the isthmus at first day after ovulation (Masson trichrome staining)

As progesterone activity begins to increase towards the end of the third day of the ovum's passage through the tube, the rate of beating of the ciliated cells increases by 20% (Borell et al., 1957) and there is a sharp decline in secretion in the isthmus. Both these factors encourage the ovum to enter the isthmus.

This mechanism is backed up by a morphological change of the tubal epithelium. Under the influence of estrogens, the secretory

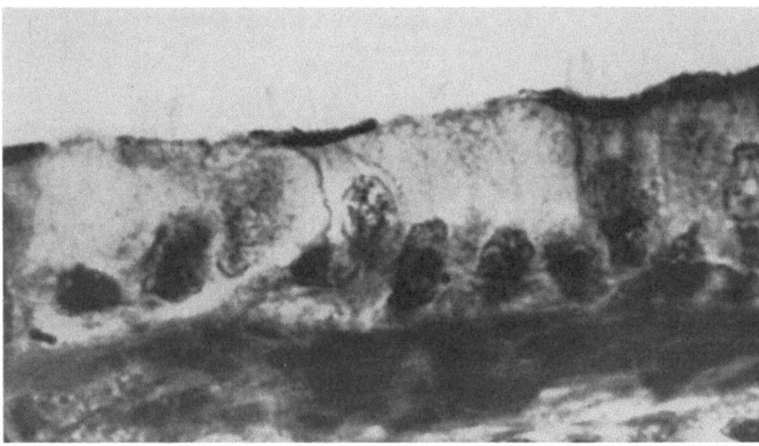

Fig. 28. Epithelium of the isthmus at third day after ovulation (Masson trichrome staining)

cells project above the ciliated cells. In the ampulla (Fig. 26) this is of little importance since there are only a few secretory cells. However, in the isthmus (Fig. 27) it has a mechanical effect, since here there are very few ciliated cells, and these are deep down between the tall secretory cells. The tips of the cilia hardly reach the tubal lumen and thus exert no effect.

In the second phase of ovum transport, the secretory cells drop back under the influence of progesterone, leaving the ciliated cells free (Fig. 28). The cilia then reach far into the tubal lumen of the isthmus and they are in a position to drive the ovum forwards.

It can, thus, be confirmed that ovum transport is controlled both by the secretory process in the tube and by the effects of hormone on tubal secretion, and also by a blockade mechanism at the junction between isthmus and ampulla, but not by muscular tubal activity. This does not occur as a result of one of the many suggested closure mechanisms at this point, but is evidently governed by the balance between ciliary beats and the counter-flow of secretion, the rate of which is determined by the different widths of the tubal lumen and by production of secretion in the isthmus, which is controlled hormonally. This mechanism which regulates migration of the ovum is dependent on closure of the utero-tubal junction (BLACK and ASDELL, 1958) which is determined by a specific estrogen-progesterone ratio (HAFEZ, 1962). Obviously, it can work when the utero-tubal junction is closed. The rapidly changing ratio of the two hormones after ovulation appears to be the fundamental factor in controlling migration of the ovum, since ovum transport through the tube occurs normally only during the first three to four days after ovulation, during which progesterone production rises sharply.

Thus, migration of the ovum is hormonally controlled, and hence can be disrupted hormonally. The results of many animal experiments have shown that ovum transport can be influenced by hormones. Both oestrogens and also progestogens are capable of delaying or accelerating ovum transport, depending on the species, dose, time and duration of administration (ELGER, 1969). Many of the observations appear to be contradictory, particularly if muscular activity of the tube is assumed to be the main mechanism responsible for transport. However, if we consider the secretory picture within the tube, it provides a possible explanation for some of the observations reported in the literature. For example, GREENWALD (1957, 1959, 1961, 1963) was able to produce retention of the ova for many days at the ampullary-isthmical junction by giving rabbits a single high dose of a depot-estrogen (250 γ estradiol-cyclopentyl-propionate) (Fig. 29). The same thing happened with daily injections of low doses of estrogen (10 γ estradiol benzoate). However, after a single dose of only 25 γ of a depot-estrogen, the opposite result was obtained. In rats and rabbits, ovum transport was accelerated and many ova appeared in the uterus prematurely (i.e. within 48 hours) and were discharged into the vagina. Similar observations were made in guinea pigs by DEANESLY (1963), CHANG and YANAGAMACHI (1965) and in rabbits and hamsters by NOYES,

Adams, and Walton (1959). In rabbits, with high doses of estrogen they were able to cause retention of ova in the tube, while low doses accelerated the passage of the ova and hence their entry into the uterus. Using similar doses of hormone, Harper (1961) was able to achieve the same effect with radioactively-labelled glass beads of the size of ova. This qualitatively-different, dose-related

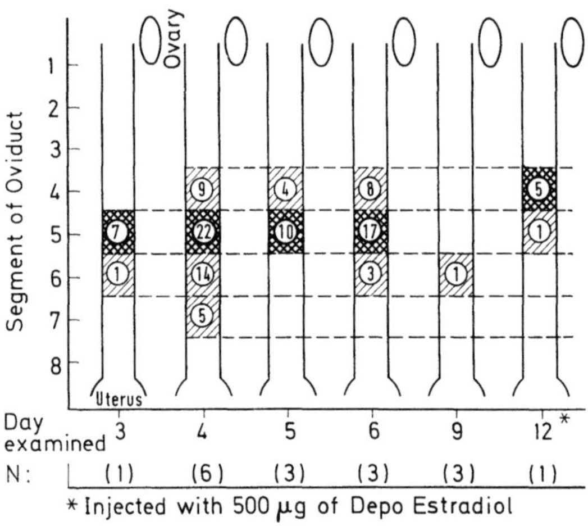

Fig. 29. Distribution of ova in the oviduct after a single injection of 250 μg of Depo Estrodiol (G. S. Greenwald, 1961)

effect had already been described by many authors (Allen and Corner, 1929; Allen, 1932; Burdick and Pincus, 1935; Pincus and Kirsch, 1936 etc.).

Looking at these findings in the light of secretory function, which is different in individual segments of tube and which can be influenced by hormones, we have the following explanation: with high doses of estrogen, there is a high output of secretion in the isthmus, as compared with the ampulla, and the utero-tubal junction remains closed, so that ova cannot move into the isthmus. Whilst the hormone is exerting its effect, ova are retained in the middle of the tube by the consistently high level of secretion and the accompanying morphological changes of the epithelial cells.

With a single low dose of estrogen, the estrogen-progesterone ratio which is necessary for closure of the utero-tubal junction

(HAFEZ, 1962) is evidently disrupted, and the ova can enter the uterus prematurely through the open utero-tubal junction. Premature opening of the tube to the uterus also alters the direction of flow of the secretion, so that ova are not retained at the junction of the ampulla and isthmus.

Reports in the literature of changes of ovum transport after treatment with progesterone also accord with a functional mechanism mediated by tubal secretion. By administering progesterone or other progestational compounds, it is possible to prevent ova from being arrested at the ampullary-isthmical junction, so that the ova enter the isthmus prematurely, although often their entry into the uterus is only a little premature (GREENWALD, 1961; ALLEN and CORNER, 1929; WU and ALLEN, 1959; WISLOCKI and SNYDER, 1933; BLACK and ASDELL, 1959). This has also been observed in the cow (DOWLING, 1949) and in sheep (ROBINSON, 1951). In oophorectomized estrogen-treated rabbits, it is again possible to prevent arrest of the ova in the middle of the tube by concurrent administration of progesterone (HAFEZ, 1952).

These observations can be explained by the reduction of secretory activity, particularly of the isthmus, under the influence of progesterone, since, when this happens, the cilia begin to exert a more powerful effect on the counter-current of secretion.

Oophorectomy will also prevent arrest of ova at the ampullary-isthmical junction, because of the resultant sharp reduction of secretion, and the migration of the ova will not be delayed very much even though tubal motility is disturbed; however, the ova will not develop properly in the uterus because of the absence of progesterone (KING, COLLINS, and PETERSON, 1932; ADAMS, 1958). This degeneration can be prevented by giving corpus luteum extract (ALLEN and CORNER, 1928) or progesterone (PINCUS and WERTHESSEN, 1938; ADAMS, 1958). Passage of the ovum can be disrupted by giving natural sex hormones and also by synthetic steroid hormones used for inhibition of ovulation (CHANG and YANAGAMACHI, 1965).

Non-steroid estrogens, prostaglandins and also compounds such as, for example, reserpine, chlorpromazine and tetrabenazine, can have an effect on ovum transport (references in BENNETT, 1969). They either cause acceleration of the ova in the tube or retention ("tube locking"), usually at the junction of ampulla and isthmus.

These findings are of practical significance. Ova in the tube can develop further only if they pass out of the tube into the uterus at a specific point in time (specific for each species of animal). If the ova pass into the uterus a few hours too early, they may perish there because the endometrium will not be properly prepared for them. On the other hand, if the ova are delayed in the tube for more than one-and-a-half days, the already developing ova will degenerate and will not implant in the uterus (Adams, 1956; Averill and Rowson, 1958; Beatty, 1951; Chang, 1948, 1955, 1951, 1964; Dowling, 1959; Doyle, Gates, and Noyes, 1963; Hafez, 1962; Hunter, Adams, and Rowson, 1955; McLaren and Michie, 1956; Noyes and Dickmann, 1960, 1961; Pincus and Kirsch, 1936).

In view of the complicated nature of ovum transport and the hormonal regulation of this process, it is astonishing that the mechanism usually works without any trouble. We certainly have here one of the intangible causes of female infertility. But above all, the mode of action of hormonal contraceptives which do not inhibit ovulation, for which there is as yet no satisfactory explanation, may well be sought in the effect they have on the interaction between muscular activity, sphincter system, ciliary current and secretion of the oviduct, and in disrupting thereby the complicated mechanism and synchronization of ovum transport.

References

Adams, C. E.: A study of fertilization in the rabbit: The effect of postcoital ligation of the Fallopian tube or uterine horn. J. Endocr. **13**, 296 (1956).

— Egg development in the rabbit: The influence of postcoital ligation of the uterine tube and of ovariectomy. J. Endocr. **16**, 283 (1958).

Alden, R. H.: The periovarial sac in the albino rat. Anat. Rec. **83**, 421 (1942a).

— The oviduct and egg transport in albino rat. Anat. Rec. **84**, 137 (1942b).

— Aspects of the egg-ovary-oviduct relationship in the albino rat. I. Egg passage and development following ovariectomy. J. exp. Zool. **90**, 159 (1942c).

Allen, W. M., Corner, G. W.: Physiology of the corpus luteum. III. Normal growth and implantation of embryos after early ablation of the ovaries under the influence of extracts of corpus luteum. Amer. J. Physiol. **88**, 340 (1929).

Anderes, E.: Zur Frage des Eitransportes durch die Tube. Schweiz. med. Wschr. **71**, 364 (1941).

ANDERSEN, D. H.: The rate of passage of the mammalian ovum trough the various portions of the Follopian tube. Amer. J. Physiol. **82**, 552 (1927).
— Comparative anatomy of the tubo-uterine junction. Amer. J. Anat. **42**, 255 (1928).
AUSTIN, C. R., BRADEN, A. W. H.: Passage of the sperm and the penetration of the egg in mammals. Nature (Lond.) **170**, 919 (1952).
AVERILL, R. L. W., ROWSON, L. E. A.: Ovum transfer in sheep. J. Endocr. **16**, 326 (1958).
BEATTY, R. A.: Transplantation of mouse eggs. Nature (Lond.) **168**, 995 (1951).
BENNET, J. P.: The effect of drugs on egg transport. Advances in the Biociences 4 (1970). Oxford-London-Edinburgh-New York-Toronto-Sidney-Paris-Braunschweig: Pergamon Press-Vieweg.
BISHOP, D. W.: The tubal secretions of the rabbit oviduct. Anat. Rec. **125**, 125 (1956a).
— Oxygen concentrations in the rabbit genital tract. Proc. III. Internat. Congr. on Animal Reproduction Cambridge, Engl., June 1956b, Sect. I.
— Active secretion in the rabbit oviduct. Amer. J. Physiol. **187**, 347 (1956c).
BJÖRK, L.: Cineradiographic studies on the Fallopian tube in rabbits. Acta radiol. Suppl. **176** (1959).
BLACK, D. L., ASDELL, S. A.: Transport through the rabbit oviduct. Amer. J. Physiol. **192**, 63 (1958).
— — Mechanism controlling entry of ova into rabbit uterus. Amer. J. Physiol. **197**, 1275 (1959).
— DAVIS, J.: A blocking mechanism in the cow oviduct. J. Reprod. Fertil. **4**, 21 (1962).
— DUBY, R. T., RIESEN, J.: Apparatus for the continuous collection of sheep oviduct fluid. J. Reprod. Fertil. **6**, 257 (1963).
BORELL, U., NILSSON, O., WESTMAN, A.: Ciliary activity in the rabbit Fallopian tube during estrus and after copulation. Acta obstet. gynec. scand. **36**, 22 (1957).
BRADEN, A. W. H., AUSTIN, C. R.: The number of sperms about the eggs in mammals and its significance for normal fertilization. Aust. J. biol. Sci. **7**, 543 (1954).
BURDICK, H. O., PINCUS, G.: The effect of oestrin injection upon the developing ova of mice. Amer. J. Physiol. **111**, 201 (1935).
CAFFIER, P.: Studien zum Eitransport beim Menschen. I. Mitteilung: Der Eiabnahmemechanismus. Zbl. Gynäk. **60**, 1873 (1936).
CHANG, M. C.: Probability of normal development after transplantation of fertilized rabbit ova stored at different temperatures. Proc. Soc. exp. Biol. (N.Y.) **68**, 680 (1948).
— Fertility and sterility as revealed in the study of fertilization and development of rabbit eggs. Fertil. Steril. **2**, 205 (1951a).
— Fertilization in relation to the number of spermatozoa in the Fallopian tubes of rabbits. Ann. Ostet. Ginec. **73**, 918 (1951b).
— Fertilization and normal development of follicular oocytes in the rabbit. Science **121**, 867 (1955).

Chang, M. C.: Effects of certain antifertility agents on the development of rabbit ova. Fertil. Steril. **15**, 97 (1964).
— Yanagamachi, R.: Effect of estrogens and other compounds as oral antifertility agents on the development of rabbit ova and hamster embryos. Fertil. Steril. **16**, 281 (1965).
Clewe, T. H., Mastrianni, L.: Mechanism of ovum pickup. I. Functional capacity of rabbit oviducts ligated near the fimbria. Fertil. Steril. **9**, 13 (1958).
Deanesly, R.: Further observations on the effect of oestradiol on tubal eggs and implantation in the guinea pig. J. Reprod. Fertil. **5**, 49 (1963).
Dowling, D. F.: Problems of the transplantation of fertilized ova. J. agric. Sci. **39**, 374 (1949).
Doyle, L. L., Gates, A. H., Noyes, R. W.: Asynchronous transfer of mouse ova. Fertil. Steril. **14**, 215 (1963).
Dubreuil, G.: Le réseau lymphatique fonctionnel des franges et des plis de la trompe, son rôle dans la captation et le cheminement de l'oeuf chez la femme. Gynéc. et Obstét. **44**, 397 (1944).
Dyroff, R.: Der Mechanismus der Eiabnahme beim Menschen und seine Störungen. Zbl. Gynäk. **56**, 2987 (1932).
Edgar, D. G., Asdell, S. A.: The valve like action of the utero-tubal junction of the ewe. J. Endocr. **21**, 315 (1960).
Elert, R.: Der Mechanismus der Eiabnahme im Laparoskop. Zbl. Gynäk. **69**, 38 (1947).
Elger, W.: Einfluß von Gestagenen auf Befruchtung, Eiernährung, Eitransport und Schwangerschaftsverlauf. Handbuch der experimentellen Pharmakologie, Band XXII/2. Berlin-Heidelberg-New York: Springer 1969.
Erb, H.: Zur hormonalen Regulation der Tubenmotilität. Basel-New York: S. Karger 1969.
Fawcett, D. W., Porter, K. R.: A study of fine structure of ciliated epithelia. J. Morph. **94**, 221 (1954).
Friz, M., Mey, R.: Die Bedeutung der Eileitersekrete für die Physiologie der Fortpflanzung. Geburtsh. Frauenheilk. **19**, 706 (1959).
Greenwald, G. S.: Interruption of pregnancy in the rabbit by the administration of estrogen. J. exp. Zool. **135**, 461 (1957).
— The comperative effectiveness of estrogens in interrupting pregnancy in the rabbit. Fertil. Steril. **10**, 155 (1959).
— Tubal transport of ova in the rabbit. Anat. Rec. **133**, 386 (1959).
— A study of the transport of ova through the rabbit oviduct. Fertil. Steril. **12**, 80 (1961).
— In vivo recording of intramural pressure changes in the rabbit oviduct. Fertil. Steril. **14**, 666 (1963).
— Interruption of early pregnancy in the rabbit by a single injection of oestradiolcyclopentyl-proprionate. J. Endocr. **26**, 133 (1963).
Gregory, P. W.: The early embryology of the rabbit. Contr. Embryol. Carneg. Inst. **21**, 141 (1930).

HAFEZ, E. S. E.: Tubo-ovarian mechanism and ova reception in mammals. Cornell Vet. **49**, 459 (1959).

— Procedures and problems of manipulation, selection, storage, and transfer of mammalian ova. Cornell Vet. **6**, 299 (1961).

— In vivo and in vitro studies on rabbit ova; non-surgical ova transfer as a target. Proc. IVth Intern. Congr. Animal Reprod. The Hague 1961.

— Endocrine control of reception, transport, development and loss of rabbit ova. J. Reprod. Fertil. **3**, 14 (1962).

— Pressure fluctuations during uterotubal kymographic insufflation in pregnant rabbits. Fertil. Steril. **13**, 426 (1962).

— The uterotubal junction and the luminal fluid of the uterine tube in the rabbit. Anat. Rec. **145**, 7 (1963).

HARPER, M. J. K.: The mechanism involved in the movement of newly ovulated eggs through the ampulla of the rabbit Fallopian tube. J. Reprod. Fertil. **2**, 522 (1961).

— The effects of constant doses of oestrogen and progesterose on the transport of artificial eggs through the reproductive tract of ovariectomized rabbits. J. Endocr. **30**, 1 (1964).

— BENNET, J. P., BOURSNELL, J. C., ROWSON, L. E. A.: An autoradiographic method for the study of egg transport in the rabbit Fallopian tube. J. Reprod. Fertil. **1**, 249 (1960).

HARTMAN, C. G.: Ovulation, fertilization and the transport and viability of eggs and spermatozoa. In: Sex and Internal Secretions, ed. E. ALLEN. Baltimore: Williams & Wilkins Comp. 1939.

HASELHORST, G.: Weibliche Sterilität. Arch. Gynäk. **161**, 81 (1936).

HORSTMANN, E.: Über die Bedeutung der Epoophoronmuskulatur. Arch. Gynäk. **182**, 314 (1952).

HUMPHREY, K. W.: Mechanisms involved in egg transport. In: Advances in the Biosciences, Vol. 4. Oxford-London-Edinburgh-New York-Toronto-Sidney-Paris-Braunschweig: Pergamon Press-Vieweg 1970.

HUNTER, G. L., ADAMS, C. E., ROWSON, L. E. A.: Inter-breed ovum transfer in sheep. J. Agric. Sci. **46**, 143 (1955).

KING, J. L., COLLINS, M., PETERSON, H. E.: Experimental interference with rabbit embryos in the early stages of their development. Amer. J. Physiol. **102**, 375 (1932).

KIPFER, K.: Die Muskulatur der Tuba uterina als funktionelles System. Acta Anat. **9**, 35 (1950).

KNEER, M.: Anatomie und Funktion der Muskulatur des menschlichen Eileiters. Arch. Gynäk. **176**, 156 (1948/49).

— CLESS, H.: Flimmerung und Strömung im menschlichen Eileiter. Geburtsh. Frauenheilk. **11**, 233 (1951).

KOESTER, H.: Tierexperimentelle Untersuchungen zur Physiologie der Eileitersekretion während der Zeit der Eipassage. Bibl. Gynaecologica (Basel) (im Druck).

KOK, F.: Bewegung des muskulösen Rohres der Fallopischen Tube. Arch. Gynäk. **127**, 384 (1926).

Lee, F. C.: The tubo-uterine junction of various mammals. Johns Hopk. Hosp. Bull. **42**, 335 (1928).

Leonard, S. L., Ferlman, P. L.: Conditions affecting the passage of spermatozoa through the utero-tubal junction of the rat. Anat. Rec. **104**, 89 (1949).

Martius, H.: Grundlagen der Gynäkologie. Stuttgart: Thieme 1950.

Mastroianni, L., Beer, F., Shah, U., Clewe, T. H.: Endocrine regulations of oviduct secretions in the rabbit. Endocrinology **68**, 92 (1961).

— Ehteshamzadeh, J.: Corona cell dispersing properties of rabbit tubal fluid. J. Reprod. Fertil. **8**, 145 (1964).

— Shah, U., Abdul-Karim, R.: Prolonged volumetric collection of oviduct fluid in the rhesus monkey. Fertil. Steril. **12**, 417 (1961).

— Wallach, R. C.: Effect of ovulation and early gestation on the oviduct secretion in the rabbit. Amer. J. Physiol. **200**, 815 (1961).

McClean, D., Rowlands, I. W.: The role of hyaluronidase in fertilization. Nature (Lond.) **150**, 627 (1942).

McLaren, A., Michie, A. D.: Studies on the transfer of fertilized mouse eggs to uterine foster mothers. J. exp. Biol. **33**, 394 (1956).

Mikulicz-Radecki, F. von: Experimentelle Untersuchungen über Tubenbewegungen. Arch. Gynäk. **128**, 318 (1926).

— Diskussionsbemerkung zu Westmann: Studien über den Bewegungsmechanismus des Eileiters.

— Der Eiauffangmechanismus bei der Frau und seine Bedeutung für die Sterilität. Schriften der Königsberger gelehrten Gesellschaft 13 (1937), Heft 6. Halle (Saale): Max Niemeyer.

Noyes, R. W., Adams, C. E., Walton, A.: The transport of ova in relation to the dosage of oestrogen in ovariectomized rabbits. J. Endocr. **18**, 108 (1959).

— — — The passage of spermatozoa through the genital tract of female rabbits after ovariectomy and oestrogen treatment. J. Endocr. **18**, 165 (1959).

— Dickmann, Z.: Relationship of ovular age ot endometrial development. J. Reprod. Fertil. **1**, 186 (1960).

— — Survival of ova transferred into the oviduct of the rat. Fertil. Steril. **12**, 67 (1961).

Parker, G. H.: The passage of sperma and eggs through the oviduct in terrestrial vertebrates, Phil. Trans. Roy. Soc. B **219**, 381 (1931).

Pincus, G., Kirsch, R. E.: The sterility in rabbits produced by injections of oestrone and related compounds. Amer. J. Physiol. **115**, 219 (1936).

— Werthessen, N. T.: The sterility in rabbits produced by injections of oestrone and related compounds. Amer. J. Physiol. **115**, 219 (1938).

Rauscher, H.: Discussion remark. Arch. Gynäk. **207**, 181 (1969).

Robinson, T. J.: Reproduction in the ewe. Biol. Rev. **26**, 121 (1951).

ROCKER, I.: The anatomy of the utero-tubal junction area. Proc. roy. Soc. Med. **57**, 707 (1964).

SCHRÖDER, R.: Der Eileiter: In: Handbuch der mikroskopischen Anatomie des Menschen, von v. MÖLLENDORF, Bd. VII, Teil 1. Berlin: Springer 1930.

SECKINGER, D.: Spontaneous contractions of the Fallopian tube of the domestic pig with reference to the oestrous cycle. Johns Hopk. Hosp. Bull. **34**, 236 (1923).

— CORNER, G.: Cyclic variations on the spontaneous contractions of the Fallopian tube of the macacus rhesus. Anat. Rec. **26**, 299 (1923).

— SYNDER, F. F.: Cyclic variations in the spontaneous contractions of the human Fallopian tube. Proc. Soc. exp. Biol. (N.Y.) **21**, 519 (1924).

— — Cyclic changes in the spontaneous contractions of the human Fallopian tube. Johns Hopk. Hosp. Bull. **39**, 371 (1926).

SIEGLER, S. L.: Fertility in women. Philadelphia: Lippincott 1944.

SQUIER, R.: The living egg and early stages of its development in the Guinea-pig. Contrib. Embryol. Carneg. Instn **21**, 223 (1932).

STANGE, H. H.: Vergleichende morphologische Untersuchungen an der menschlichen Tube in extremen Funktionszuständen zur Klärung der Frage: „Gibt es einen Sphincter infundibuli?" Zbl. Gynäk. **74**, 1176 (1952).

— Zur funktionellen Morphologie des Fimbrienendes der menschlichen Tube und des Epoophoron. Arch. Gynäk. **182**, 77 (1952).

STEGNER, H. E.: Elektronenmikroskopische Untersuchungen über die Sekretionsmorphologie des menschlichen Tubenepithels. Arch. Gynäk. **197**, 351 (1962).

SWYER, G. I. M.: A tubal factor concerned in the denudation of rabbit ova. Nature (Lond.) **159**, 873 (1947).

WALTON, A., HAMMOND, J.: Observations on ovulation in the rabbit. Brit. J. exp. Biol. **6**, 190 (1928).

WESTMAN, A.: A contribution to the question of the transit of the ovum from ovary to uterus in rabbits. Acta obstet. gynec. scand. **25**, 475 (1926).

— Studien über den Bewegungsmechanismus der Eileiter. Z. Geburtsh. Gynäk. **95**, 189 (1929).

— Investigations into the transit of ova in man. J. Obstet. Gynec. Brit. Emp. **44**, 821 (1937).

WIMSATT, W. A., WALDO, C. M.: The normal occurence of a peritoneal opening in the bursa ovarii of the mouse. Anat. Rec. **93**, 74 (1945).

WIMPFHEIMER, S., FERESTEN, M.: The effect of castration in tubal contraction of the rabbit as determined by the Rubin test. Endocrinology **25**, 91 (1939).

WINTENBERGER, S.: Recherches sur les relation entre l'oeuf et le tractus maternel chez les mammiféres. Etudes de la traversé de l'oviducte par l'oeuf fécondé de brebis. Ann. Zootech. **2**, 269 (1953).

Wintenberger-Torrés, S.: Mouvements des trompes et progression des oeufs chez la brebis. Ann. Biol. anim. **1**, 121 (1961).

Wislocki, G. B., Guttmacher, A. F.: Spontaneous persitalsis of the excised whole uterus and Fallopian tubes of the sow with reference to the ovulation cycle. Johns Hopk. Hosp. Bull. **35**, 246 (1924).

— Synder, F. F.: The experimental acceleration of the rate of transport of ova through the Fallopian tube. Johns Hopk. Hosp. Bull. **52**, 379 (1933).

Wu, D. H., Allen, W. M.: Maintenance of pregnancy in castrated rabbits by 17-alpha-hydroxy-progesterone and by progesterone. Fertil. Steril. **10**, 439 (1959).

Acknowledgement:

This presentation is an excerpt from „Bibliotheca Gynaecologica" (in press) Karger, Basel, New York, and printed with friendly premission of the Karger Verlag.

Metabolism of the Ovum Between Conception and Nidation*

R. L. BRINSTER

Laboratory of Reproductive Physiology, School of Veterinary Medicine, University of Pennsylvania, Philadelphia, Pennsylvania

With 5 Figures

Introduction

During the past ten years there has been a considerable increase in the interest in the metabolism of the early mammalian embryo, which has resulted in a gradual accumulation of information about the biochemistry and physiology of these stages in development. Despite this increase in information, our understanding of the events which occur between ovulation and implantation is rather fragmentary and far from complete. We are well aware of the more obvious events such as fertilization, cleavage, blastocyst formation, and early attachment of the embryo to the uterus, but we know almost nothing of the genetic changes and biochemical processes which underlie these obvious morphological events. At the present time we have the most information about energy metabolism and next most about protein metabolism; outside these two areas very little is known. Consequently, much of the following discussion will deal primarily with energy and protein metabolism and information which relates to these two areas.

Although a good deal of the interest concerning the developing zygote is in the human embryo and perhaps in the embryos of the economically important domestic animal species, even the casual observer in the field knows at once that the type of study required to obtain sufficient information on human embryos can never be

* I thank Excerpta Medica for permission to reprint here substantial parts of my chapter entitled "Developing Zygote" from their book, *Reproductive Biology*, edited by H. BALIN and S. GLASSER, 1970.

performed solely or predominantly on human embryos. There simply will never be sufficient human embryos available to conduct the detailed biochemical studies required, and, furthermore, many studies on human embryos would be precluded for moral reasons. Fortunately, there is increasing evidence that many similarities exist among the zygotes and early embryos of various Eutherian mammals. Consequently, it should be possible to conduct detailed and extensive studies on laboratory animals, thereby establishing a firm foundation of information concerning the early developmental stages. Then by the execution of a few well planned experiments, employing primate embryos and human embryos, it should be possible to make very accurate estimations of the conditions that exist in the development of the early human embryo. In fact, by employing this type of approach, we shall eventually understand early development in all mammalian embryos, including the human. However, at the present time most of the information that we have is about the metabolism of developing mouse and rabbit embryos, and consequently these are the species which will be discussed most in the following account.

Nutritional Requirements

A good place to begin a discussion of the metabolism of the mammalian embryo is with the requirements the embryo has for survival and development. Some of these we know quite well; others we know only very little about. The ionic requirements of the developing zygote is one of those areas in which we have some understanding. Most of what we know has been derived from studies made *in vitro*. Whitten (1956) found that the omission of calcium, magnesium, or potassium from the medium prevented growth and that development was delayed without phosphate. He also found that growth was possible in a medium with the osmolarity reduced to 0.09 osmols. Brinster (1964) showed that calcium concentration affected *in vitro* development of the mouse embryo but that magnesium concentration did not. He also demonstrated that the optimum osmolarity for *in vitro* development of the two-cell mouse embryo was 0.276 (Brinster, 1965a). Whitten and Biggers (1968) have obtained cleavage of one-cell mouse embryos to blastocysts in a medium with an osmolarity of 0.256 osmols. This has only occurred in a very few hybrid strains, and the increase in albumin contained in the medium appears to be the critical characteristic (BIGGERS and WHITTINGHAM, personal communication).

The effect of hydrogen ion concentration on development of the mouse embryo was also examined by WHITTEN, who found that mouse embryos developed between pH 6.9 and 7.7 (WHITTEN, 1956). BRINSTER (1965a) found that development of two-cell mouse embryos occurred between pH 5.87 and 7.78 in an atmosphere of five percent CO_2 and air. Although development occurred over a wider pH range, BRINSTER demonstrated that the optimum pH for development of the mouse embryo was dependent on the concentration of pyruvate or lactate in the medium. The higher the pH, the higher the concentrate of substrate necessary to obtain optimum development, suggesting that uptake of these compounds is related to the amount of the compound in the acid form. Thus, very little can be said about pH requirements without considering the energy source substrates in the culture medium, and we must await further investigation before we know what the optimum pH is for the development of the zygote. However, these results on the interaction of pH and energy substrate do suggest that the membrane of the developing embryo is able to show selective permeability to substrates (BRINSTER, 1965b).

An additional problem in studying the effect of pH on the embryo is that the studies involve the use of bicarbonate buffers containing CO_2. It has been known for a number of years that the bicarbonate buffer system containing CO_2 offered the best conditions for embryo development (BRINSTER, 1964, 1969d), but it has not been possible to show a specific effect of CO_2 on development by varying CO_2 content of the medium (BRINSTER, 1970a). Recently, WALES, QUINN, and MURDOCH (1969) have shown that CO_2 from the environment is incorporated by the embryo probably by the carboxylation of pyruvate to oxaloacetate. However, oxaloacetate plus another buffer cannot substitute for the bicarbonate in the medium, so there must be a further function of the bicarbonate in embryo metabolism.

Work on the oviductal secretions in a variety of species, including the rabbit (MASTROIANNI and WALLACH, 1961; HAMNER and WILLIAMS, 1965), the sheep (RESTALL, 1966), and the monkey (MASTROIANNI, URZUA, AVALOS, and STAMBAUGH, 1969) suggests that the ionic composition of the fallopian tube is similar to the ionic composition of blood serum. Whether there are minor variations within the fluid which are important is not definitely known, but at the present time there is no evidence for this. Hydrogen ion concentration is extremely difficult to determine *in vivo*. Nonetheless, a number of attempts have been made to determine the pH of reproductive tract fluids. BLANDAU, JENSEN, and RUMMERY (1958) found that the pH of the reproductive tract fluids of the rat was significantly higher than the pH of the peritoneal fluid. Likewise, VISHWAKARMA (1962) found that the mean pH of the fluids from the ligated rabbit uterus was 7.86 and that from the fluids from the ligated oviduct was 7.91. However, HAMNER and WILLIAMS (1965) found that bicarbonate concentration of rabbit oviductal fluid was 1.8 mg/ml, which is close to the concentration found in blood serum. Similar values have been found by RESTALL (1966) for sheep oviductal fluid. Therefore, if the CO_2 tension in the oviductal fluid is close to that of blood, which seems likely, then the pH would be approximately 7.4 for these fluids. It is not inconceivable that the difficulties involved in measuring pH in small volumes as for *in vivo* studies could have led to somewhat elevated pH estimations in other studies.

The carbohydrate or energy source requirements of zygotes developing *in vitro* have been studied in considerable detail, especially in the mouse embryo, and from these studies we have learned a great deal about the nutrition of the early embryo. HAMMOND (1949) and WHITTEN (1956) showed that the eight-cell mouse embryo would develop in a medium containing glucose and a simple protein. WHITTEN (1956, 1957) demonstrated that a number of other compounds, such as mannose, lactate, pyruvate, and malate could provide energy for the eight-cell embryo. However, compounds such as fructose, lactose. maltose, acetate, propionate, citrate and glycine could not provide the energy for development. He further demonstrated that two-cell mouse embryos developed into blastocysts in the presence of calcium lactate. BRINSTER (1965c) extended WHITTEN'S studies on energy sources for mouse embryos. He examined a large group of compounds and found only four, pyruvate, lactate, oxaloacetate, and phosphenolpyruvate, which would support development of the two-cell embryo to a blastocyst. He also determined optimum concentrations for these compounds and demonstrated that pyruvate occupied a central position in the support of development of the early mouse embryo (BRINSTER, 1965d). BIGGERS, WHITTINGHAM, and DONAHUE (1967) examined the energy source requirements of the oocyte and one-cell fertilized mouse embryo and found that of the five compounds studied (the above four and glucose), only pyruvate and oxaloacetate would allow development of these early stages. However, if cumulus cells were included with the embryos, then the other three compounds would allow development of the oocyte and one-cell fertilized ovum. COLE and PAUL (1965) employed feeder layers of HeLa cells to support early stages of the mouse embryo and considered that one of the contributions of these feeder layers of HeLa cells was to supply lactate and pyruvate. Perhaps a variety of cells is able to produce pyruvate and lactate, which can be utilized by the embryos. Recently, DONAHUE and STERN (1968) have shown biochemically that the cumulus cells are capable of producing pyruvate and lactate from glucose. At the eight-cell stage of development, a variety of compounds is able to support development of the mouse embryo into a blastocyst (BRINSTER and THOMSON, 1966). Table 1 shows the change which occurs with age in the energy source requirements of the developing mouse embryo.

Table 1. *Energy sources which will allow development of the mouse embryo. Taken from* BRINSTER *(1965c);* BIGGERS, WHITTINGHAM, *and* DONAHUE *(1967);* BRINSTER *and* THOMSON *(1966)*

Substrate	Oocyte	One-cell	Two-cell	Eight-cell
Pyruvate	+	+	+	+
Oxaloacetate	+	+	+	+
Lactate	—	—	+	+
Phosphoenolpyruvate	—	—	+	+
Glucose	—	—	—	+
α Ketoglutorate	—	—	—	+
Malate	—	—	—	+
Acetate	—	—	—	+
Succinate	—	—	—	—
Glucose-6-phosphate	—	—	—	—
Ribose	—	—	—	—

+ means development.
— means no development.

Energy substrates necessary for the rabbit embryo have not been worked out in great detail. However, DANIEL and OLSON (1968) have shown that pyruvate, lactate, and phosphoenolpyruvate are beneficial for development of the cleavage stages of the rabbit embryo. BRINSTER (1969a) has shown that the optimum concentration of pyruvate and lactate for the rabbit embryo is very similar to the optimum concentrations found for the mouse embryo, thus indicating, here again, that there are similarities between mammalian zygotes. This important similarity exists despite the morphological differences between the rabbit embryo and the mouse embryo and despite the large difference in lactic dehydrogenase found in these embryos (see later). However, it has been shown that the rabbit embryo is not as dependent upon exogenous energy substrates as the mouse embryo, thus indicating a difference in this characteristic, which may suggest the existence of internal energy stores that can be utilized at an earlier stage in the rabbit embryo than in the mouse embryo (BRINSTER, 1969a).

How closely the energy source requirement *in vitro* resembles the energy source available *in vivo* is not completely known. However, it has been shown (BISHOP, 1956; MASTROIANNI and WALLACH, 1961; and HOLMDAHL and MASTROIANNI, 1965) that the glucose concentration is very low in the rabbit oviduct, whereas the lactate and pyruvate concentrations are high. Similar observations have been made on the fallopian tube of the sheep (RESTALL, 1966) and the monkey (MASTROIANNI et al., 1969). LUTWAK-MANN (1962) has shown that in the uterus of the rabbit the glucose concentration is low and lactate concentration is high.

Studies on the *in vitro* requirements of the preimplantation embryo have demonstrated that amino acids and proteins are important to the development of the zygote. Mouse embryos younger than the eight-cell stage will die if the medium does not contain amino acids or protein. This requirement for an amino nitrogen source has been found to exist in the zygotes of a number of mammalian species. Even the rabbit embryo, which can survive and cleave *in vitro* without an energy sourve, will die when the amino nitrogen source is absent from the medium. The cleavage stages of developing embryos appear to be less particular about the form of the amino nitrogen source than is true of most other cell lines grown *in vitro*. For instance, in the two-cell mouse embryo it is not possible to demonstrate a requirement for a group of essential amino acids (BRINSTER, 1965e), and, in fact, it has been shown that the two-cell mouse embryo will develop into a blastocyst when the only amino nitrogen source is glutathione (BRINSTER, 1968a). Of the amino acids, cystine seems to be the one whose supply is most critical to the early cleavage stages of the mouse.

DANIEL and OLSON (1968) in studies on the early cleavage stages of the rabbit embryo identified six amino acids as essential for cleavage. However, the studies of MAUER, HAFEZ, EHLERS, and KING (1968) with five amino acids, including one of those described as essential by DANIEL and OLSON, did not show an essential requirement of the two-cell rabbit embryo for these five amino acids. Furthermore, BRINSTER (1969a) found that the early rabbit embryo would show normal cleavage for two days when the only amino nitrogen source in the medium was oxidized glutathione or such single amino acid as alanine or glutamine. Therefore, it appears that under certain conditions there is no essential amino acid requirement for the preblastocyst rabbit embryo, a condition which is similar to the situation found in the mouse embryo.

It should be emphasized that some amino nitrogen source is required in the media at all preimplantation stages for both the mouse and the rabbit embryo, and also for the rat (BRINSTER, unpublished). In addition, during blastocyst formation and expansion there is no doubt that amino acids and protein are beneficial both to the mouse and to the rabbit embryo. GWATKIN (1966) has shown that a macromolecule in the medium is required for blastocyst outgrowth of the mouse embryo *in vitro*, and furthermore he was able to identify a group of amino acids which were essential for outgrowth to occur.

The protein concentration of the fallopian tube fluid varies from one species to another. In the sheep it is 10—20 mg/ml (RESTALL, 1966), whereas in the rabbit it is 2—3 mg/ml (HAMNER and WILLIAMS, 1965). The fallopian tube contains a variety of amino acids, several of which are in high concentration. GREGOIRE, GONGSAKDI, and RAKOFF (1961) showed that the following amino acids were present in the fallopian tube of the rabbit: alanine, glycine, glutamic acid, threonine, and serine. The studies thus far made on the tubal and uterine fluid would suggest that the amino acids and proteins present are adequate for zygote development as far as nutrient requirement is concerned.

An extremely interesting question is whether there are specific inducer proteins or macromolecules in the fallopian tube which would act as stimulators of the zygote. The work of BEIER (1968) on uteroglobulin and KRISHNAN and DANIEL (1967) on blastokinin suggests that specific proteins do exist in the secretions of the uterus. KRISHNAN and DANIEL have postulated a specific biological role for blastokinin, which is the induction of blastocyst cavitation and expansion. However, the biological proof of this role has not been convincing to all. This is particularly true in view of the fact that ONUMA, MAUER, and FOOTE (1968) claim to have obtained rapid blastocyst expansion in the absence of this specific uterine secretion.

Since mammalian embryos have been grown outside the body in a variety of media, some of which are relatively simple, there is considerable question whether the embryos need specific growth factors, vitamins or nucleic acid precursors. COLE and PAUL (1965) reported that they found a beneficial effect of nucleic acid precursors when employed in the culture media for mouse embryos. However, TENBROECK (1968) has shown that the presence of nucleic acid precursors in the culture environment, either singly or in combination, does not enhance development of the preimplantation mouse embryo. Since we know that labeled nucleosides are incorporated by the early mouse embryo (MINTZ, 1964), the lack of effect cannot be due to impermeability of the embryos to the compound. Therefore, it appears likely that the early embryo either contains an adequate endogenous supply or is able to synthesize the components required for the formation of RNA and DNA. DANIEL (1967) found that the five-day rabbit blastocyst had a requirement for certain vitamins and growth factors. He found that development was improved by thiamine, riboflavin, niacin, pyridoxine, folic acid, inositol, and hypoxanthine. BRINSTER and BRUNTON (unpublished), using the techniques described by BRINSTER (1963, 1965a) could find no beneficial effect of thiamine, riboflavin, niacin, pyridoxine, panthothenic acid, biotin, choline, inositol, B-12, or ascorbic acid on the development of two-cell mouse embryos. Each vitamin was employed singly in concentrations from 10^{-3} M to 10^{-7} M. Thus, the zygote as observed from *in vitro* culture experiments appears to have rather simple but specific requirements for development.

Stored Nutrients

Embryos such as the invertebrate sea urchin and the amphibian frog are able to and, in fact, must be able to store considerable amounts of nutrient in order to live in a relatively neutral or hostile environment. The mammalian embryo, although "free living" during the preimplantation period, nonetheless lives in an environment which contains a variety of nutrients. However, it is obvious even from morphological evidence that the embryo contains endogenous stores of nutrients. Some embryos (carnivores) appear to contain larger amounts of lipoprotein globules than other embryos. However, we cannot be sure of this until adequate biochemical measurements are made. LOEWENSTEIN and COHEN (1964) found that the one-cell mouse embryo contained 20 mμgrams of protein, 8 mμgrams of carbohydrate, and 4 mμgrams of fat. The total wet weight was estimated to be 318 mμgrams. LEWIS and WRIGHT (1935) found that the volume of the mouse embryo actually decreased from the one cell stage to the early morula. BRINSTER (1967a) found that the protein content of the mouse embryo decreases by 25 percent from ovulation to the morula stage. After blastocyst formation protein content begins to increase. It seems a general characteristic of preimplantation embryos that there is little or no increase in mass before blastocyst formation, and then during blastocyst formation there is an increase in size and tissue mass up to and through implantation. The increase in blastocyst size varies, depending on the species; a substantial increase in mass occurs in the rabbit embryo during blastocyst formation but quite frequently a much smaller increase in mass occurs in other species, such as the mouse, rat, and human before implantation.

REAMER (1963) has demonstrated a considerable store of both DNA and RNA in the early stages of the mouse embryo. At ovulation the mouse embryo contains 45 μμgrams of DNA per embryo. This is considerably more than the 2.5 μμgrams one would expect to be present in a haploid mouse cell. Figure 1 shows the level of DNA in the early cleavage stages of the mouse embryo. As cleavage goes on, the DNA content per cell gradually decreases until at the 16-cell stage only about 20 μμgrams of DNA are present per cell. The DNA per cell must continue to decrease through succeeding cleavage stages in order to reach the 5 μμgrams per cell which is found in postimplantation embryonic cells. The nature of this excess DNA in the newly ovulated ovum is not known, and in fact its very existence has been questioned (MINTZ, 1964), because the number of polar bodies included in the assays and the state of DNA synthesis within the embryo were not determined. However, regardless of these conditions, the measurements still indicate excess DNA in the embryo. It has been suggested that the excess

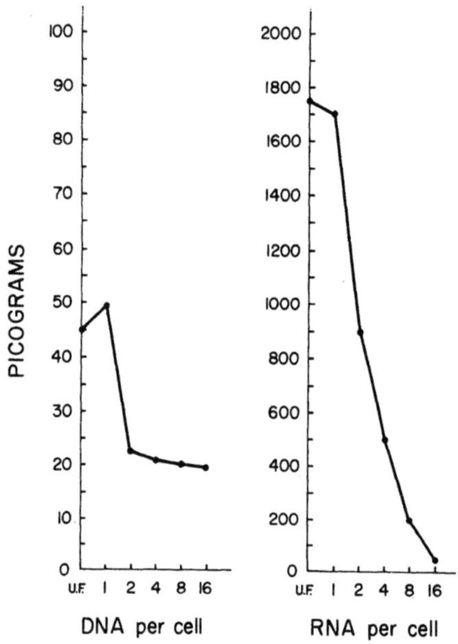

Fig. 1. DNA and RNA content of the preimplantation mouse embryo (REAMER, 1963)

represents an extrachromosomal pool of DNA or DNA precursors upon which the embryo may rely during the initial period of cell replication. Whether some of this DNA is in the form of mitochondrial DNA or associated with yolk spherules, as in the sea urchin embryo (TYLER, 1967), is not known, but ultrastructural studies indicate that the number of mitochondria present in the early mouse embryo is small, and the internal cristae of these mitochondria are poorly developed (CALARCO and BROWN, 1969). Certainly, additional studies will be required to resolve the existence and the nature of the DNA in the early embryo.

The RNA content of the mouse embryo at ovulation is extremely high (REAMER, 1963). The average postimplantation embryonic cell RNA content is 20 $\mu\mu$grams, whereas at ovulation there are 1750 $\mu\mu$grams of RNA per ovum. The RNA per embryo rises slightly and then decreases steadily to the 16-cell stage. The

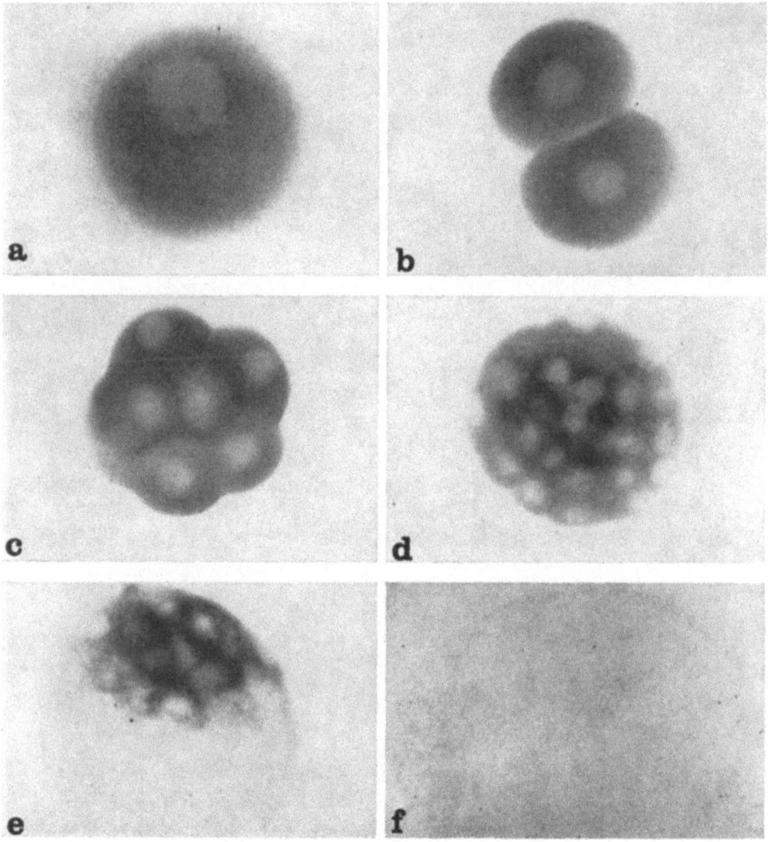

Fig. 2a

Fig. 2a and b. Glycogen content of the preimplantation mouse embryo. a. Histochemical studies. The dark color indicates the presence of glycogen. The developmental stages are: (*a*) 1-cell, (*b*) 2-cell, (*c*) 8-cell, (*d*) Morula, (*e*) Early blastocyst, (*f*) Blastocyst just prior to implantation. (THOMSON and BRINSTER, 1966)

RNA content per cell falls rapidly from the one-cell to the 16-cell stage (see Fig. 1). At the 16-cell stage the RNA per individual cell is only about 50 μμgrams. Again, the significance of this RNA is not known. It has been suggested that the RNA represents a pool of precursors from which specific RNA species are formed later in embryonic life. Certainly the ultrastructural studies on early

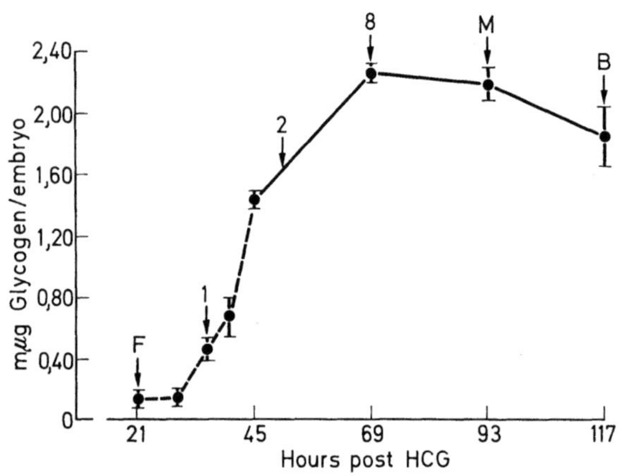

Fig. 2b. Biochemical studies. Ovulation occurs approximately 12 hours after HGC injection. (*F:* Fertilized 1-cell egg; *1:* approximate time of first cleavage; *2:* approximate time of second cleavage; *8:* eight-cell eggs; *M:* morulae and early blastocysts; *B:* late blastocysts). (STERN and BIGGERS, 1968)

mammalian embryos do not suggest that there are large quantities of RNA present in the form of polyribosomes. Generally, the ribosomal constituents of the embryo after ovulation and during the first few cleavages are relatively small in number and are not formed into polyribosomes (CALARCO and BROWN, 1969; ENDERS and SCHLAFKE, 1967; SZOLLOSI, 1967; ZAMBONI, MISHELL, BELL and BACA, 1966). It would certainly be desirable to have supportive evidence or additional determinations on the nucleic acid content of the early mammalian embryo.

A possible source of energy for the embryo would be endogenous stores of glycogen. In the mouse embryo histochemical studies (THOMSON and BRINSTER, 1966) indicate substantial amounts of glycogen are present at ovulation and that the amount increases until the eight-cell stage of development. After blastocyst formation a major part of the glycogen is utilized (see Fig. 2a). The histochemical studies also indicate that the pathway for glycogen breakdown is only operating at a very low level before blastocyst formation, since it has been shown that the stored glycogen is not available to the embryo for utilization. Histochemical studies on the rat (GREGOIRE, personal communication) and the rabbit (THOMSON and BRINSTER, unpublished) also indicate glycogen present in the early cleavage stages. Biochemical studies (STERN and BIGGERS, 1968) show that the embryo synthesizes almost all of its glycogen between the one-cell and the eight-cell stage of development and then uses a small portion of this glycogen during blastocyst formation (see Fig. 2b). Thus, the biochemical studies suggest more synthesis early in development and less utilization later in development than do the histochemical studies. However, a recent investigation on incorporation of pyruvate and glucose into the mouse embryo between the one-cell and eight-cell stage indicates that only very small quantities of carbon from these substrates are incorporated into the glycogen of the embryo in the early cleavage stages. Incorporation of substrate into the polysaccaride fraction is much greater at the time of blastocyst expansion, thus suggesting that the glycogen turnover at the blastocyst stage is high. Some of the incorporation could be into a polysaccaride of unknown chemical structure which is synthesized in large quantities by the mouse blastocyst (PIKÓ, 1970). In general, the incorporation findings are more in agreement with the histochemical than with the biochemical studies. It would be very interesting to determine if the embryo growing *in vitro* shows the same glycogen accumulation from the one-cell to the eight-cell stage as was found in the biochemical studies.

Oxygen Consumption and Metabolic Rate

Fundamental to understanding the development of the zygote is a knowledge of its basic metabolic rate, because the metabolic rate of the cell or group of cells will determine the maximum ability to do work. In short, the metabolic rate gives us a very good idea of the synthetic capability of the tissue. Because of the small size of the mammalian embryo it has proven difficult to make accurate measurements of oxygen consumption and some parameter of tissue mass, such as dry weight, wet weight, or protein content. Cartesian diver techniques have proven useful in measuring oxygen consumption of a variety of embryos, and some of these results are summarized in Table 2. Unfortunately, very few definitive studies have been made on tissue mass of the preimplanted embryo. In most cases we must depend on volume measurements and estimations to convert these volumes to protein content or dry weight.

Table 2. *Oxygen uptake of preimplantation embryos*

Stage of Development	Uptake in mμl/embryo/hour					
	Mouse[a]	Rabbit[b]	Rat[c]	Cow[d]	Pig[e]	Sea Urchin[f]
Follicular	—	—	1.11	23.0	4.31	—
Unfertilized	0.16	—	—	—	0.82	0.40
Fertilized	0.16	0.64	0.56	—	—	0.60
Day 2	0.15	0.60	—	—	—	—
Day 3	0.19	0.84	1.23	—	—	—
Day 4	0.41	5.45	1.40	—	—	—
Day 5	0.54	53.00	2.00	—	—	—
Day 6	—	214.20	—	—	—	—

[a] Mills and Brinster (1967).
[b] Fridhandler (1961).
[c] Sugawara and Umezu (1961).
[d] Dragiou, Benetato, and Opreau (1937).
[e] Sugawara (1962).
[f] Monroy (1965).

If we examine the values found by Fridhandler (1961) for oxygen uptake of the rabbit embryo and those found by Mills and Brinster (1967) for the mouse embryo, we see that these are in good agreement for the preblastocyst stages of the embryo. The rabbit embryo is approximately 3.5 times the volume of the mouse embryo, and the oxygen consumption of the rabbit embryo likewise is approximately 3.5 times the oxygen consumption of the mouse embryo. After blastocyst formation the rabbit embryo undergoes an unusually rapid expansion and increase in tissue mass, and therefore the oxygen consumption rises rapidly. Consequently, comparisons made between the embryos of the two species after blastocyst formation can have very little meaning.

Since the protein content of the mouse embryo has been determined, it is possible to estimate very accurately the dry mass of the embryo. Using the oxygen consumption and the dry mass, we can then calculate the Q_{O_2} for the embryonic tissue at various stages of development from ovulation to implantation. We find that the newly ovulated mouse ovum has a Q_{O_2} of 3.7 µl/mg dry weight, which is a relatively low metabolic rate, comparable to skin, whereas the blastocyst at the time of implantation is metabolically very active and has a Q_{O_2} of 16.3, which is comparable to whole brain. Thus, there is a dramatic increase in the overall metabolic rate of the mouse embryo from the time of ovulation to implantation. Furthermore, a rather large part of this increase in metabolic rate is found at about the time of blastocyst formation.

Energy Metabolism

The utilization of substrates by the embryo has recently been investigated, using radioactive techniques. The findings of these studies have materially increased our knowledge of the pathways utilized by the embryo during development in the breakdown of substrates and the formation of energy. The first study performed on substrate oxidation demonstrated that the mouse embryo was able to oxidize only a very small amount of glucose during the first one or two days after ovulation (BRINSTER, 1967b). However, the embryo's ability to form CO_2 from glucose increased one hundredfold during the preimplantation period (see Table 3). Major increases in CO_2 occurred at fertilization and particularly at the time of blastocyst formation. Studies on the oxidation of pyruvate and lactate (BRINSTER, 1967c) demonstrated that both of these substrates were very actively oxidized to CO_2 by all stages of the mouse embryo (see Table 3). Pyruvate was oxidized to a significantly greater degree than lactate. At the time of ovulation pyruvate oxidation could account for 100 percent of the oxygen utilized by the embryo, whereas glucose oxidation could account for only two percent. By the time of implantation glucose appeared to be oxidized as readily as pyruvate. Either substrate alone could account for approximately 65 percent of oxygen consumption. The unaccounted-for 35 percent of oxygen consumption must be used for oxidation of other substrates, both endogenous and exogenous. A probable endogenous substrate is stored glycogen, whereas exogenous substrates could be bovine serum albumin or amino acids.

Studies on rabbit embryos likewise indicate that glucose is only poorly oxidized by the newly ovulated zygote (BRINSTER, 1968b). However, at later stages, particularly in the late morula and blastocyst, glucose is oxidized in substantial amounts (Table 3).

Table 3. *Comparison of carbon dioxide production from glucose, pyruvate and lactate by the preimplantation mouse and rabbit embryo. Taken from* BRINSTER *(1967b, c; 1968b; 1969b)*

Stage of development	Glucose		Pyruvate		Lactate	
	mouse	rabbit	mouse	rabbit	mouse	rabbit
Unfertilized	0.13	0.61	7.24	16.68	3.09	14.71
Fertilized	0.68	0.81	6.95	14.47	3.31	11.28
Day 2	1.19	5.98	6.03	13.42	2.77	22.11
Day 3	2.16	13.71	7.25	22.31	4.54	23.06
Day 4	8.84	50.33	11.96	201.19	11.20	178.20
Day 5	14.69	1335.15	15.73	2032.91	15.06	655.20
Day 6	—	4904.98	—	9882.36	—	3844.20

All values are μμmoles of CO_2 produced/embryo/hour.

The substrate concentration was for glucose, 1 mg/ml; for pyruvate, 5×10^{-4} M; and for lactate, 5×10^{-2} M.

Carbon dioxide is formed from pyruvate quite readily by all stages of the rabbit embryo and accounts for 50 to 100 percent of oxygen uptake throughout the preimplantation period (BRINSTER, 1969b). Lactate is also oxidized very well by the preimplantation rabbit embryo (BRINSTER, 1969b). Therefore, despite the fact that the rabbit embryo has only about 1/70th the lactate dehydrogenase activity (see later) that the mouse embryo contains, both lactate and pyruvate are utilized very well by the rabbit embryo. It should be emphasized that in the early stages of development, oxidation of all three substrates, glucose, pyruvate, and lactate, is comparable for the embryos of the two species. Both embryos show a marked preference for pyruvate in the early stages of development, whereas glucose is hardly oxidized at all by the newly ovulated and recently fertilized embryo. The rabbit embryo oxidizes about three times the amount of substrate that the mouse embryo oxidizes, and this is in agreement with their comparable sizes.

Recent investigations on the oocyte of the rhesus monkey indicate that pyruvate is oxidized to a greater extent than glucose (BRINSTER, 1970b). However, the ratio of CO_2 formed from pyruvate to CO_2 formed from glucose by the oocyte of the rhesus monkey is 20 : 1 instead of 50 : 1 as is found in the mouse oocyte. The data indicate that the difference in the ratio results from a relatively greater ability of the rhesus monkey oocyte to oxidize glucose.

The small amount of glucose which is oxidized during the first day or two of life is handled differently by the mouse and rabbit embryo. Studies employing specifically labelled glucose have shown that CO_2 formed from carbon-one of glucose is approximately equal to the CO_2 formed from carbon-six of glucose by the mouse embryo throughout the preimplantation period (BRINSTER, 1967b). Since the C_1 to C_6 ratio is near one, this suggests that oxidation is occurring primarily through the Krebs cycle and not by the pentose shunt. In the rabbit embryo FRIDHANDLER (1961) showed that the preblastocyst rabbit embryo oxidized carbon-one to a much greater extent than carbon-six. After blastocyst formation the ratio was approximately one. BRINSTER (1968b) has found ratios very similar to FRIDHANDLER'S, thus ruling out the possibility that the different ratios for mouse and rabbit are due to different techniques. Therefore, it appears that the pentose shunt is active in glucose oxidation in the preblastocyst rabbit embryo but not in the blastocyst; whereas there is no evidence that the pentose shunt is very active at any stage of development in the mouse embryo.

Recently, WALES (1969) has been able to measure lactate formation from glucose during the preimplantation period in the mouse. In the newly ovulated ovum equal molar amounts of glucose are utilized for oxidation and for lactate formation. However, only about one-third as much glucose is oxidized as is converted to lactate by the mouse blastocyst. An interesting observation is that lactate formation by the mouse embryo parallels closely the oxidation of glucose, that is, lactate formation from glucose increases approximately one hundredfold during the preimplantation period (see Table 4). This is additional evidence for the hypothesis (BRINSTER, 1967b) that the oxidation of glucose is low in the early stages of development because the Embden-Meyerhof pathway operates only very slowly. Table 4 summarizes glucose utilization by the

Table 4. *Regulation of glucose utilization in the preimplantation mouse embryo. Taken from* BRINSTER *(1967b, 1968d, 1969c), and* WALES *(1969)*

Stage of development	Hexokinase activity μμmoles of NADP reduced/hr/embryo	CO_2 produced[a]	Lactate produced[a]	Carbon incorporated[b]	Total[a]
Unfertilized	1.23 (7.38)	0.13	0.15	—	—
Fertilized	1.76 (10.56)	0.68	0.72	—	—
Two-cell	1.70 (10.20)	1.19	2.10	1.65	4.94
Eight-cell	2.24 (13.44)	2.16	3.75	1.72	7.63
Morula	5.63 (33.78)	6.73	27.96	—	—
Blastocyst	7.94 (47.64)	10.94	33.30	7.01	41.70
Late blastocyst	9.40 (56.40)	14.69	42.30	20.01	77.00

( ) Indicate maximum intracellular glucose carbon available to the embryo.

[a] Values are μμmoles of carbon from glucose per hour per embryo.

[b] Values are average of previous 24 h.

mouse embryo during the preimplantation period. Wales found that in the mouse morula and blastocyst, pyruvate and acetate could also be identified as products of the metabolism of glucose. Lactate, pyruvate and acetate were found approximately in the ratio of 90 : 10 : 1. In the rabbit embryo FRIDHANDLER, WASTILA, and PALMER (1967) have measured lactate production from glucose by the five-day, six-day, and seven-day rabbit blastocyst. Their studies indicated that most of the glucose uptake of the six-day rabbit blastocyst was accounted for by lactate production even under aerobic conditions. These findings are at variance with those of BRINSTER (1968b), who found substantial amounts of CO_2 formed from glucose by the six-day rabbit blastocyst. Although the methods of FRIDHANDLER and BRINSTER are different, a rough comparison of the data suggests that the six-day rabbit blastocyst probably converts glucose to equal moles of CO_2 and lactate. This would be similar to the mouse blastocyst just prior to implantation.

Studies have recently been made on the incorporation of pyruvate and glucose carbon into the mouse embryo over relatively long periods of time (BRINSTER, 1969c). These studies show that incorporation of glucose carbon is greater than pyruvate carbon for all developmental stages. This is in marked contrast to substrate oxidation where pyruvate carbon is oxidized much better than glucose carbon up to the blastocyst stage, after which both substrates seem to be oxidized equally well. These studies show that pyruvate is oxidized more than it is incorporated and that glucose is incorporated more than it is oxidized throughout the preimplantation period.

The studies on metabolism so far show that there is progressive development or change in the importance of the major energy pathways. This is particularly true in the case of the Embden-Meyerhof pathway. We see very little glycolytic ability, either in the mouse, rabbit, or rhesus monkey ovum, at or near the time of ovulation. Glycolytic ability gradually increases as development proceeds with a marked increase at about the time of blastocyst formation. The Krebs cycle appears to be a major energy supplying pathway throughout the preimplantation period. This is evidenced by the high rate of oxidation of pyruvate, both in the mouse and rabbit embryo. The evidence from the rhesus monkey oocyte indicates that pyruvate oxidation is also important in this species. Statements which have been made in the past indicating or suggesting little or no Krebs cycle activity and low oxidative ability of the developing zygote have not been borne out by critical experiments. The pentose shunt appears to be very active in the preblastocyst rabbit embryo but not in the blastocyst of the rabbit embryo. There is no evidence in the mouse embryo that pentose shunt activity is very high. It seems probable that changes in enzyme activity play an important role in altering the capability or activity of the different pathways.

Protein and Nucleic Acid Metabolism

MINTZ (1962) has studied, using autoradiography, the uptake of nucleic acids and protein precursors by the preimplantation mouse embryo grown *in vitro*. Although the embryos will develop without exogenous precursors of DNA in the medium, when tritiated thymidine is present, it is incorporated into the nuclei of the cleavage stages and is still present in the nuclei when the embryos reach the blastocyst stage. Likewise, when tritiated uracil is present, it is taken up by the cells of the embryo throughout the preimplantation period. However, in the case of radioactive uracil there is little radioactivity seen in sections of the very early embryo. It is not until true nucleoli are formed, after the second cleavage, that much tritiated uracil is taken up by the embryo. In the late morula and blastocyst, the autoradiographs indicate very heavy labelling.

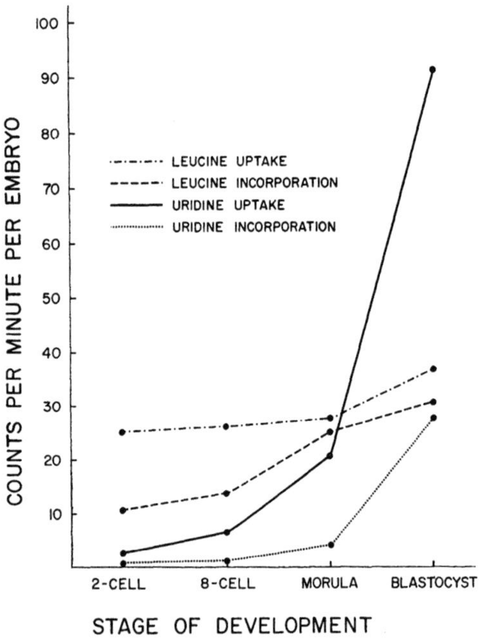

Fig. 3. Uptake and incorporation of C^{14}-leucine and H^{3}-uridine into protein and RNA of mouse embryos (data taken from TASCA, 1969)

The label appears first in the nucleus, then in the nucleoli, and finally in the cytoplasm. The pattern of incorporation would suggest that the RNA may be ribosomal in type. When tritiated leucine is included in the culture medium in which the embryos are grown, all developmental stages show autoradiographic evidence of uptake. Again, very low levels are seen in the one-cell and two-cell stages, but incorporation increases at the eight-cell stage and is high in the blastocyst.

A number of workers have tried to assess quantitatively the incorporation or uptake of labelled precursors by the developing zygote. MONESI and SALFI (1967) found that the uptake of tritiated lysine, leucine, and uridine was relatively level until about the eight-cell stage, and then there was a dramatic increase in uptake by the mouse embryo. TASCA (1969) measured the uptake and incorporation separately for both leucine and uridine. His findings are shown in Fig. 3 and provide an example of the results found by others in this field. In general, both the uptake and incorporation of leucine and of uridine increases with the age of the embryo; however, the greatest rate of increase seems to at the morula stage.

The effect of inhibitors of nucleic acid and protein metabolism on the *in vitro* development of the preimplantation mouse embryo has been examined, and it was found that these compounds will inhibit development of the embryo as well as the uptake of precursors. For instance, it was found that the dose of actinomycin (which inhibits DNA dependent RNA synthesis) required to prevent cleavage and development of the mouse embryo is about one-millionth the dose required to inhibit cleavage of the sea urchin embryo. This suggests that the mouse embryo relies much more than the sea urchin on DNA dependent RNA synthesis during the early cleavage stages.

TASCA (1969) found that actinomycin inhibited one-quarter to one-half of the uridine incorporation in mouse embryos. Surprisingly, the actinomycin did not affect leucine incorporation during the three-hour *in vitro* incubation period. If leucine incorporation continued for three hours in the absence of new RNA synthesis, protein synthesis at the translational level must continue in the absence of new RNA synthesis. This suggests that the embryo depends during the early cleavage stages to some extent upon previously synthesized RNA. In the sea urchin embryo synthesis during the early cleavage stages up to gastrulation depends almost entirely upon preovulatory synthesized RNA (GROSS and COUSINEAU, 1963, 1964). However, the protein synthesis in the mouse embryo seems to be more dependent on new RNA than is protein synthesis in the sea urchin embryo. This is suggested from experiments which show that both incorporation and cleavage of the mouse embryo are inhibited with much lower levels of actinomycin than are required to inhibit protein synthesis and cleavage in the sea urchin embryo (THOMSON and BIGGERS, 1966; MINTZ, 1964; GROSS and COUSINEAU, 1964). However, recently MANES and DANIEL (1969) on the basis of incorporation studies in the rabbit embryo suggested that the protein synthetic apparatus necessary to carry the embryo through cleavage is totally present in the zygote and is merely parcelled out to each blastomere with cleavage division. They interpret their data as indicating that the embryonic genome is not utilized for protein synthesis which occurs during cleavage. MANES and DANIEL feel that the actinomycin inhibitory effect seen in a number of other studies is a "side effect" of actinomycin and that its depression of RNA synthesis is not the cause of the inhibition of embryonic development. Although it is appealing to believe that the mammals are similar to the invertebrates relative to their nucleic acid synthesis during the preimplantation period, considerably more evidence certainly will be needed to substantiate this hypothesis, since the bulk of the work to date suggests that important differences do exist between the embryos of invertebrates and mammals.

TASCA's study showed that cycloheximide inhibits both leucine incorporation and uridine incorporation into the early cleavage stages of the mouse embryo. One would expect cycloheximide, which has been shown to inhibit leucine incorporation into other cell systems, to inhibit leucine incorporation into the early mouse embryo. However, the inhibition of uridine incorporation by cycloheximide was unexpected and suggests that the inhibitor is blocking the formation of a protein which is necessary for RNA synthesis and uridine incorporation. This protein could very well be RNA polymerase.

ELLEM and GWATKIN (1968) have identified the specific nucleic acid fractions into which RNA is incorporated. They were not able to detect incorporation before the eight-cell stage of development, but from the eight-cell stage through the blastocyst stage, they found a gradual increase in incorporation into soluble and messenger RNA. This suggests a gradual increase in the number of genes being transcribed during this period. They found a very marked increase in incorporation of labelled uridine into ribosomal RNA from the eight-cell stage to the morula and blastocyst stages. Table 5 shows the percentage distribution of isotope among the nucleic acids after exposure to uridine. The preponderance of incorporation into ribosomal RNA is striking. ELLEM and GWATKIN found that actinomycin severely inhibited ribosomal RNA and soluble RNA incorporation but did not affect incorporation into messenger RNA.

Table 5. *Percentage distribution of isotope among nucleic acids after 5 hours exposure to H^3-uridine. Modified from* ELLEM *and* GWATKIN *(1968)*

Nucleic acid	Embryonic stage			Diploid[a] cells *in vitro*
	8-cell	Morula	Blastocyst	
Soluble RNA	18.2	9.8	14.6	5.0
Ribosomal RNA	63.9	82.0	68.7	44.9
Messenger RNA	17.9	8.2	16.7	50.1

[a] Percentage distribution of H^3-uridine among fractions of nucleic acids from monolayers of human diploid cells (W138) after one hour exposure to isotope.

WOODLAND and GRAHAM (1969) have examined in detail the species of RNA into which labelled uridine is incorporated. Fig. 4 shows the gradual change in incorporation pattern from the two-cell stage to the eight-cell stage. At the early two-cell stage most of the activity is incorporated into low molecular weight RNA (5S). During succeeding cleavage stages incorporation shifts into high molecular weight RNA. By the eight-cell stage of development the radioactive peaks appear at the same position as 28S and 16S ribosomal RNA. The low molecular weight RNA can be identified in the four-cell stage as 4S (soluble) RNA. These studies show very clearly the initiation of synthesis of specific types of RNA and indicate that mice are quite different from most other embryos which have been studied (sea urchin and frog) in the extremely early stages of development, at which they begin to synthesize 4S RNA and ribosomal RNA. It seems quite probable that the genes for 4S RNA and ribosomal RNA both become activated at the four-

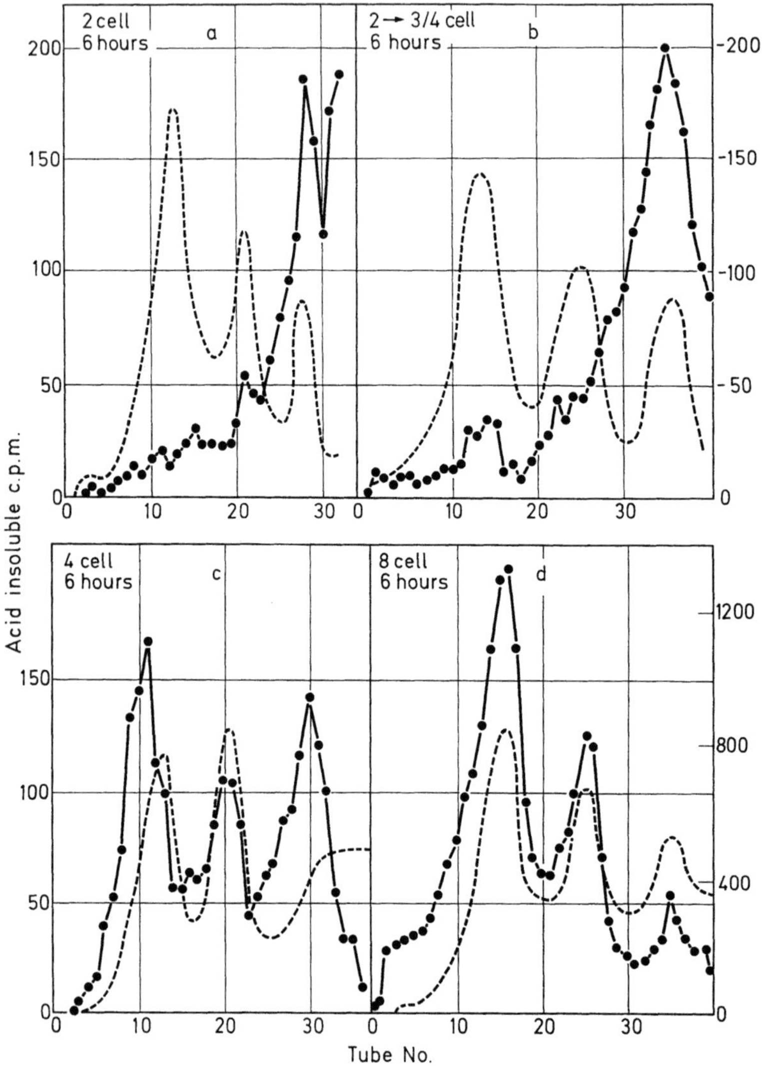

Fig. 4a—d. Sucrose-density-gradient centrifugation of RNA extracted from mouse embryos labelled with 5-^{3}H-uridine for 6 h. (a) 947 two-cell embryos. (b) 641 two-cell embryos which had reached the three or four cell stage at the end of incubation. (c) 384 four-cell embryos labelled for the first 6 h of the four-cell stage (some three-cell at the start of incubation). (d) 100 six to twelve cell embryos, average 8.3. The OD trace represents added mouse liver RNA, which was used as a marker for the types of RNA. (●) Acid soluble c.p.m.; ... OD 254 mμ (arbitrary scale). Low molecular weight RNA sediments near tube 30, 16S RNA near tube 20, and 28S RNA near tube 10 in these examples

cell stage and during the same cell cycle. This is quite different than in the sea urchin where 4S RNA synthesis begins a considerable time before ribosomal RNA synthesis can be detected.

Recently PIKÓ (1970) has confirmed and extended some of the observations of WOODLAND and GRAHAM. From his studies, PIKÓ estimated that the half life of the 39S RNA (precursor of ribosomal RNA) was fifteen minutes, and the half-life of heterodisperse RNA (probably containing messenger RNA and precursors of messenger RNA) was ten minutes or less in the blastocyst just prior to implantation. He also estimated that the minimum rate of production of ribosomes was 6×10^5 ribosomes per minute in a blastocyst of 60 cells. This minimum rate of ribosome production is comparable on a per cell basis to that estimated for HeLa cells (MADEN, 1968).

The distribution of the RNA contained in the newly ovulated mouse embryo to succeeding cells of the cleavage stages results in a cellular content of RNA of approximately 50 $\mu\mu$grams at 16-cell stage (REAMER, 1963). Since this is not too different from what is found in the normal postimplantation embryonic cell, we would expect to see an increase in the embryo's ability to synthesize RNA in order to prevent the depletion of cellular RNA. This appears to be the situation. Beginning at about the four-cell stage the embryo's ability to synthesize RNA increases gradually, and then at about the 16-cell stage of development there is a marked increase in RNA synthesis. The primary species of RNA being synthesized seems to be soluble RNA and ribosomal RNA. Since the newly ovulated and early mouse embryo cleavage stages do not contain very many polyribosomes or ribosomes, it seems logical that ribosomal RNA should be one of the first types of RNA synthesized by the embryo. Also, since we know that protein synthesis is occurring in the mouse embryo from fertilization onwards and that protein synthesis increases markedly at about the morula stage of development, we can expect an increase in the need of the embryo for soluble RNA. The embryo seems to meet this need by synthesizing this species of RNA beginning at the four-cell stage. Unfortunately, some very interesting questions are left unanswered. One of the questions is the form of the RNA present in the ovulated egg which was measured by REAMER in her studies. A second and perhaps even more important question which may be related to the first question is what type of messenger RNA is present in the newly ovulated egg and when does the embryo begin synthesizing its own messenger RNA. Perhaps these questions will be answered in the next few years.

There is little known about the details of DNA synthesis during the preimplantation period. It has been shown autoradiographically that tritiated thymidine is incorporated into the mouse embryo at all the cleavage stages (MINTZ, 1964), indicating a synthesis of DNA throughout this period. However, it has also been shown that the embryos develop quite well without an exogenous source of DNA precursors (TENBROECK, 1968), thus suggesting adequate precursors are available in the embryo and that the embryo is capable of whatever synthetic steps are required to make DNA from its precursor material. Perhaps some of the high DNA content measured by REAMER (1963) represents precursor material. Thymidine incorporation into DNA, at least during the blastocyst stage, is proportionate to the number of nuclei produced during the incubation period (PIKÓ, 1970). The exogenous precursor (thymidine) in these experiments was shown to be rapidly incorporated, and it was estimated that 65 % of the thymidine residues in the DNA of the nuclei produced during the labelling period were from labelled precursor. It appears that little or no mitochondrial DNA is synthesized by the cleaving mouse embryo (PIKÓ, 1970). This is similar to the findings in sea urchin embryos (GROSS, 1964).

Enzyme Activity

The activity of several important enzymes has been measured in the preimplantation embryo. In the mouse embryo lactate dehydrogenase, malate dehydrogenase, glucose-6-phosphate dehydrogenase, and hexokinase have been measured. In the rat and rabbit embryo, lactate dehydrogenase has been measured. Lactate dehydrogenase is extremely interesting for several reasons. First, it is closely associated with lactate and pyruvate, two of the most important substrates for the early mammalian embryo and, second, because there is a marked change in the isoenzyme pattern with development. It has been found that the mouse embryo has a lactate dehydrogenase activity ten times higher than skeletal muscle, which is the most active tissue in the adult mouse body (BRINSTER, 1965f). The activity falls off fairly rapidly through the preimplantation period so that at implantation the activity is only about one-tenth that found in the newly ovulated ovum. This high activity was at first thought to be related to the need of the embryo for lactate, but it has now been found that there is a whole range of lactate dehydrogenase activity among the oocytes of various species (BRINSTER, 1968c). The rabbit and the human have only about 1/70th the lactate dehydrogenase activity of the mouse embryo, but we know that lactate and pyruvate are as important to the energy metabolism of the rabbit as they are to the mouse embryo (see previous sections). Furthermore, it can be calculated that there is 10,000 times as much lactate dehydrogenase present in the mouse ovum as is necessary to convert lactate to pyruvate for pyruvate oxidation. One possible explanation is that the lactate dehydrogenase found in the newly ovulated ovum may be more closely related to ovarian oocyte metabolism than to the metabolism of the cleaving embryo. An interesting characteristic of the lactate dehydrogenase in the newly ovulated ovum is that the electrophoretic isozyme pattern is a single band, of type H (AUERBACH and BRINSTER, 1967). The electrophoretic pattern does not change throughout cleavage and, in fact, cannot be altered by a variety of experimental techniques. However, at implantation new electrophoretic bands of lactate dehydrogenase appear, which indicates the synthesis of type M LDH (see Fig. 5). In addition, it appears that not only M type synthesis but also H type synthesis is turned on at implantation (AUERBACH and BRINSTER, 1968).

The activity of glucose-6-phosphate dehydrogenase is approximately three percent of that found for lactate dehydrogenase (BRINSTER, 1966a). The activity is very high compared to other tissues, but again since the pentose shunt is not very active in the early mouse embryo, it is difficult to ascribe an exact role for this enzyme during cleavage. Perhaps here again the enzyme may be most essential during oocyte development. Malate dehydrogenase has a relatively level activity during the preimplantation but shows an increase at the time of blastocyst formation when overall oxidative ability is increasing (BRINSTER, 1966b). Perhaps the most interesting enzyme which has been measured in the preimplantation mouse embryo is hexokinase, since it has been shown to be a regulatory enzyme in other cell lines. In the mouse embryo it has a very low activity at the time of ovulation, but the activity increases about sevenfold during the preimplantation period (BRINSTER, 1968d). Of significant importance, however, is that the carbon available from glucose, which must come through the hexokinase reaction, at each stage of development, is very close to the sum of glucose carbon

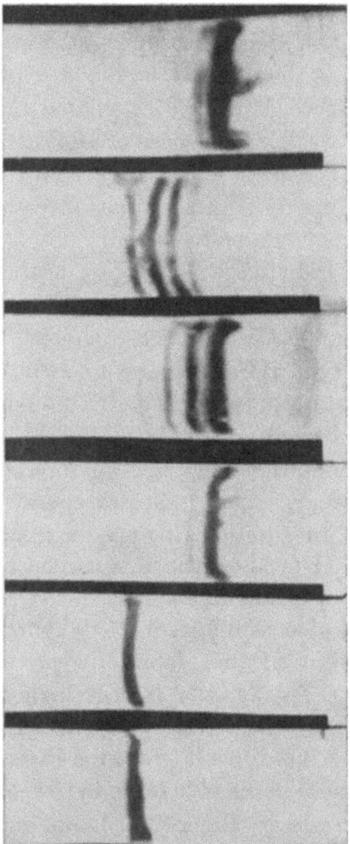

Fig. 5. Lactate dehydrogenase isozymes in the preimplantation mouse embryo. From top to bottom, the first two strips are samples of one-cell ova, third strip nine-day fetus, fourth strip nine-day decidua, fifth strip adult heart, sixth strip adult skeletal muscle. The start is marked by the edge of the black paper and migration is toward the right (anode). The first four strips were run at equal total LDH activities. From left to right the bands represent LDH-5 (M_4), LDH-4 (M_3H), LDH-3 (M_2H_2), LDH-2 (MH_3), and LDH-1 (H_4). (Auerbach and Brinster, 1967)

which can be accounted for by CO_2 produced, lactate formed, or carbon incorporation into the embryo (see Table 4). This suggests that the hexokinase reaction is important in regulating glucose availability to the early mouse embryo. Another regulatory enzyme in the glycolytic pathway is phosphofructokinase. Preliminary measurements of the activity of this enzyme indicate two things: first, that the activity is relatively level throughout the preimplantation period and, second, that the activity found is three to thirty times higher than that found for hexokinase (BRINSTER, unpublished). This would argue in favor of hexokinase as the primary regulator of glucose metabolism in the preimplantation mouse embryo.

Permeability

Changes in metabolic pathway capability or changes in key enzyme activity have been advanced as the mechanism by which the changes in patterns of energy metabolism within the embryo are brought about. An alternative explanation might lie in changes of the embryo's permeability to substrates. Perhaps the first indication of selective permeability of the zygote's membrane was the interaction between hydrogen ion concentration and substrate concentration (BRINSTER, 1965a). In these experiments it was shown that the optimum concentration of the substrate was influenced by the pH of the medium. The higher the pH of the medium, the higher the optimum concentration of the substrate required. It appeared that the embryo responded to the amount of the substrate in the acid form.

More recently, direct studies on permeability have been made. WALES and BRINSTER (1968) studied the uptake of glucose by two- and eight-cell mouse embryos. They found no significant difference in the permeability at these two stages despite the fact that glucose is able to support development of the eight-cell but not the two-cell embryo. Likewise, there was no difference in the permeability of one- and two-cell mouse embryos to pyruvate or lactate despite the fact lactate will support the two-cell embryo but not the one-cell embryo (WALES and WHITTINGHAM, 1967). However, in the case of malate there was a sevenfold increase in the permeability of the embryo to malate between the two- and the eight-cell stage (WALES and BIGGERS, 1968). This suggests that a difference in permeability may be the basis for the ability of malate to support the development of the eight-cell embryo but not the two-cell embryo. However, caution must be exercised in interpreting results of this series of experiments, since it was not possible to determine the intracellular intermediates. Without knowing to what form the substrate had been converted, it is not possible to rule out the possibility that the substrate is equally permeable at different stages of development but that the enzymes within the cell convert the substrate to other intermediates more rapidly at one developmental stage than at another. However, in the case of malate where we know there is little change in malate dehydrogenase activity during the period studied, it seems quite possible that a change in cell permeability accounts for the change in the embryo's response to malate in the culture medium.

Summary

There appears to be a morphological, biochemical and developmental similarity among the preimplantation stages of the Eutherian mammals. This is particularly true for the newly ovulated ovum and during the first two to three days of development of the embryo. The embryos require an amino nitrogen source for development and in most cases require an energy substrate. Pyruvate appears to be the central energy substrate in those species where nutritional requirements of the zygote have been examined in detail. During the first day or two of the embryo's life, theEmbden-Meyerhof pathway has a very low capability, but after blastocyst formation there is a sharp increase in glycolytic ability. The Krebs cycle activity is the main source of energy throughout the preimplantation period. Although changes in permeability have been suggested as a possible cause of metabolic changes within the embryo, there is considerable evidence that differential rates of synthesis and/or activity of enzymes are the basis for variation in the capability of different synthetic or metabolic pathways.

There is evidence that protein and nucleic acid synthesis occurs in the mammalian embryo from the time of fertilization onward, but the rate is slow during the early cleavage stages. The largest increase in RNA synthesis during the preimplantation period is in the ribosomal RNA fraction, but there is also a considerable increase in soluble RNA synthesis. The increase in RNA synthesis occurs at a time when the RNA content in the embryo has declined to a low level and at a time just before net protein synthesis increases. It may be said that the embryo begins to produce ribosomes and other RNA in preparation for blastocyst development, implantation and differentiation.

Acknowledgments

I thank Miss MILDRED COMBS for assistance in preparation of the manuscript. Financial support for the author's research is currently from PHS Research Grant No. 03071 from the National Institute of Child Health and Human Development.

References

AUERBACH, S., BRINSTER, R. L.: Lactate dehydrogenase isozymes in the early mouse embryo. Exp. Cell Res. **46**, 89 (1967).

— — Effect of oxygen concentration on the development of two-cell mouse embryos. Nature (Lond.) **217**, 465 (1968).

BEIER, H. M.: Uteroglobin: A hormone-sensitive endometrial protein involved in blastocyst development. Biochim. Biophys. Acta (Amst.) **160**, 289 (1968).

BIGGERS, J. D., WHITTINGHAM, D. G., DONAHUE, R. P.: The pattern of energy metabolism in the mouse oocyte and zygote. Proc. nat. Acad. Sci. (Wash.) **58**, 560 (1967).

BISHOP, D. W.: Active secretion in the rabbit oviduct. Amer. J. Physiol. **187**, 347 (1956).

BLANDAU, R., JENSEN, L., RUMMERY, R.: Determination of the pH values of the reproductive tract fluids of the rat during heat. Fert. and Steril. **9**, 207 (1968).

BRINSTER, R. L.: A method for *in vitro* cultivation of mouse ova from two-cell to glastocyst. Exp. Cell Res. **32**, 205 (1963).

— Studies on the development of mouse embryos *in vitro*. Ph. D. Thesis, University of Pennsylvania, Philadelphia 1964.

— Studies on the development of mouse embryos *in vitro*. I. The effect of osmolarity and hydrogen ion concentration. J. exp. Zool. **158**, 49 (1965a).

— Studies on the development of mouse embryos *in vitro*: Energy metabolism. In: Ciba Foundation Symposium on Preimplantation Stages of Pregnancy. Ed. by G. W. WOLSTENHOLME. London: J. & A. Churchill 1965b.

— Studies on the development of mouse embryos *in vitro*. II. The effect of energy source. J. exp. Zool. **158**, 59 (1965c).

— Studies on the development of mouse embryos *in vitro*. IV. Interaction of energy sources. J. Reprod. and Fertil. **10**, 227 (1965d).

— Studies on the development of mouse embryos *in vitro*. III. The effect of fixed-nitrogen source. J. exp. Zool. **158**, 69 (1965e).

Brinster, R. L.: Lactic dehydrogenase activity in the preimplanted mouse embryo. Biochim. Biophy. Acta (Amst.) **10**, 439 (1965f).
— Glucose 6-phosphate dehydrogenase activity in the preimplantation mouse embryo. Biochem. J. **101**, 161 (1966a).
— Malic dehydrogenase activity in the preimplantation mouse embryo. Exp. Cell Res. **43**, 131 (1966b).
— Protein content of the mouse embryo during the first five days of development. J. Reprod. and Fertil. **13**, 413 (1967a).
— Carbon dioxide production from glucose by the preimplantation mouse embryo. Exp. Cell Res. **47**, 271 (1967b).
— Carbon dioxide production from lactate and pyruvate by the preimplantation mouse embryo. Exp. Cell Res. **47**, 634 (1967c).
— Effect of glutathione on the development of two-cell mouse embryos *in vitro*. J. Reprod. and Fertil. **17**, 521 (1968a).
— Carbon dioxide production from glucose by the preimplantation rabbit embryo. Exp. Cell Res. **51**, 330 (1968b).
— Lactate dehydrogenase activity in the oocytes of mammals. J. Cell Biol. **35**, 17A (1968c).
— Hexokinase activity in the preimplantation mouse embryo. Enzymologia (Basel) **34**, 304 (1968d).
— Culture of two-cell rabbit embryos to morulae. J. Reprod. and Fertil. (in press). (1969a).
— Radioactive carbon dioxide production from pyruvate and lactate by the preimplantation rabbit embryo. Exp. Cell Res. **54**, 205 (1969b).
— Incorporation of carbon from glucose and pyruvate into the preimplantation mouse embryo. Exp. Cell Res. **58**, 153 (1969c).
— Mammalian embryo culture. In: The Mammalian Oviduct, ed. by E. S. E. Hafez and R. J. Blandau. Chicago: University of Chicago Press (1969d).
— *In vitro* culture of the embryo. In: Pathways to Conception, (Harold C. Mack Symposium on the Physiology and Pathology of Reproduction), Springfield, Ill.: Charles G. Thomas Publ. Comp. (in press) (1970a).
— Carbon dioxide production from pyruvate and glucose by the monkey oocyte. J. Reprod. and Fertil. (in press) (1970b).
— Thomson, J. L.: Development of eight-cell mouse embryos *in vitro*. Exp. Cell Res. **42**, 308 (1966).

Calarco, P. G., Brown, E. H.: An ultrastructural and cytological study of preimplantation development of the mouse. J. Exp. Zool. **171**, 253 (1969).

Cole, R. J., Paul, J.: Properties of cultured preimplantation mouse and rabbit embryos, and cell strains derived from them. In: Preimplantation Stages of Pregnancy, ed. by G. E. W. Wolstenhome and M. O'Connor. London: J. & A. Churchill Ltd. 1965.

Daniel, J. C., Jr.: Vitamins and growth factors in the nutrition of rabbit blastocysts *in vitro*. Growth **31**, 71 (1967).
— Olson, J. D.: Amino acid requirements for cleavage of the rabbit ovum. J. Reprod. and Fertil. **15**, 453 (1968).

Donahue, R. P., Stern, S.: Follicular cell support of oocyte maturation: Production of pyruvate *in vitro*. J. Reprod. and Fertil. **17**, 395 (1968).

DRAGIOU, I., BENETATO, G., OPREAU, R.: Recherches sur la respiration des ovocytes des mammiferes. Compt. Rend. Soc. Biol. **126**, 1044 (1937).

ELLEM, K. A. O., GWATKIN, R. B. L.: Patterns of nucleic acid synthesis in the early mouse embryo. Develop. Biol. **18**, 311 (1968).

ENDERS, A. C., SCHLAFKE, S.: A morphological analysis of the early implantation stages in the rat. Amer. J. Anat. **120**, 185 (1967).

FRIDHANDLER, L.: Pathways of glucose metabolism in fertilized rabbit ova at various preimplantation stages. Exp. Cell Res. **22**, 303 (1961).

— WASTILA, W. B., PALMER, W. M.: The role of glucose in metabolism of the developing mammalian preimplantation conceptus. Fertil. and Steril. **18**, 819 (1967).

GREGOIRE, A. T., GONGSAKDI, D., RAKOFF, A.: The free amino acid content of the female rabbit genital tract. Fertil. and Steril. **12**, 322 (1961).

GROSS, P. R.: The immediacy of genomic control during early development. J. exp. Zool. **157**, 21 (1964).

— COUSINEAU, G. H.: Synthesis of spindleassociated proteins in early cleavage. J. Cell Biol. **19**, 260 (1963).

— — Macromolecule synthesis and the influence of actinomycin on early development. Exp. Cell Res. **33**, 369 (1964).

GWATKIN, R. B. L.: Amino acid requirement for attachment and outgrowth of mouse blastocyst *in vitro*. J. Cell Physiol. **68**, 335 (1966).

HAMMOND, J., JR.: Recovery and culture of tubal mouse ova. Nature (Lond.) **163**, 28 (1949).

HAMNER, C. E., WILLIAMS, W. L.: Composition of rabbit oviduct secretions. Fertil. and Steril. **16**, 170 (1965).

HOLMDAHL, T. H., MASTROIANNI, L., JR.: Continuous collection of rabbit oviduct secretions at low temperature. Fertil. and Steril. **16**, 587 (1965).

KRISHNAN, R. S., DANIEL, J. C., JR.: Blastokinin: Inducer and regulator of blastocyst development in the rabbit uterus. Science **158**, 490 (1967).

LEWIS, W. H., WRIGHT, E. S.: On the development of the mouse. Carnegie Inst. Washington, Contribution to embryology **25**, 113 (1935).

LOEWENSTEIN, J. E., COHEN, A. I.: Dry mass, lipid content and protein content of the intact and zona-free mouse ovum. J. Embryol. exp. Morphol. **12**, 113 (1964).

LUTWAK-MANN, C.: Glucose, lactine acid and bicarbonate in rabbit blastocyst fluid. Nature (Lond.) **193**, 653 (1962).

MADEN, B. E. H.: Ribosome formation in animal cells. Nature (Lond.) **219**, 685 (1968).

MANES, C., DANIEL, J. C., JR.: Quantitative and qualitative aspects of protein synthesis in the preimplantation rabbit embryo. Exp. Cell Res. **55**, 261 (1969).

MASTROIANNI, L., JR., WALLACH, R. C.: Effect of ovulation and early gestation on oviduct secretions in the rabbit. Amer. J. Physiol. **200**, 815 (1961).

— URZUA, M., AVALOS, M., STAMBAUGH, R.: Some observations on fallopian tube fluid in the monkey. Amer. J. Obstet. Gynec. **103**, 703 (1969).

MAUER, R. E., HAFEZ, E. S. E., EHLERS, M. H., KING, J. R.: Culture of two cell rabbit eggs in chemically defined media. Exp. Cell Res. **52**, 293 (1968).

MILLS, R. M., JR., BRINSTER, R. L.: Oxygen consumption of preimplanted mouse embryos. Exp. Cell Res. **47**, 337 (1967).

MINTZ, B.: Incorporation of nucleic acid and protein precursors by developing mouse eggs. Amer. Zool. **2**, 432 (1962).

— Synthetic processes and early development in the mammalian egg. J. exp. Zool. **157**, 85 (1964).

MONESI, V., SALFI, V.: Macromolecular syntheses during early development in the mouse embryo. Exp. Cell Res. **46**, 632 (1967).

MONROY, A.: Biochemical aspects of fertilization. In: The Biochemistry of Animal Development. Vol. 1., ed. by R. WEBER. New York: Academic Press 1965..

ONUMA, H., MAUER, R. E., FOOTE, R. H.: *In vitro* culture of rabbit ova from early cleavage stages to the blastocyst stage. J. Reprod. Fertil. **16**, 491, (1968).

PIKÓ, L.: Synthesis of macromolecules in early mouse embryos cultured *in vitro*: RNA, DNA, and a polysaccharide component. Develop. Biol. **21**, 257 (1970).

REAMER, G. R.: The quantity and distribution of nucleic acids in the early cleavage stages of the mouse embryo. Ph. D. Thesis, Boston University, Boston 1963.

RESTALL, B. J.: The fallopian tube of the sheep. III. The chemical composition of the fluid from the fallopian tube. Austr. J. Biol. Sci. **19**, 687 (1966).

STERN, S., BIGGERS, J. D.: Enzymatic estimation of glycogen in the cleaving mouse embryo. J. exp. Zool. **168**, 61 (1968).

SUGAWARA, S.: Metabolism of the mammalian ova. Jap. J. Tootechn. Sci. **33**, 1 (1962).

— UMEZU, M.: Studies on metabolism of the mammalian ova. II. Oxygen consumption of the cleaved ova of the rat. Tohoku J. Agricultural Res. **12**, 17 (1961).

SZOLLOSI, D.: Modification of the endoplasmic reticulum in some mammalian oocytes. Anat. Rec. **158**, 59 (1967).

TASCA, R. J.: RNA synthesis and protein synthesis in preimplantation stage mouse embryos. Ph. D. Thesis, Temple University, Philadelphia 1969.

TENBROECK, J. T.: Effect of nucleosides and nucleoside bases on the development of preimplantation mouse embryos *in vitro*. J. Reprod. Fertil. **17**, 571 (1968).

THOMSON, J. L., BIGGERS, J. D.: The effect of inhibitors of protein synthesis on the development of mouse embryos *in vitro*. Exp. Cell Res. **41**, 411 (1966).

— BRINSTER, R. L.: Glycogen content of preimplantation mouse embryos. Anat. Rec. **155**, 97 (1966).

TYLER, A.: Masked messenger RNA and cytoplasmic DNA in relation to protein synthesis and processes of fertilization and determination in embryonic development. Develop. Biol. Supl. **1**, 170 (1967).

VISHWAKARMA, P.: The pH and bicarbonate-ion content of the oviduct and uterine fluids. Fertil. and Steril. **13**, 481 (1962).

WALES, R. G.: Accumulation of carboxylic acids from glucose by the preimplantation mouse embryo. Aust. J. Biol. Sci. **22**, 701 (1969).
— BIGGERS, J. D.: The permeability of two and eight-cell mouse embryos to L-malic acid. J. Reprod. Fertil. **15**, 103 (1968).
— BRINSTER, R. L.: The uptake of hexoses by mouse embryos. J. Reprod. Fertil. **15**, 415 (1968).
— QUINN, P., MURDOCH, R. N.: The fixation of carbon dioxide by the eight-cell mouse embryo. J. Reprod. Fertil. **20**, 541 (1969).
— WHITTINGHAM, D. G.: A comparison of the uptake and utilization of lactate and pyruvate by one- and two-cell mouse embryos. Biochim. Biophys. Acta (Amst. **148**, 703 (1967).
WHITTEN, W. K.: Culture of tubal mouse ova. Nature (Lond.) **176**, 96 (1956).
— Culture of tubal ova. Nature (Lond.) **179**, 1081 (1957).
— BIGGERS, J. D.: Complete development *in vitro* of the preimplantation stages of the mouse in a simple chemically defined medium. J. Reprod. Fertil. **17**, 399 (1968).
WOODLAND, H. R., GRAHAM, C. F.: RNA synthesis during early development of the mouse. Nature (Lond.) **221**, 327 (1969).
ZAMBONI, L., MISHELL, D. R., JR., BELL, J. H., BACA, M.: Fine structure of the human ovum in the pronuclear stage. J. Cell Biol. **30**, 579 (1966).

Endometrial Secretion and Early Mammalian Development

HENNING M. BEIER, GERHARD PETRY, and WOLFGANG KÜHNEL
Anatomisches Institut II, Philipps-Universität, Marburg/Lahn, Germany

With 14 Figures

Endometrial Transformation

The maternal organism provides essential conditions for preimplantation development of the mammalian embryo by means of tubal and uterine proliferation and secretion. The uterine mucosa of a sexually mature rabbit shows the typical picture of an oestrous endometrium, moderate proliferation of the cavum and glandular epithelium (Fig. 1a). The surface of the endometrium carries a

a

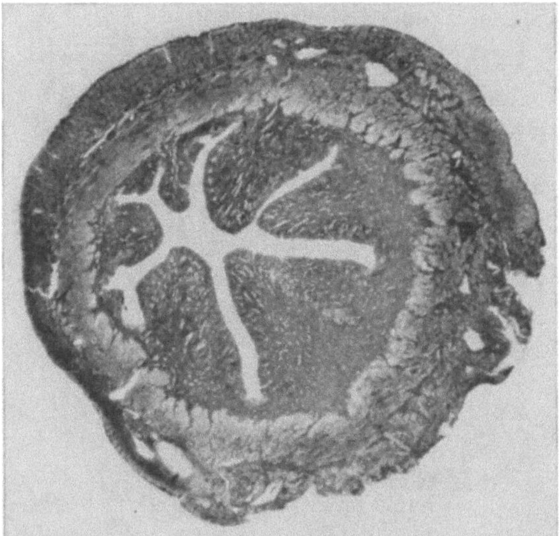

Fig. 1a—c. Cross section of rabbit uterus on oestrus. (a) Typical picture of moderate endometrial proliferation. Azan staining; 15×

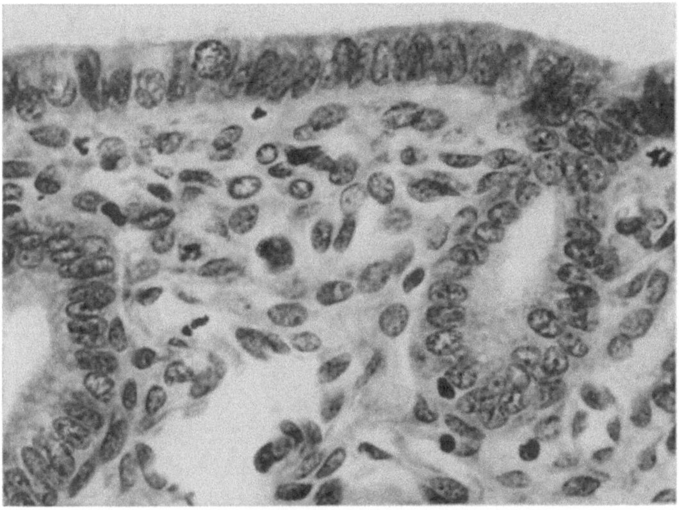

b

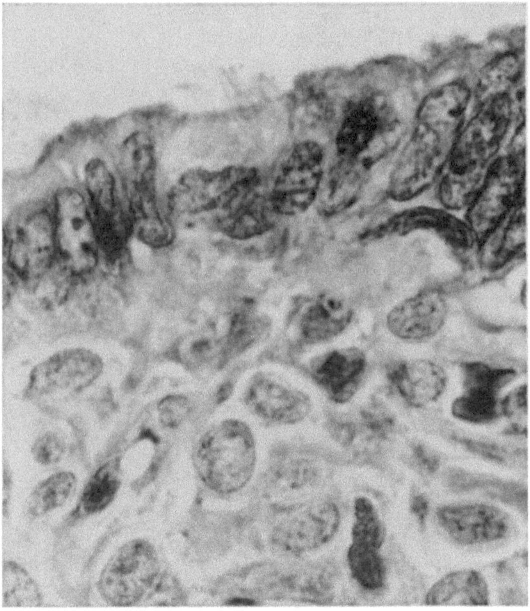

c

Fig. 1b und 1c. (b) Endometrium with cylindrical cavum epithelium and short crypt. van Gieson staining; 560×. (c) Cylindrical cavum epithelium with cilia cells. Masson trichrome staining; 1400×

a

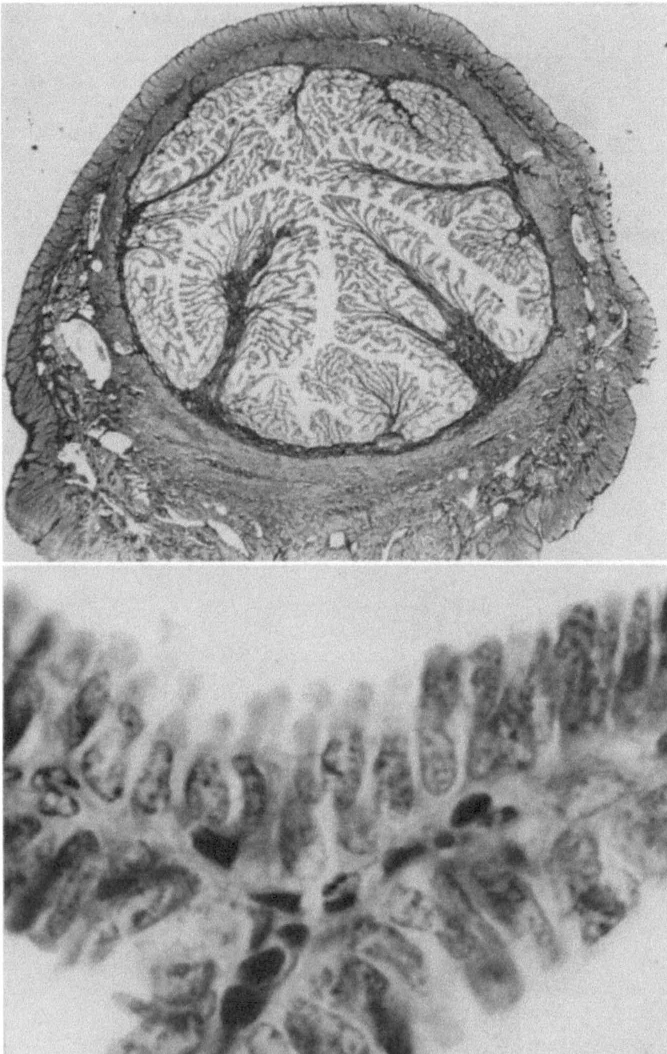

b

Fig. 2a and b. Cross section of rabbit uterus 6 d p.c. (a) Transformation of endometrium into filigreelike, branched folds. Azan staining; 15 ×. (b) Mucosa fold with typical apocrine secretion. Masson trichrome staining; 1400 ×

single-layered, cubic to cylindrical epithelium, areas of which are covered with cilia. The cavum epithelium sinks down into short, narrow crypts, the uterine glands, in the tunica propria. There are no signs of typical secretory activity (Fig. 1 b, c).

Within the first 3 days of pregnancy, the uterus increases in length and volume and the endometrium experiences a drastic transformation. The bulges in the endometrium, originally plump, are transformed by the fourth day p.c. into filigree-like, branched folds from which numerous villi project (Fig. 2a). The cells of the cavum epithelium become multi-nucleic between the fifth and sixth days p.c., and cell borders are difficult to determine at this point by photomicroscopy. During this transformation the height of the epithelium increases; furthermore, apical cytoplasmic convex folds develop, which fulfill the morphological criteria for an apocrine secretion (Fig. 2b). The glandular tubes, whose form varies considerably, become wider and deeper. Their epithelium also increases in height, but it remains mono-nucleic.

a

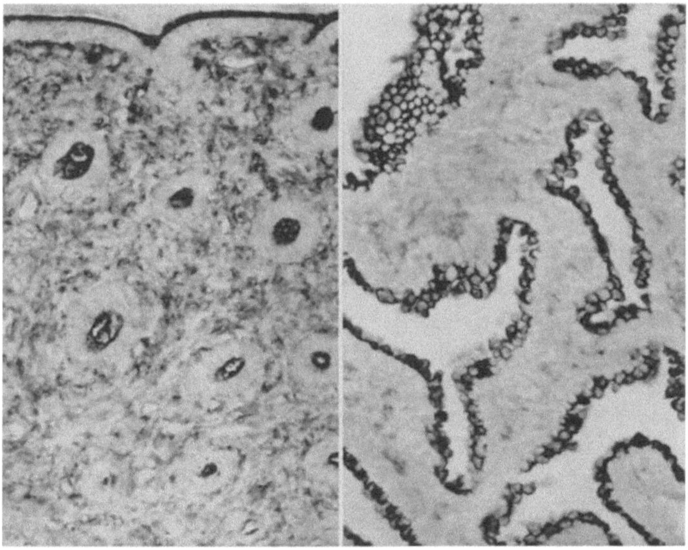

b

Fig. 3a and b. Rabbit endometrium, Hale reaction. (a) Oestrus. (b) 6 d p.c.; 350 ×

The transformation in the epithelium should be understood on the one hand as a preparation of the endometrium for the implantation of the blastocyst. On the other hand, we must see the basis of the secretory function of this organ in the morphologically measurable endometrium transformation.

The intracellular metabolism, which represents the prerequisite for delivery of secretion components into the cavum uteri, can be traced with histochemical and enzyme topochemical methods (Fig. 3–6). A histochemical analysis of large molecular carbohydrate substances results in the finding, for example, that during the entire preimplantation period in the rabbit, there is no evidence of glycogen in the cavum epithelium. Shortly after the seventh day p.c. glycogen granula appear for the first time in the central sections of the crypta; they appear in the surface epithelium on the eighth day p.c. Apparently glycogen is provided to the rabbit embryo by the mother as a source of energy beginning at the point in time when mesometral implantation occurs (the point at which placentation actually begins).

Enzymes Involved in Endometrial Secretory Process

The enzyme histochemical distribution patterns develop in a characteristic way during preimplantation and show such multifarious deviations from the oestrous endometrium during the first days of pregnancy, that it would lead us too far afield to discuss our findings in a more detailed form here[1]. However, the topochemistry of the adenosine triphosphatase, non-specific esterase, and leucine aminopeptidase should be discussed in close connection with the results to be explained here.

Adenosine triphosphatase converts ATP to ADP and simultaneously releases energy. This energy is required for syntheses, secretory activities, active nutrient transport and cilia movement. ATPase elicits a positive reaction on the surfaces of the cavum and glandular epithelium during oestrus, as well as a strong reaction from all blood vessels (Fig. 4). The apical epithelium reacts much more strongly during the oestrus than during preimplantation. By way of contrast, the colouring of the capillary and subepithelial cell

1 Methods and detailed results on histochemical investigations are published elsewhere [19].

layer increases more and more from oestrus to nidation. What we see here in the behaviour of ATPase in the endometrium during preimplantation is apparently a hormone-dependent process. The production and secretion of oestrous uterine fluid is regulated

a 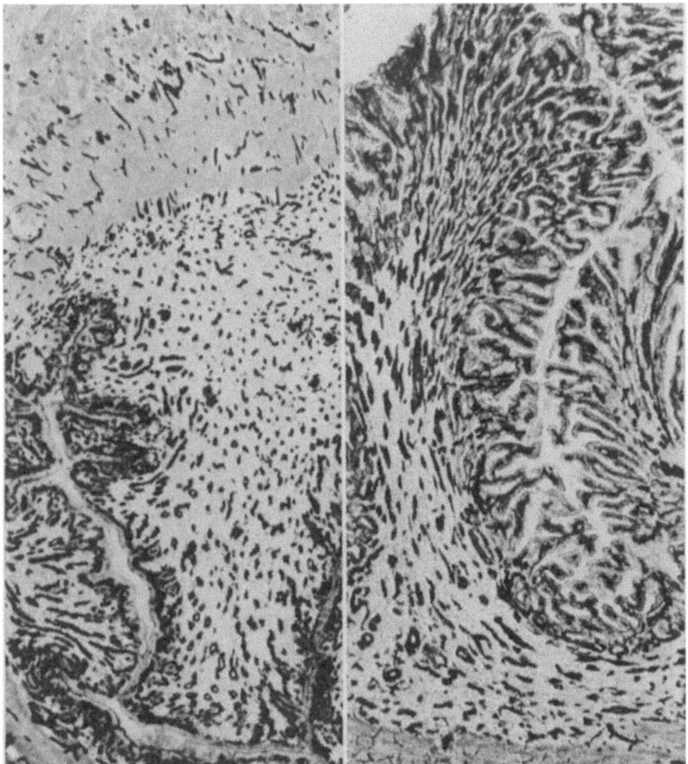 b

Fig. 4a and b. Rabbit endometrium ATPase reaction (pH 9,2). (a) Oestrus. (b) 6 d p.c.; 90×

primarily by the cavum epithelium, while the secretion located separately in the uterine lumen during the first days of pregnancy represents a product of the endometrial glands as well. Because ATPase plays an important role in nutrient transfer through membranes, it would not be far-fetched to look at this enzyme in connection with the process of secretion. ATPase activity increases with stimulation by oestrogens, as shown by Henzl et al. [11]. We may conclude from this that, taking the permeability of the intrauterine epithelium and endometrial capillaries into consideration, hormone-dependent changes do occur.

Non-specific esterase can be found in fairly high activity in the oestrous endometrium epithelium. At 6 days p.c. this enzyme has

a 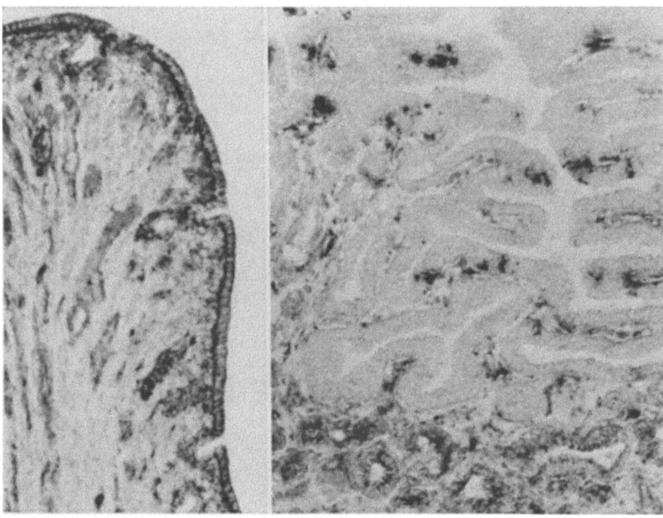 b

Fig. 5a and b. Rabbit endometrium, 4-Chloro-5-bromo-indoxylacetate esterase reaction (non-specific esterase). (a) Oestrus; (b) 6 d p.c.; 140 ×

been disappeared from epithelial cells and reveals only positive reactions in the stroma (Fig. 5). Evidence has been demonstrated earlier by Beier [3], that non-specific esterase is part of the endometrial secretion delivered at 6 d p.c. So far it remains speculative if this enzyme is essential for preimplantation metabolism and involved in the development of blastocysts. Decreased activities of non-specific esterase have also been observed during the postovulatory part of the human menstrual cycle [9]. This observation, together with results of our experiments on hormonal contraceptives [17], provide evidence of progesterone on effects non-specific esterase activity.

Leucine aminopeptidase (LAP) is a membrane enzyme and is responsible for the hydrolysis of the peptide binding of an end-state leucine with free amino group. LAP shows striking behaviour during the preimplantation phase in the rabbit endometrium (Fig. 6).

a 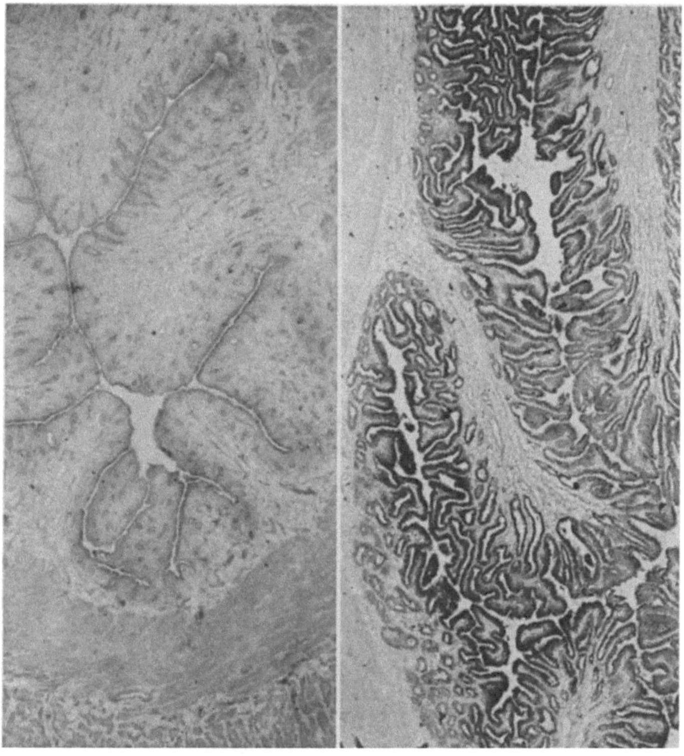 b

Fig. 6a and b. Rabbit endometrium, leucine aminopeptidase reaction. (a) Oestrus; (b) 6 d p.c.; 35 ×

During oestrus it cannot be found in the endometrium, but its activity becomes clearly positive at the point where the embryo enters the uterus (4 days p.c.) and it reaches maximum activity on the sixth day p.c., as shown also by DENKER [8]. This increase in activity corresponds in time to the formation of blastocysts and to the phase in which only uterine secretion is available to the embryo as a source of nutrition and energy. We may conclude from this that LAP is very important for intracellular protein metabolism and for the provision of these proteins to the uterine secretion.

Endometrial Secretion Proteins

The mammalian egg possesses no supply of nutrient for its journey from ovulation up to the time of implantation. Before that point has been reached, the embryo must fill its energy requirements from the reservoir of nutrients contained in the tubal and uterine secretions.

As may be shown by analytical investigations, transplantations and in vitro experiments, the dependence of the blastocysts on their surrounding milieu does not end with their nutritional reqirements [2, 3, 5, 12, 13, 16]. On the contrary, some uterine secretion proteins seem to have specific influence on the development of the rabbit blastocyst. Their role as so-called extrinsic factors produced by the maternal organism must be compared and contrasted to the endogenous intrinsic factors of the blastocyst.

We subjected the intrauterine protein milieu to an analytic investigation with disc-electrophoresis. We used 6% polyacrylamide as a separation gel with tris glycine buffer (pH 9.0). The amount of protein applied to each column was 250–350 μg. We performed photometric interpretation of the electrophoreses with the supplementary densitometer equipment to the Zeiss-PMQ II according to Koch [15].

The electrophoresis picture of the uterine secretions after fertilization contains points of concentration different from those in the serum spectrum. The fraction most in evidence are prealbumin and uteroglobin, neither of which occur in serum (Fig. 7).

The secretion protein pattern during oestrus is, however, not too different from that of serum: albumin and transferrin appear as the predominant fractions. The strong macroglobulin and lipoprotein bands were not found in oestrous uterine secretion.

In normal pregnancy, a pregnancy-specific protein pattern forms. The uterus secretion patterns for the sixth and seventh days p.c. show considerable changes compared to the oestrous secretions: a high, sharp prealbumin peak towers above the following albumin peak, which is lower and somewhat broader. With regard to serum we found comparatively little albumin. The band in the next position behind albumin corresponds to a pregnancy-specific post-albumin. Uteroglobin is the strongest fraction of the endometrium secretion and is far above the other individual components. Very little transferrin could be identified before implantation time, only

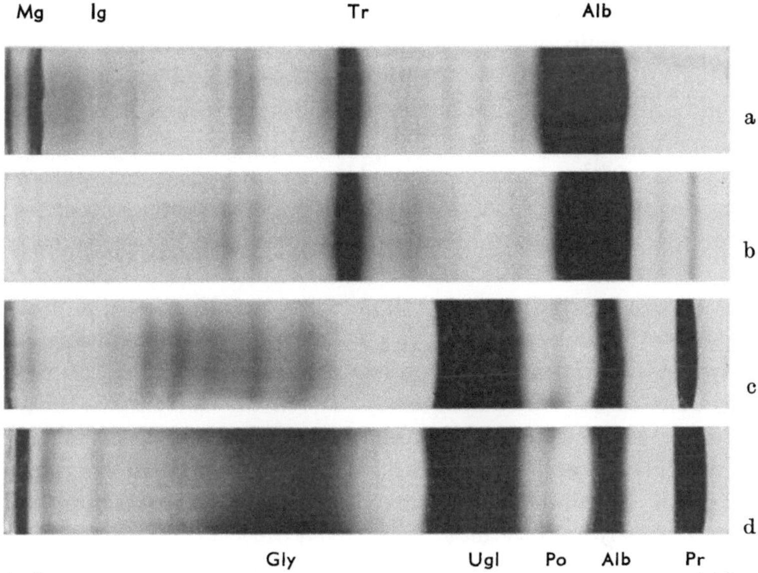

Fig. 7a—d. Disc-electrophoretic protein patterns of rabbit serum (a), rabbit oestrous uterine fluid (b), uterine secretion from the 6th day p.c. (c), and 7th day p.c. (d). Alb Albumin, Gly β-Glycoproteins, Ig Immunoglobulins, Mg Macroglobulins, Pr Prealbumin, Po Postalbumin, Tr Transferrin, Ugl Uteroglobin. (Modified from PETRY, KÜHNEL, and BEIER, 1970)

a weakly indicated band shows the position of this protein. The broad fraction of the β-glycoproteins and larger molecular macroglobulins is indicated on the sixth day p.c. by a large number of individual, less prominent bands. On the seventh day p.c., the day on which implantation of the rabbit blastocyst begins, the broad fraction of the β-glycoproteins experiences a considerable change, and it appears in the form of a prominent peak.

The results indicate that the secretion protein pattern in the cavum uteri during oestrus shows no meaningful variation from that of blood serum as regards its main components albumin and transferrin. However, within the first three days p.c., as the embryos are migrating through the Fallopian tube, a specific uterine secretion begins, the results of which may be seen in the protein patterns for the sixth and seventh days p.c. Uterus-specific proteins which cannot be found immunologically in serum, are secreted from the endometrium in greater and greater amounts: uterus prealbumin and postalbumin, uteroglobin, β-glycoprotein, as well as β-U-macroglobulin. Other proteins, identical to those found in serum, are found only in very small comparative concentrations in secretion (albumin, transferrin, immunoglobulin IgG). This reveals to us the function of the endometrium: biosynthesis of specific uterus proteins, selection of a few particular serum proteins, and the production of secretion adapted and coordinated in time to early pregnancy.

Apparently during oestrus the selective uptake of some proteins from blood plays the more important role, without causing an important quantitative change in the relative percentage of these fractions to occur simultaneously. It remains for enzyme histochemical methods to explain this process of selection on the level of epithelial and capillary permeability. As we discussed at the beginning, we have seen indications that changes in epithelium and capillary permeability may be triggered especially by the activity of adenosine triphosphatase and alcaline phosphatase. Under the increasing influence of progesterone the endometrium takes over production of more secretion up to the nidation of the blastocysts: the absolute protein content of the uterine secretion climbs continuously. In addition with regard to proteins of serum origin, a quantitatively differentiating selection process develops out of the qualitative selection which occurs in oestrus. Combining the findings of earlier experiments [5, 20] with those of this experiments, we may characterize the temporal sequence of typical uterine secretion protein patterns as the "developmental specific protein pattern dynamics" of the preimplantation phase.

Some Biochemical and Biological Parameters of Uterine Proteins

Up to this point we have identified nine different proteins in the endometrium secretion. Of these, six have the characteristics of glycoproteins. As clearly shown by KIRCHNER [14], the high percentage of N-acetyl-neuraminic acid (NANA) in the uterus prealbumin fraction is remarkable here, as well as the glucose content of the postalbumin fraction. The β-globulin complex contains a large quantity of galactose. Also, the percentage of total carbohydrates in this protein is considerably higher than that in the other known uterus glycoproteins.

The majority, that is about 55 to 60 % of all uterus secretion protein is relatively small-molecule protein (25,000 to 70,000). Only a very small percentage may be definitely classified as large-molecule protein (α-macroglobulin, β-U-macroglobulin) with a molecular weight between 600,000 and 1,000,000 (Table).

Table. *Uterine secretion proteins*

	Sedimentation coefficient	Molecular weight
Prealbumin	2.00	ca. 50,000
Albumin	4.20	69,000
Postalbumin	—	ca. 30,000
Uteroglobin	1.38	27— 30,000
Transferrin	4.32	80— 90,000
β-Glycoprotein	2.26	70—100,000
Immunoglobulin IgG	6.07	150—160,000
α-Macroglobulin	16.60	700—800,000
β-U-Macroglobulin	—	>800,000

The most important uterus-specific secretion protein is thought to be uteroglobin (BEIER, [1, 4]). We have found experimental evidence that this protein is dependent on progesterone for its occurence. Uteroglobin has a molecular weight between 27 and 30,000. It makes up the strongest protein fraction not only in the uterine fluid during the preimplantation phase, but also in the blastocyst fluid from 5- and 6-day-old blastocysts, as has been shown previously [1–3, 10, 13].

KRISHNAN and DANIEL [16] isolated a protein fraction from rabbit uterine fluid, which seems to be identical to uteroglobin. In an in-vitro experiment, this protein fraction supported the growth of 4-day-old blastocysts. As a result, the authors named it "blastokinin". The decisive finding of this experiment lies in the fact that the development of the embryo into the blastocyst stage is apparently a power residing in the embryo itself, or rather in its cleavage stage, a power which cannot be released without the presence of certain endometrial factors. This protein fraction is to be assumed to promote mitotic activity in animals having delayed implantation as the armadillo and mink [6].

Human Genital Tract Protein Patterns

The findings of studies of the normal rabbit uterus to date almost force the question about comparable conditions in the *human uterus*.

This led us to cooperative work with two hospitals [2] in which we investigated operational tissue suited for our analysis (uterus myomatosus, descensus uteri). The usable Fallopian tube and endometrium secretion samples were prepared and handled in the same way as rabbit tissue after flushing as described earlier [3, 19]. Disc electrophoresis patterns are shown in Fig. 8.

The proteins of human uterine secretion represent a secretion product which is produced according to the same principles of tubal and endometrial glands and epithelium as in the rabbit. Thus it is not a simple transsudat of the blood plasma. The intra-tubal and intra-uterine protein pattern apparently develops not only in the rabbit but in the human through selection of individual plasma proteins on the one hand and through biosynthesis of *uterus-specific* proteins on the other.

In the tubal and uterine secretion of the human, we have not yet been able to locate a *quantitatively dominating* protein fraction which can be described as uterus-specific, such as uteroglobin in the rabbit. There probably exist still more specific tube and uterus proteins comparable to the relatively weak prealbumin fraction, which migrates between the serum prealbumin and the albumin; however, we have not yet been able to demonstrate these immunologically as was the case with, for example, the serum-identical proteins prealbumin, orosomucoid, albumin, α_1-antitrypsin, transferrin and the three immunoglobulins IgG, IgA, IgM.

---

2 Kreiskrankenhaus Homberg (Bez. Kassel), Chefarzt Dr. C. Hellwig; Stadtkrankenhaus Kassel, Gynäk. Abtlg., Chefarzt Prof. Dr. P. Pfau.

Because we had to limit our initial investigation to human secretion samples from the 12th to 15th and from the 26th and 27th days of the cycle, our comparative analysis must be limited to comment on the principle of the secretion process.

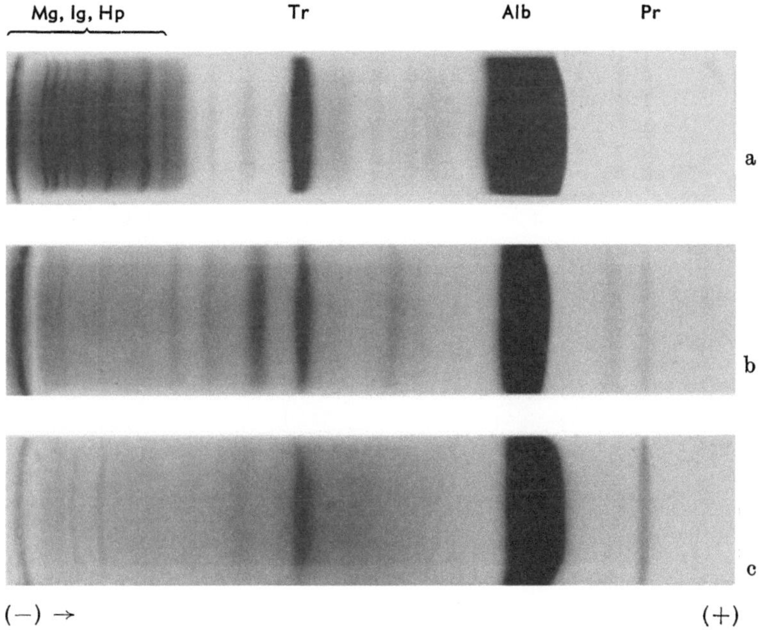

Fig. 8a—c. Disc-electrophoretic protein patterns of human blood serum (a), human endometrial secretion from the 14th day of the menstrual cycle (b), and flushed secretion of the human Fallopian tube from the 15th day of the menstrual cycle (c). Specimens of serum and Fallopian tube from same patient. Alb Albumin, Hp Haptoglobins, Ig Immunoglobulins, Mg Macroglobulins, Pr Prealbumin, Tr Transferrin.

Ovarian Hormone Effects on the Rabbit Uterine Protein Pattern

Oestrogens and gestagens introduce and maintain optimal conditions in tube and uterus for migration, development and implantation of the young embryo. This fundamental endocrinological insight has been part of our knowledge since the point at which we learned to influence early mammalian development to any extent using exclusion experiments on endocrine organs (such as ovariectomy) and subsequent substitution of hormones [7, 18]. The possibility to perform protein-biochemical investigations in endometrial secretion now offers us a new level for the analysis of hormonal control during preimplantation.

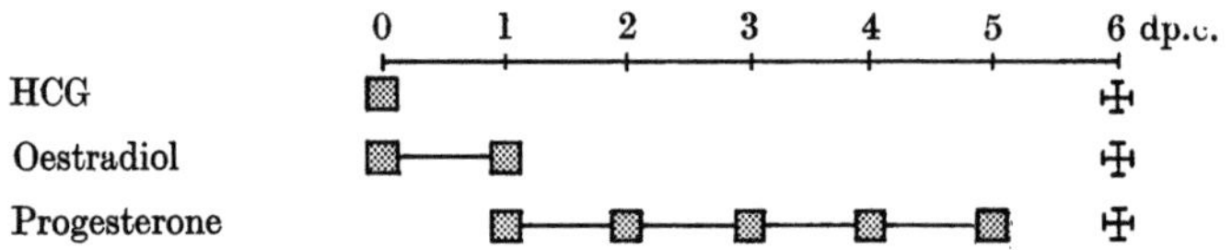

Experim.	♀ + ♂	HCG	Oestradiol	Progesterone
A	—	—	—	—
B	—	—	100 and 150 µg	—
C	—	—	100 and 150 µg	5×5 mg
D	—	20 I.U.	—	—
E	—	—	—	5×5 mg
F	♀♂	—	—	—
G	♀♂	—	—	5×5 mg
H	♀♂	—	100 and 150 µg	5×5 mg
I	♀♂	—	100 and 150 µg	—

Fig. 9. Design of experiments with hormones

According to the scheme outlined in Fig. 9 we triggered exogenous imbalances in the ovarian hormone level. The effects of these treatments on the endometrial secretion and on the development of the embryos were investigated.

Qualitative and quantitative interpretation of the disc-electrophoreses of endometrium secretion taken from the 6-days pregnant (or pseudopregnant) rabbit can be best made along the lines of Fig. 10. This table lists the different treatments and states of pregnancy. The arrows are meant to visualize the considerable variation that existed between the different test states and the normal progress of early pregnancy. Beginning at oestrus (uterus in normal state before copulation), the straight arrow from left to right symbolises the progress of normal pregnancy, which is accompanied by completely normal blastocyst development.

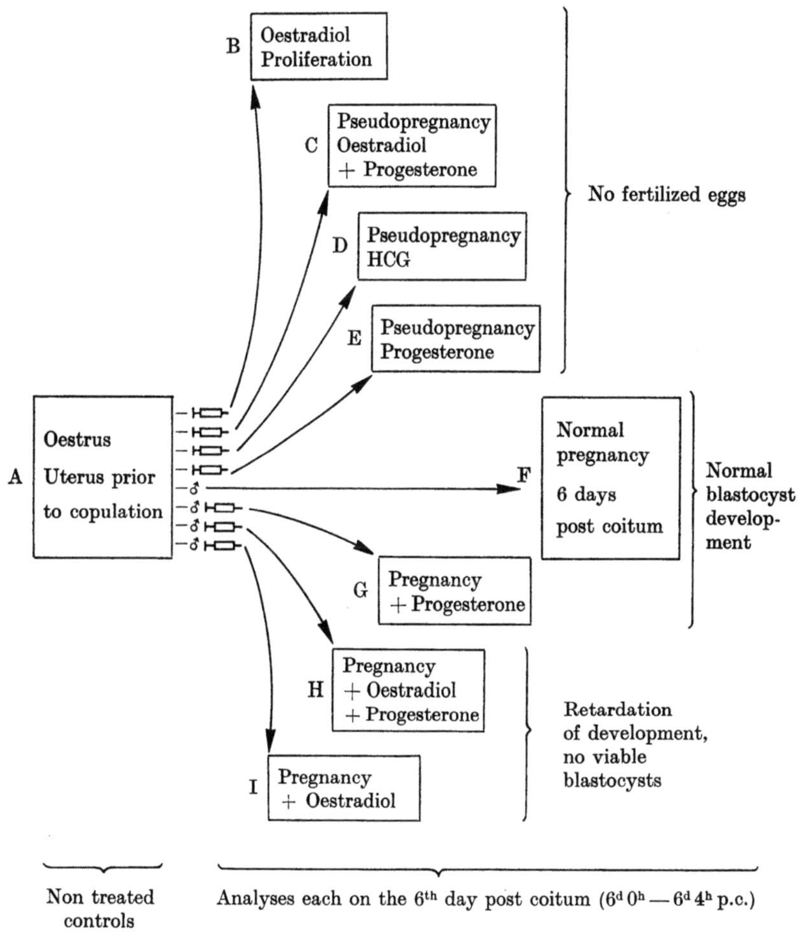

Fig. 10. Ovarian hormones and the uterine environment in pseudopregnant and pregnant rabbits

The four arrows above this "normal axis" correspond to the results of experimental hormone treatments, which led, in one case, to an oestradiol-dependent proliferation (B) and in the three other cases to pseudopregnancy (C, D, E).

The three arrows below the normal axis correspond to the results of hormonal influence on normal early pregnancies (G, H, I).

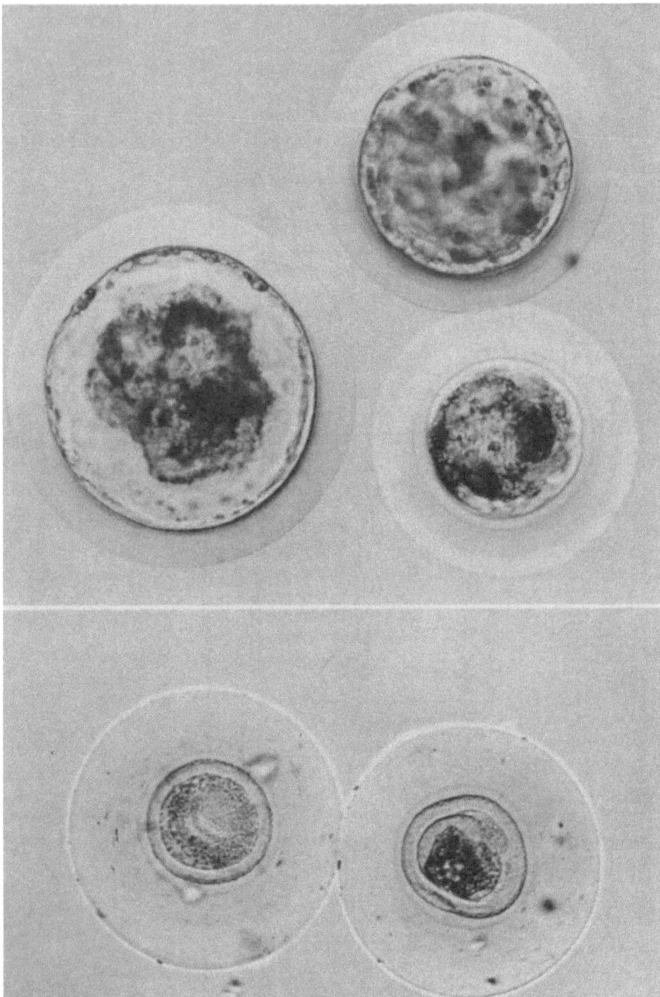

a

b

Fig. 11a and b. Rabbit blastocysts recovered 6 d p.c. from uteri of hormone treated animals, experiments *H* and *I*. (a) Combined treatment with oestradiol and progesterone revealed retardation of development, failure of blastocyst expansion, and abnormal mucin layers. (b) Administration of oestradiol to the mother led to increased retardation in egg development, degeneration of the cytoplasm, and damage to the zona pellucida

Here, the hormone injections were preceded in each case by normal copulation with a fertile male. There was no suppression of ovulation and fertilization here; in all cases we found blastocysts in different stages of development. But it seemed that in only a few cases was normal development assured (G). On the contrary, the majority of the embryos showed retarded development of the most various kinds (Fig. 11).

As shown in Figs. 12–14, all oestrogen dominated uteri show protein patterns with more reference to the serum protein pattern, especially regarding albumin and transferrin. The more progesterone activity is increased, the more an uterus specific pattern accumulates. Uteroglobin and β-glycoprotein seem to show clearly a progesterone dependent secretion rate. This is true, when progesterone is injected into oestrous and pregnant animals (Experiments E, G), and even when progesterone activity is triggered by injection of HCG into oestrous rabbits (Experiment D). The β-glycoprotein fraction at 6 days p.c. is stimulated by both HCG and progesterone in an extend which can be compared with the β-glycoprotein secretion of the 7th day p.c. during normal pregnancy (Fig. 7 and 13–14).

Conclusions

Our results present evidence that the biochemically analyzed parameters of the endometrium depend greatly on the balance of the mother's hormone supply. When we experimentally affect the autonomous control of the mother's hormone regulation system during preimplantation, decompensation of these sensitive regulation mechanisms can easily be the result. The effect of the interference can be gathered from its results on the target organ, which must prepare an optimum milieu for the early development of the young embryos. Endometrial secretion is disturbed, to an extent dependent upon the dosage and frequency of the exogenously applied ovarial hormones, so that a protein pattern which is appropriate according to time sequence and stage of pregnancy has no chance to develop. Parallel to this distorted secretion, we see various forms of retarded blastocyst development. Both of these findings are so concrete, that the question about a causal relationship between them automatically arises.

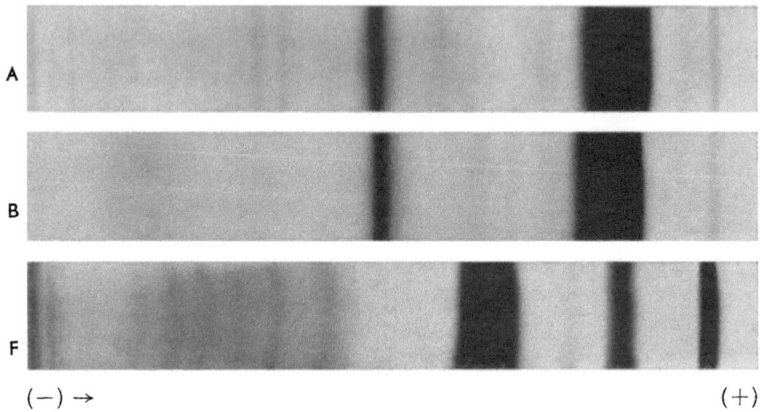

Fig. 12. Disc-electrophoretic protein patterns of rabbit uterine secretion. Hormone treatment experiment (*B*), compared with oestrous pattern (*A*) and normal preimplantation pattern of 6th day p.c. (*F*)

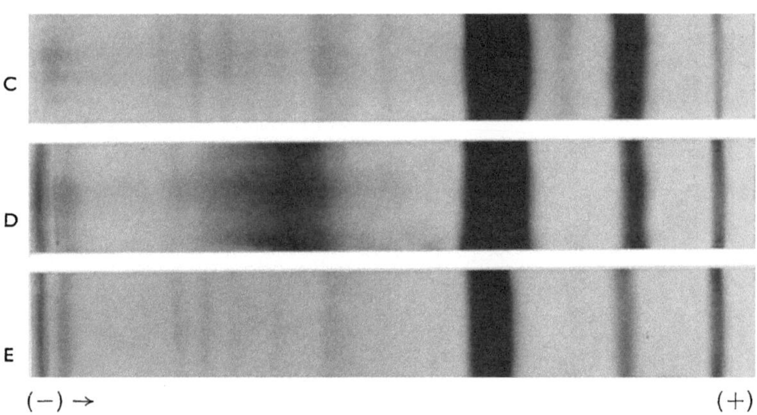

Fig. 13. Disc-electrophoretic protein patterns of uterine secretion 6 d p.c. from pseudopregnant rabbits. Hormone treatment experiments (*C*), (*D*), and (*E*)

Our findings allow us to conclude that normal blastocyst development in the uterus before implantation is insured only if the uterine milieu corresponds, in terms of time and substance sequences, to the specific development stage and physiological requirements of the embryo. Nevertheless, the question as to

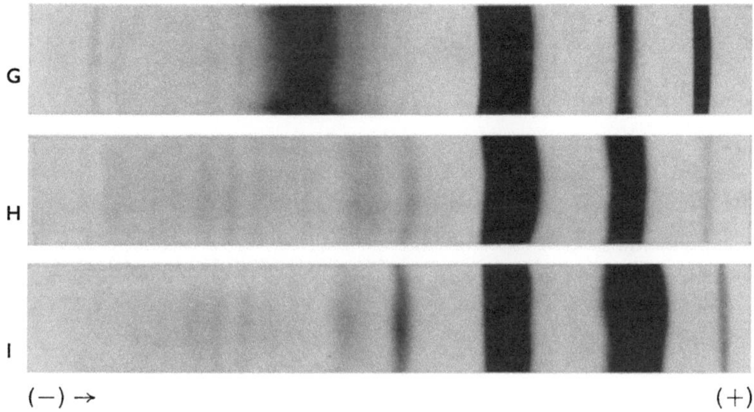

Fig. 14. Disc-electrophoretic protein patterns of uterine secretion 6 d p.c. from pregnant rabbits. Hormone treatment experiments (*G*), (*H*), and (*I*)

causal relationship between protein pattern dynamics and possibilities of blastocyst development cannot, on the basis of our experience to date, be answered conclusively. Far more extensive investigations in the field of hormonal regulation of preimplantation are necessary.

In order to answer the causality question, we must gain enough knowledge to be able to say how the regulatory circuit of interdependency between mother and embryo functions. We must find out whether the feedback mechanisms have their source in the mother's system or in the embryo's. This knowledge is basic and, not least, meaningful in a practical sense, especially as regards *human reproduction* and ways to influence it by steroidal hormones anti-conceptionally.

Summary

The maternal organism provides the basis for early embryonic development in mammals via tubal and uterine proliferation and secretion. The morphological and biochemical adaptations of the rabbit uterus during the preimplantation period were examined by histological, histochemical, and electrophoretic methods.

Biochemical and biological parameters of the proteins of endometrial secretion seem to be of special significance. The majority of them were classified as glycoproteins of comparatively low molecular weight (25–70,000). Among them enzymes and enzyme-inhibitors were found.

The "uteroglobin" is regarded as the most important secreted uterusspecific protein (BEIER, 1966).

The pattern of intra-uterine proteins is set by single, selected plasma proteins and by biosynthesis of uterine-specific proteins secreted by endometrial epithelium and glands, respectively. The same seems to be true of the secretions of the human genital tract.

The effect of experimental imbalance of ovarian hormones on endometrial secretion pattern and embryonic development was investigated. Morphological and biochemical studies revealed characteristic changes of the uterine quantities of distinct protein fractions and the inability in a proper time schedule. These alterations were found to be accompanied by a disturbed development or death of the early embryos.

The causal relation of these phenomena remains to be elucidated. Both maternal and embryonic organism are not autonomous entities and are likely to exist in mutual dependence controlled by feedback mechanisms. The investigation of these controls seems to be most relevant for an understanding and discussion of human reproductive physiology and contraception.

Acknowledgment

This investigation was supported by grant Pe 48/11 of the Deutsche Forschungsgemeinschaft, Germany.

References

1. BEIER, H. M.: Das Proteinmilieu in Serum, Uterus und Blastocysten des Kaninchens vor der Nidation. Coll. Biochemie d. Morphogenese. Konstanz, 1966.
2. — Veränderungen am Proteinmuster des Uterus bei dessen Ernährungsfunktion für die Blastocyste des Kaninchens. Verh. dtsch. zool. Ges. Heidelberg, **31**, 139—148 (1967).
3. — Biochemisch-entwicklungsphysiologische Untersuchungen am Proteinmilieu für die Blastocystenentwicklung des Kaninchens (Oryctolagus cuniculus). Zool. Jb. Anat. **85**, 72—190 (1968).
4. — Uteroglobin: A hormone-sensitive endometrial protein involved in blastocyst development. Biochim. biophys. Acta (Amst.) **160**, 289—291 (1968).

5. Beier, H. M.: Protein patterns of endometrial secretion in the rabbit. In: P. O. Hubinont et al. (Eds.): Ovo-implantation, human-gonadotropins and prolactin, pp. 157—163. Basel-München-New York: S. Karger 1970.
6. Daniel, J. C., Krishnan, R. S.: Studies on the relationship between uterine fluid components and the diapausing state of blastocysts from mammals having delayed implantation. J. exp. Zool. **172**, 267—282 (1969).
7. Deanesly, R.: The endocrinology of pregnancy and foetal life. In: A. S. Parkes (Ed.): Marshall's physiology of reproduction, Vol. 3, p. 891—1063. London: Longmans 1966.
8. Denker, H. W.: Zur Enzym-Topochemie von Frühentwicklung und Implantation des Kaninchens. Inaug.-Diss. Med. Fakultät, Marburg 1969.
9. Garcia-Bunnuel, R., Brandes, D.: Lysosomal enzymes in human endometrium. A histochemical study of acid phosphatase, nonspecific esterase and E-600 resisteant esterase. Amer. J. Obstet. Gynec. **94**, 1045—1055 (1966).
10. Hamana, K., Hafez, E. S. E.: Disc electrophoretic patterns of uteroglobin and serum proteins in rabbit blastocoelic fluid. J. Reprod. Fertil. **21**, 557—560 (1970).
11. Henzl, M. R., Smith, R. E., Magoun, R. E., Hill, R.: The influence of estrogens on rabbit endometrium. An ultrastructural-cytochemical and biochemical study. Fertil. Steril. **19**, 914—935 (1968).
12. Kirby, D. R. S.: The role of the uterus in the early stages of mouse development. In: G. E. W. Wolstenholme (Ed.): Preimplantation stages of pregnancy, pp. 325—344. Ciba Found. Symp., London 1965.
13. Kirchner, Ch.: Untersuchungen an uterusspezifischen Glycoproteinen während der frühen Gravidität des Kaninchens (Oryctolagus cuniculus). Roux' Arch. EntwMech. **164**, 97—133 (1969).
14. — Glycoproteine im Uterussekret des Kaninchens während der frühen Entwicklung der Blastocyste. Verh. dtsch. zool. Ges., Würzburg 1969 (in press).
15. Koch, P.: Disc-Elektrophorese: Ein Photometerzusatz zur Auswertung von Gelsäulen. Zeiss-Mitt. **4**, 397—403 (1968).
16. Krishnan, R. S., Daniel, J. C., Jr.: Blastokinin: Inducer and regulator of blastocyst development in the rabbit uterus. Science **158**, 490—492 (1967).
17. Kühnel, W., Petry, G., Amon, H.: Der Einfluß sogenannter Ovulationshemmer auf die Esteraseaktivität des Endometriums. Experientia (Basel) **24**, 715 (1968).
18. Parkes, A. S., Deanesly, R.: The ovarian hormones. In: A. S. Parkes (Ed.): Marshall's physiology of reproduction, Vol. III, p. 570—828. London: Longmans 1966.
19. Petry, G., Kühnel, W., Beier, H. M.: Untersuchungen zur hormonellen Regulation der Präimplantationsphase der Gravidität. I. Histologische, topochemische und biochemische Analysen am normalen Kaninchenuterus. Cytobiologie **2**, 1—32 (1970).
20. Seidel, F.: Entwicklungspotenzen des frühen Säugetierkeimes. Arbeitsgem. Forsch. Nordrhein-Westf. 193. Köln-Opladen: Westdeutscher Verlag 1969.

Immunology of Reproduction

H. KRIEG

Universitäts-Frauenklinik, 8700 Würzburg, Josef-Schneider-Str. 4

With 19 Figures

Among the different topics in the field of immunoreproduction, immunologic factors of human sterility and infertility are of special clinical interest.

As early as 1899 the antigenicity of spermatozoa was discovered by LANDSTEINER [54] and METCHNIKOFF [64]. In the following years numerous papers with rather conflicting results were published on the topic – namely whether or not antibodies against semen may influence fertility in humans and other mammals.

In 1954, WILSON [99] and RÜMKE [74] independently reported the presence of spermagglutinins in the seminal plasma and blood serum of sterile human males. During the next few years RÜMKE [75] examined the sera of 2015 sterile males: 3.2% of whom possessed spermagglutinins with a titer of 1 : 32 or higher, while none of the sera of 415 fertile males gave positive reactions of this titer using the macroagglutination technique of KIBRICK [45].

The serum factor, which is responsible for the agglutination of spermatozoa in high titers, has been identified as a true antibody: the sperm agglutinating activity remains unchanged after heating of the sera at 56° C for 30 min, it can be completely removed by absorption with seminal spermatozoa and is only present in the gamma-globulin fraction. Sperm agglutinating antibodies are mainly of the 7S type [26], they may, however, partly belong to the class of IgM antibodies [81]. COOMBS, EDWARDS, and GURNER [14] found recently IgG and IgM antibodies to spermatozoa in sera, and IgG and IgA antibodies in seminal plasma of sterile human males. Since IgA, which is evidently synthesized at mucus surfaces, could only be detected in seminal plasma, EDWARDS [23] raised the question of a local synthesis of potent spermagglutinins in the male genital tract. This concept is supported by occasional reports on higher spermagglutinin titers in seminal plasma than in serum, generally the seminal plasma titers are however much lower than the serum titers. According to RÜMKE [77] spermagglutinins transudate within the prostate, from the blood into the seminal fluid. Normally there is a direct relationship between the serum spermagglutinin titer and the extent of the agglutination of spermatozoa in the ejaculate: in males with high serum titers the spermagglutination in the ejaculate is usually strong and often complete shortly after ejaculation. In cases of low serum titers agglutination if it occurs, commences very slowly and remains incomplete.

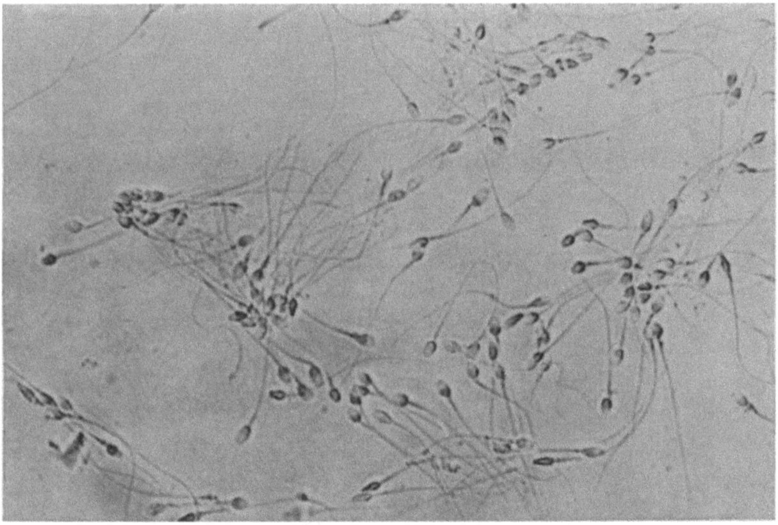

Fig. 1. Spontaneous head-to-head agglutination of spermatozoa in an ejaculate from a man with sperm antibodies

There are various types of immunoagglutination of spermatozoa: the head-to-head type (Fig. 1) can be distinguished from the tail-to-tail type. Furthermore there is a mixed agglutination type with either both the head-to-head and the tail-to-tail type or head-to-tail agglutination.

FJÄLLBRANT [26] found in a series of 36 infertile males a close correlation between spermagglutinin levels in the serum and the ability of spermatozoa to penetrate cervical mucus in vitro, using postcoital tests: the mucus penetration of spermatozoa diminished with rising spermagglutinating titers. Probably, only complete agglutination of spermatozoa in the ejaculate may cause sterility. It is likely that in men, who are fertile, in spite of high titers of circulating spermagglutinins, antibodies fail to enter seminal plasma [77].

In a series of about a hundred cases of azoospermia due to impaired spermatogenesis RÜMKE and HELLINGA [75] found only one case with spermagglutinins in the serum: it is therefore unlikely that the presence of spermagglutinating antibodies in man is related to impaired spermatogenesis.

Spermagglutinating antibodies are often accompanied by immobilizing antibodies. According to FJÄLLBRANT [26] spermimmobilisins play a major role as an immunologic cause of infertility.

The formation of spermagglutinins in men is mainly the result of an excessive resorption of spermatozoa in the genital tract. Several authors reported the presence of spermatozoa in the interstitium of the epididymis and in the blood- and lymph vessels in cases of granulomatous epidymitis or obstruction of the efferent ducts [15, 29, 46]. Prostatovesiculitis may also be responsible for the insufficient drainage and extravasation of the seminal material, followed by the initiation of an immunologic response [26]. According to RÜMKE [76] 75 % of the males with obstruction, however, do not develop spermagglutinins in spite of excessive resorption. The following factors may be responsible for the variable incidence of spermagglutinins [77]:

1. immunotolerance induced by continuous resorption of seminal antigens;

2. blocking of the formation of spermagglutinins by other antibodies;

3. the different predisposition of individuals to form antibodies;

4. and finally the observation of PHADKE [68] that in many cases, phagocytosis of spermatozoa takes place in the lumen of the efferent ducts, and normally this does not induce antibody formation.

Therapeutic trials with ACTH or corticosteroids in infertile men due to autoagglutination of spermatozoa, remained unsuccessful [76]. BANDHAUER [3] reported the disappearance of spermagglutinins in two cases, following the excision of a cyst of the epididymis in one, and a tuberculous epididymis in the second. SCHOYSMAN [80] treated 17 infertile males with spermagglutinin titers of 1 : 64 or higher with testosterone (250 mg, once every two weeks). The spermatogenesis and the resorption of spermatozoa could be suppressed in all patients. In 14 of these cases the antibody titer dropped to below 1 : 32 within one year. After testosterone-treatment was discontinued, spermatozoa reappeared in the ejaculate and five of these males became fertile.

Isolation and purification of seminal antigens are absolutely essential. The antigenic composition of human semen, studied by means of immunoelectrophoresis and the double gel diffusion test of OUCHTERLONY [66] seems to be rather complex. The results we have hitherto are difficult to compare since different techniques and antisera were used. The number of antigens reported in human seminal plasma varies from three to thirteen [2, 6, 24, 33, 57, 70, 71]. Seminal plasma antigens may be firmly attached to the surface of spermatozoa, which in addition possess own antigens too. RAO [71] reported 3 antigens specific to spermatozoa. FELTKAMP et al. [25] found that spermagglutinating sera, when tested with the fluorescent antibody technique, reacted with testicular and not with coating antigens.

Seminal components may be immunogenic not only for the male but also for the female organism when they are deposited in the genital tract. Diffusible and soluble antigens can penetrate the tissues of the female organs and reach the sites of antibody formation. In tissue cultures REID [72] never saw spermatozoa penetrating squamous epithelium of the human cervix; columnar epithelium occasionally showed penetration of spermatozoa between adjacent epithelial cells, while 75 % of cases of metaplastic epithelium were penetrated by spermatozoa throughout the entire thickness of the epithelium.

The question whether there is local production of antibodies against seminal material in the human vagina and cervix needs further investigation. In the mucosa of the hamster's vagina, plasma cells may be found in varying numbers during the estrous cycle, and are supposed to be related to an immunological response [73]. In man however there is little or no lymphatic tissue in the vaginal wall or in the uterine cervix [21, 40]: hence major local antibody production in this area under normal physiological conditions is very unlikely.

Several authors have reported the presence of circulating antispermatozoal antibodies in women. The transmission of systemic antibodies to the vagina and other parts of the female genital tract seems to be possible. Comparative studies on anti-A and anti-B agglutinins show for instance that these isoantibodies could be detected in the cervical secretion of women with rather high anti-A and anti-B titers in the serum. The titers in the cervical fluid were always lower than the serum titers [51]. These observations may indirectly reflect the possibility of transmission of antibodies from the serum into the secretions of the female reproductive tract. However, the mechanism of this transfer is not yet clear [22].

In spite of extensive investigations there are contradictionary opinions as to whether or not circulating antispermatozoal antibodies may produce sterility in female animals as well as in women. According to Franklin and Dukes [27, 28] the presence of circulating spermagglutinins is correlated with unexplained infertility: in a series of 43 patients without demonstrable organic reasons for infertility 72.1 % had spermagglutinating antibodies, whereas in a control group of 31 fertile women only 5.7 % had spermagglutinins. 13 couples of the original 43 infertile couples used condoms for a period ranging from 2 to 6 months; following this in 10 women the spermagglutinin titer dropped to non-detectable levels; nine becoming pregnant, when they had intercourse at the time of expected ovulation.

Isojima et al. [39] found a correlation between the presence of spermimmobilisins in the serum and unexplained sterility in women. Parish and Ward [67] reported one case of an infertile female who had an IgG cytotoxic antibody in her serum and in the cervical mucus, which disrupted spermatozoa in the presence of complement. The cervical secretion of a second childless woman contained a cytotoxic antibody with an immobilizing effect on spermatozoa without causing morphological changes.

The female organism may obviously form different antibodies after contact with seminal antigens. It may, however, be stated that hitherto it has not been proved that active immunisation of females against homologous seminal material leads to sterility, even though antibodies can be demonstrated in the serum. Several theories have been postulated to explain immunological infertility in females, but these need further investigation.

Under physiological circumstances, active spermatozoa may be found in the human uterine cervix between one and a half and three minutes after ejaculation and in the oviduct 30 min after insemination [89]. The immunoagglutination or immobilisation in the vagina or in the posterior fornix, could expose spermatozoa to an unfavourable acid medium for longer intervals than occurs under normal conditions, and so probably impair fertility.

KATSH [42, 43], found in animal experiments, that isolated uteri from guinea pigs, immunised with homologous testicular material or sperm, reacted after contact with spermatozoa by producing prolonged and strong contractions. Hence, the question may be raised, whether fertility may be impaired by a strongly, sustained, contracted uterus in a previously sensitized female organism at the time of ovulation.

Spontaneous allergic sensitization to a glycoprotein in human seminal fluid was described in a 20 year old woman by HALPERN et al. [32]. The patient had a severe anaphylactic reaction, consisting of generalized urticaria, edema, asthma, uterine pain and cardiovascular collapse, immediately after coitus. The responsible antibody was shown to be reagin-like and could be classified as a γA-globulin.

In recent years, the attention of immunologists and clinicians has been drawn not only to the immunology of conception but also to the complex topic of the immunology of intrauterine growth and development, which includes the mother, the placenta and the fetus.

The mechanisms which allow a mother to tolerate the "homograft fetus" in utero for 40 weeks are not clear. The grafting of tissues of one individual to another occurs naturally only during pregnancy: the conceptus is the graft and the uterus the graft site. In the placenta, there is a direct physical contact between maternal and fetal tissues. We have only little information on the question whether immunological rejection may occur occasionally, perhaps oftener than suspected and be a factor leading to spontaneous abortion, if there is some defect in the systems of the mother, placenta or fetus.

The following hypotheses have been proposed to explain the fact that the mammalian fetus, although it has some paternal, genetic characteristics, normally survives pregnancy:

1. the fetusis is antigenically immature;
2. the uterus is an immunologically privileged site;
3. there is a weakened, immunologic reactivity of the mother during pregnancy;
4. mother and fetus are separated by a physical and immunological barrier.

Little [58] suggested in 1924, that the embryo has no individual characteristics foreign to the mother. This hypothesis is disproved since we know that the embryo possesses antigens [49]. The blood group substances A and B were detected in fetal blood and tissues at very early stages of fetal development [50, 90, 93, 94, 96]. The possibility that AB0 incompatibility between mother and fetus may cause spontaneous abortion has not been excluded to date [13, 50, 53, 60]. In addition to the A and B antigens the isoantigens of the Rhesus-, MNSs-, Duffy-, Kell- and P-systems were found on fetal erythrocytes [96].

Furthermore it is today generally accepted that the fetus possesses transplantation antigens [49]. Schlesinger [79] detected these antigens in mice by midgestation, they appeared, however, to be very weak and increased their antigenicity only after birth. The different immunogenicity of adult and embryonic tissues may be at least of some importance, for the successful development of the fetus as a homograft.

The hypothesis, that the uterus is a privileged site for grafts is no longer generally accepted. Privileged sites are for example the hamster cheek pouch [8, 9], the anterior chamber of the eye, and the brain [61]. The lack of lymphatic drainage in these sites is responsible for the absence of an immunologic response.

Schlesinger [78] studied the fate of homografts which were transplanted into the uterine lumina of laboratory rodents: in specifically, presensitized mice and rats the intrauterine homografts underwent accelerated rejection irrespective as to whether the animals were non-pregnant or pseudopregnant.

The possibility of advanced extrauterine pregnancy in humans also indicate that the uterus is immunologically not a privileged site since it is not absolutely essential for the development of the fetus.

The question, whether there is a weakened immunologic reactivity of the pregnant organism, caused by an increased adrenocortical activity during pregnancy was raised by MEDAWAR [62]. This hypothesis is supported by experiments of HESLOP et al. [34] who demonstrated that skin grafts transplanted to pregnant rabbits survive almost twice as long when the transplantation was performed in the third week of pregnancy and not at an earlier or later stage of gestation, nor in non-pregnant animals; the peak of steroid secretion in rabbits is supposed to be in the third week of pregnancy. However, these observations could not be repeated in mice and cattle [7, 63].

ANDRESEN and MONROE [1] showed that adult human skin homografts survived for longer periods of time in pregnant women than in non-pregnant hosts. According to them this delay in rejection during gestation may be due to the elevated levels of corticosteroids. Administration of cortisone or adrenocorticotrophic hormone is followed by a fall in the number of circulating lymphocytes and an involution of the tissues of the thymolymphatic system [18, 30], which occupy a key position in the production of immunologic changes. It is difficult to asess the effects of the increased production of estrogens and progesterone during pregnancy on the lymphoid structures [65]. Atrophy of the thymus and a secondary depression of the remaining lymphoid tissue could be observed after estrogen administration in rabbits [59].

The normal development of the fetus cannot be explained solely by the suppresive effect of elevated hormone levels on the immunologic responsiveness of the mother. The possibility of the formation of anti-A and anti-B immune antibodies as well as anti-D during gestation is one example, that the pregnant female is immunologically not inert. Furthermore BARDAWIL et al. [5] demonstrated that in a series of 20 pregnant women with a history of spontaneous abortions ten rejected a skin graft from the husband earlier than a skin graft from unrelated males; in eight cases there was no difference in survival time between husband's and control grafts; and in two women the husband's graft was rejected later.

Sixteen of the husband's grafts were rejected as white grafts; this observation is believed to represent a state of sensitization.

WOODRUFF [100] showed in rabbits and rats that homografts from the fetus to the mother were rapidly rejected probably as a result of a preexisting sensitization.

Finally there remains the hypothesis that there is a physical barrier in the placenta interposed between maternal and fetal tissues which has no immunologic properties and prevents sensitization of the mother to fetal antigens. The trophoblast is believed to be this barrier which behaves as an immunological "buffer zone". It is the only fetal tissue which is in direct contiguity with maternal tissues [55, 56, 101].

SIMMONS and co-workers [82–87] in a series of papers demonstrated that in mice histocompatibility antigens are absent from the trophoblast. The transplantation of embryonic, placental tissues from F_1 hybrid mice conceptuses into maternal animals of the same strain which were previously immunized to paternal antigens, showed that embryonic tissues were destroyed, while the trophoblast grew and did not provoke homograft rejection. However, on the basis of these transplantation experiments alone it cannot be stated that the trophoblast is non-antigenic; the examination of pure trophoblast free of contaminating stroma and other elements is needed [20]. KIRBY et al. [12, 47, 48] stated, that the trophoblast cells in mice possess antigens, which are covered by a layer of fibrinoid; this layer may constitute an immunological barrier. Fibrinoid was found between maternal and fetal components of the placenta in a number of species including man [4]. CURRIE and co-workers [16] injected pure trophoblast and trophoblast pretreated with neuraminidase into mice. Only the animals, which had been inoculated with neuraminidase-treated trophoblast showed accelerated rejection of donor skin grafts. It is suggested that neuraminidase, has removed the layer of fibrinoid from the surface of the antigenic trophoblastic cells.

There is no clear information available on the question as to whether the human trophoblast is antigenic or not [31]. In some of the previous investigations on the antigenicity of the human placental tissue, homogenates or extracts were used without chemical analysis and without histological separation of the different placental cell components [97, 98]. KAKU [41] isolated a polysaccharide with antigenic properties in an extract from fresh human placenta; it is not specifically mentioned, however, from which area of the placenta the tissue was obtained for extraction.

HULKA and co-workers [35, 36] reported on the appearance of anti-trophoblast antibodies from the fourth post partum day in women after normal pregnancies. In cases of toxemia, these antibodies were already detected during labor and in the immediate post partum period. The fluorescent antibody method was used on placental tissue sections. which were incubated with the fluorescein-tagged globulin fraction from the sera of the respective patients. According to the authors the tagged serum contained antitrophoblast antibodies in cases of fluorescence of the syncytiotrophoblastic cytoplasm. It has to be emphasized that in these experiments neither the antibody nor the antigen in the trophoblast were isolated and identified.

Indirect evidence for the antigenicity of human trophoblast was also obtained by CURZEN [17] in immunohistological studies. Placental sections showed specific fluorescence in the cytoplasm of the trophoblast after staining with a labelled immune mouse serum. The microsomal fraction from fresh human placenta was used for the immunisation of these mice. After incubation of tissue sections from human kidneys, with the same labelled anti-trophoblast serum, there was a strong fluorescence, especially of the tubules.

Boss [10] demonstrated by immunofluorescence studies, that placental antigens in the trophoblastic basement membranes were identical or very similar to antigens of the basement membranes of the kidney. There is a series of papers regarding the nephrotoxic action of anti-placental sera. Concerning spontaneous abortion as a possible result of immunological rejection the statement of PRESSMAN and KORNGOLD [69] is of special interest. They believe that different antibodies are responsible for the interruption of pregnancy and the production of nephrotoxicity, at least in the rat.

The antigenicity of the human trophoblast cell was the subject of our investigations.

The placenta at term is composed of the following structures from above downwards:

1. the chorionic plate; its inner surface is covered by a trophoblastic epithelium;
2. the chorionic villi, which are covered by a layer of syncytium;

3. the decidua basalis with trophoblastic epithelium on the upper surface.

The placenta can be divided into multiple units separated from each other by septae. Each unit may be regarded as one interseptal segment.

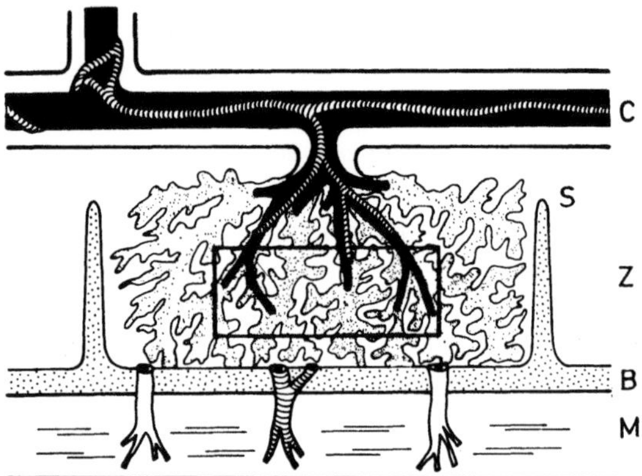

Fig. 2. Diagram of an interseptal segment. *C* = chorionic plate; *Z* = chorionic villi; *B* = decidua basalis; *S* = placental septum. The area where the tissue was obtained for antigen extraction is marked with a rectangle

Pure fetal, placental tissue, i.e. trophoblastic cells, stroma and blood vessel walls of the chorionic villi, were separated from the maternal components.

In order to eliminate the fetal erythrocytes and serum components, the fetal site of the interseptal segments from fresh placentae of mothers after normal pregnancy and delivery were perfused with physiologic saline.

After perfusion we dissected away the chorionic plate together with 3–4 mm of adjacent villous tissue. This was followed by removing the outer portion of the interseptal segments and about 3 mm from the maternal end of the placenta. The center of the interseptal segments was isolated since in this area only pure fetal placental tissue is present.

To eliminate the maternal blood components from the intervillous spaces the isolated perfused fetal tissue portions were rapidly washed two or three times in normal saline. The cut surfaces were examined histologically to confirm that the tissue was purely of fetal origin and free of erythrocytes. This was followed by homogenisation and extraction of the homogenate with a Na-phosphate buffer (pH 8; 0.033 M). The tissue precipitate was separated from the supernatant by centrifugation and the two fractions utilised for the immunisation of rabbits. Employing complete FREUND'S adjuvant, all animals produced precipitating antibodies after injection of either precipitate or supernatant.

Using the Ouchterlony double diffusion test the antisera from rabbits immunised either with the precipitate or supernatant obtained from perfused fetal placental tissue, constantly gave one line of precipitation against placental extract (Fig. 3).

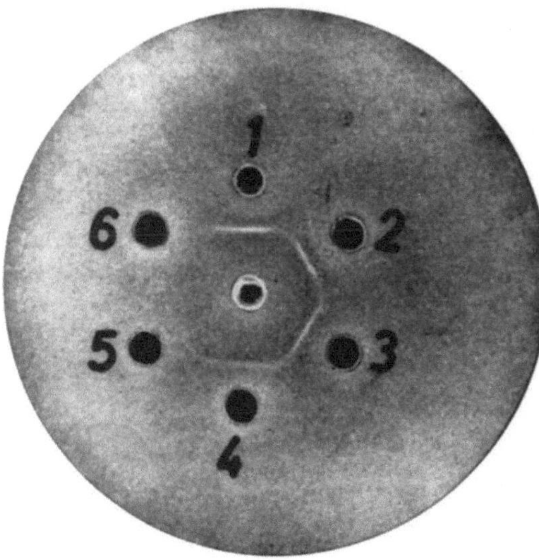

Fig. 3. Double diffusion gel precipitation pattern: undiluted anti-placenta serum (1) and dilutions of the antiserum in saline (2 = 1 : 1; 3 = 1 : 2; 4 = 1 : 4; 5 = 1 : 8; 6 = 1 : 16). Central well: extract from perfused fetal placental tissue

When the extract from perfused fetal placental tissue was tested with the rabbit-anti-placenta serum by means of immunoelectrophoresis, one constant precipitation line was seen in the α_2-

globulin area. There was no precipitation line against normal human serum (Fig. 4).

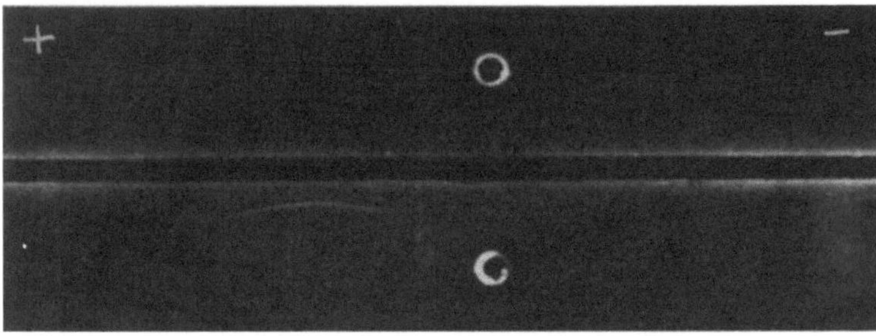

Fig. 4. Immunoelectrophoretic pattern: extract from perfused fetal placental tissue (lower well) with non-absorbed rabbit-anti-placenta serum (central trough). Upper well = normal human serum

The precipitation line in the α_2-globulin area did not disappear after absorption of the antiserum with human plasma and Rh-positive erythrocytes from AB-group donors (Fig. 5).

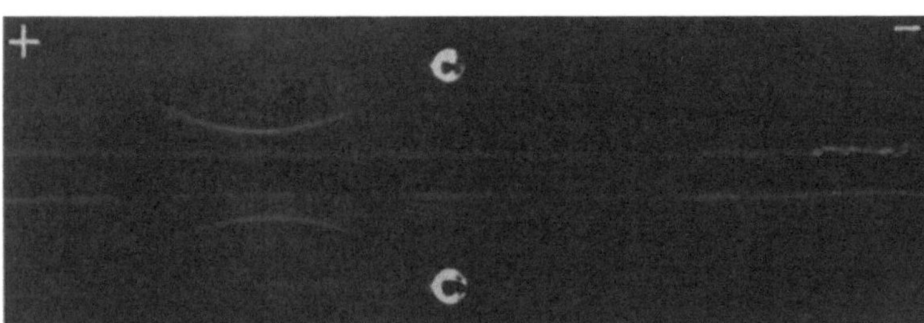

Fig. 5. Immunoelectrophoretic pattern: non-lyophilized extract from perfused fetal placental tissue (lower well) with rabbit-anti-placenta serum absorbed with human plasma and AB/Rh-positive erythrocytes (central trough). Upper well = lyophilized extract from perfused fetal placental tissue

Hence, it can be stated that fetal tissue of the term placenta possesses at least one antigen, which is not present in normal human serum.

Investigations regarding the organ-specificity of this antigen were performed with extracts from liver-, spleen-, myometrium-, kidney-, ovary- and heart muscle tissue from human adults. The rabbit-anti-placenta sera gave no precipitation line against these extracts using the gel diffusion test (Figs. 6, 7).

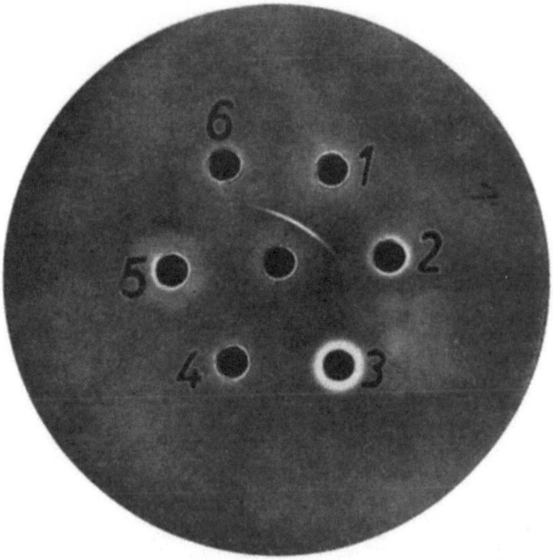

Fig. 6. Double diffusion gel precipitation pattern: central well: rabbit-anti-placenta serum absorbed with human plasma. *1* = extract from fetal placental tissue, *2* = extract from adult human liver, *3* = extract from adult human spleen, *4* = extract from adult human myometrium, *5* = normal serum of a non-pregnant woman, *6* = extract from adult human kidney

Furthermore we extracted organs from fetuses of the third, fourth, and fifth months of gestation. A precipitation line was seen only when fetal liver extract was tested with absorbed rabbit-anti-placenta serum (Fig. 7).

Another precipitation line was obtained with rabbit-anti-placenta serum against serum from fetuses of the third, fourth and fifth months of gestation.

Immunoelectrophoretically, the antigens in fetal liver tissue and fetal serum differ from each other as well as from the antigen in the fetal portion of the term placenta (Fig. 8).

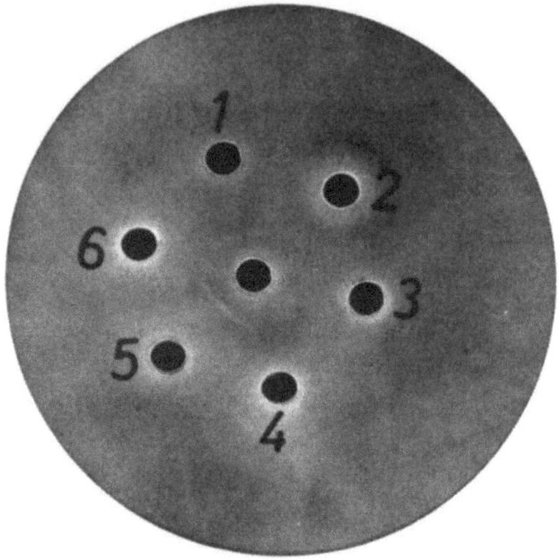

Fig. 7. Double diffusion gel precipitation pattern: central well: absorbed rabbit-anti-placenta serum. *1* = extract from perfused fetal placental tissue, *2* = normal serum of a non-pregnant woman, *3* = extract from ovarian tissue of an adult, *4* = extract from fetal kidney, *5* = extract from fetal liver, *6* = extract from heart muscle of an adult

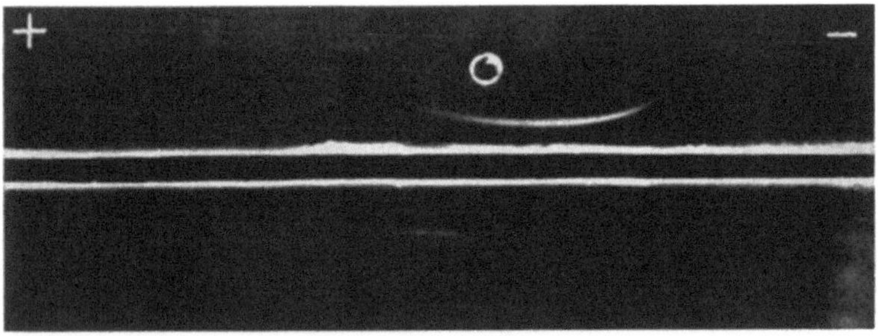

Fig. 8. Immunoelectrophoretic pattern: central trough: rabbit-anti-placenta serum absorbed with adult human plasma; upper well: extract from fetal liver; lower well: fetal serum

To determine whether, the antigen is present only in the chorionic villi and is absent from other parts of the human term placenta, we tested extracts from tissue of the decidua basalis, the upper portion of the chorion plate, the amnion and the umbilical cord, using the Ouchterlony technique; amniotic fluid was also tested; no precipitation line was seen against rabbit-anti-placenta serum (Fig. 9).

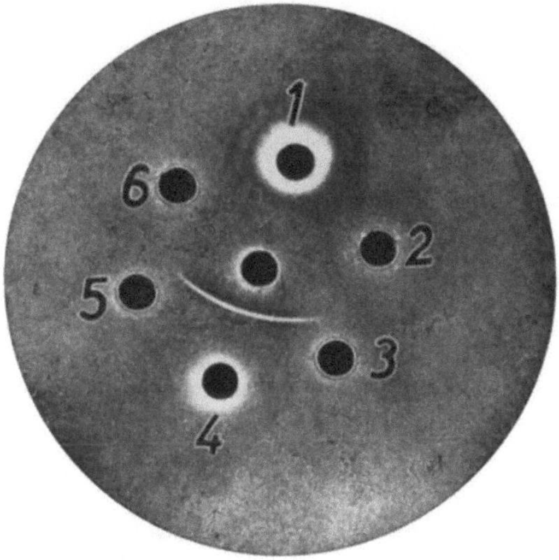

Fig. 9. Double diffusion gel precipitation pattern: central well: absorbed rabbit-anti-placenta serum. *1* = normal serum of a non-pregnant woman, *2* = extract from the umbilical cord, *3* = extract from the amnion, *4* = extract from perfused fetal placental tissue, *5* = amniotic fluid, *6* = extract from the upper portion of the chorionic plate

Besides this antigen, which is probably organ-specific and gives a precipitation line in the α_2-globulin area by immunoelectrophoresis, pure fetal villous tissue from human term placenta possesses two more antigens which differ from each other as well as from the organspecific antigen; these two antigens are in higher concentration in the fetal liver, and in the fetal serum respectively, as compared to fetal placental tissue.

In order to determine whether we had lost some antigens during the process of perfusion we immunised rabbits with extracts from non-perfused fetal placental tissue. When normal human serum was tested with the non-absorbed rabbit-anti-serum a spectrum of precipitation lines were obtained, however, extract from perfused fetal placental tissue gave the typical precipitation line only in the α_2-globulin area (Fig. 10).

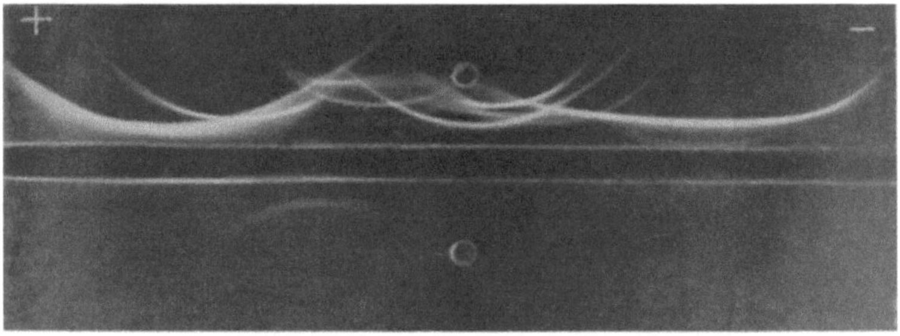

Fig. 10. Immunoelectrophoretic pattern: upper well: normal human serum; lower well: extract from perfused fetal placental tissue; central trough: non-absorbed immune serum from a rabbit immunised with extract from non-perfused fetal placental tissue

The immune serum from rabbits, immunised with extract from non-perfused fetal placental tissue, gave no precipitation lines against human serum after absorption with human plasma. The precipitation line in the α_2-globulin area remained unchanged against extract from perfused fetal placental tissue (Fig. 11).

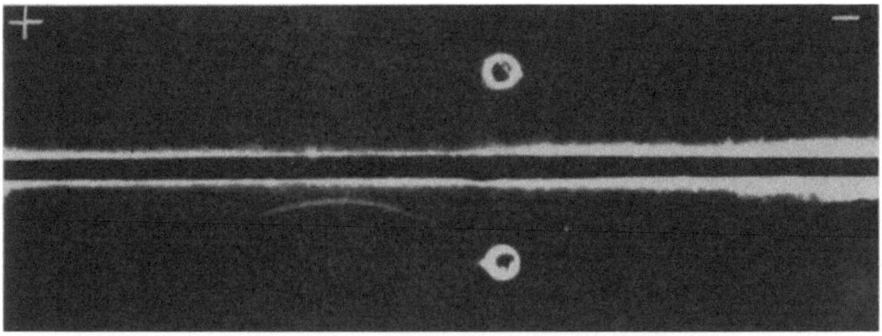

Fig. 11. Immunoelectrophoretic pattern: upper well: normal human serum; lower well: extract from perfused fetal placental tissue; central trough: with human plasma absorbed immune serum from a rabbit immunised with extract from non-perfused fetal placental tissue

Testing the extract from non-perfused fetal placental tissue by means of immunoelectrophoresis with absorbed (human plasma) immune serum from rabbits, which were immunised with extract from non-perfused fetal placental tissue, we obtained two precipitation lines: the typical line of the organspecific antigen of the chorionic villi in the α_2-globulin area and a second line in the β_1-globulin area; this latter antigen was eliminated by perfusion (Fig. 12).

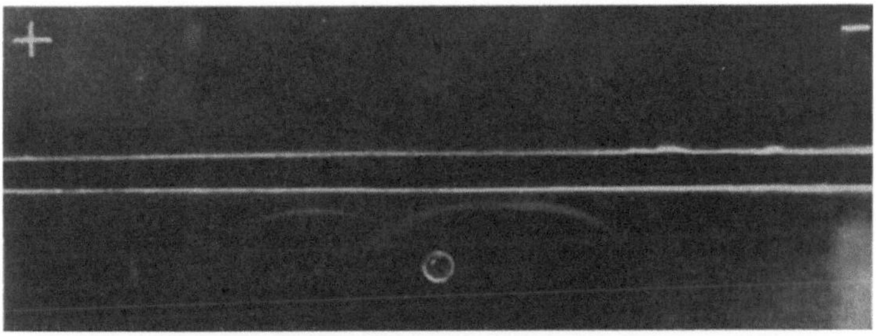

Fig. 12. Immunoelectrophoretic pattern: upper well: normal human serum; lower well: extract from non-perfused fetal placental tissue; central trough: with human plasma absorbed immune serum from a rabbit immunised with extract from non-perfused fetal placental tissue

The antigen in the fetal placental tissue, which was eliminated by perfusion, as well as the antigen in the extract of fetal liver tissue were both found in immunoelectrophoresis in the β_1-globulin area, and hence to determine whether there is an immunological identity between these two antigens, we used the double diffusion technique of OUCHTERLONY. Only a partial identity could be established by this method (Fig. 13).

The determination of the exact localisation of the organ-specific antigen in the chorionic villous tissue was the subject of further investigations. Since, it is technically impossible to isolate a sufficient amount of trophoblastic cells without stroma from the human term placenta, for the immunisation of animals, we first started with the extraction of homogenates which contained trophoblastic cells as well as stroma and fetal vessel walls. According to STRAUSS

[92] only 15 % of the total volume of the chorionic villus consists of trophoblastic cells.

For the morphological localisation of the organspecific antigen, the fluorescent antibody method was used. The crude γ-globulin portion of the absorbed precipitating rabbit-anti-placenta sera was

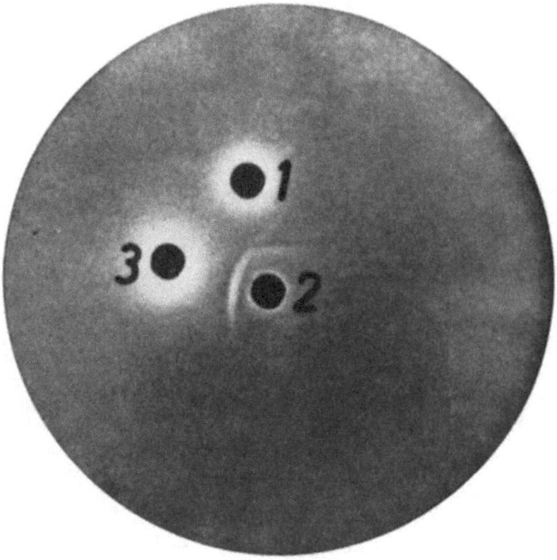

Fig. 13. Double diffusion gel precipitation pattern: *1* = extract from fetal liver tissue, *2* = with human plasma absorbed immune serum from a rabbit immunised with extract from non-perfused fetal placental tissue, *3* = extract from non-perfused fetal placental tissue

conjugated with fluorescein isothiocyanate. After conjugation of this fraction, uncoupled fluorescein radicals and impurities were removed on a sephadex column: conjugates with optimal fluorescein:protein ratios were obtained by gradient elution from a DEAE-cellulose column.

After incubation of cryostat sections from chorionic villi with the labelled antiserum, a specific, isolated fluorescence of the trophoblastic cells was noted, villous stroma, however, did not give any fluorescence (Fig. 14). Occasionally, there was a slight autofluorescence of the fetal vessel walls.

Routine immunofluorescent controls were run simultaneously.

Cryostat sections from the amnion, the decidua basalis and the upper portion of the chorionic plate showed no fluorescence with tagged anti-placenta serum. Sections of the kidney of adults too gave no specific fluorescence.

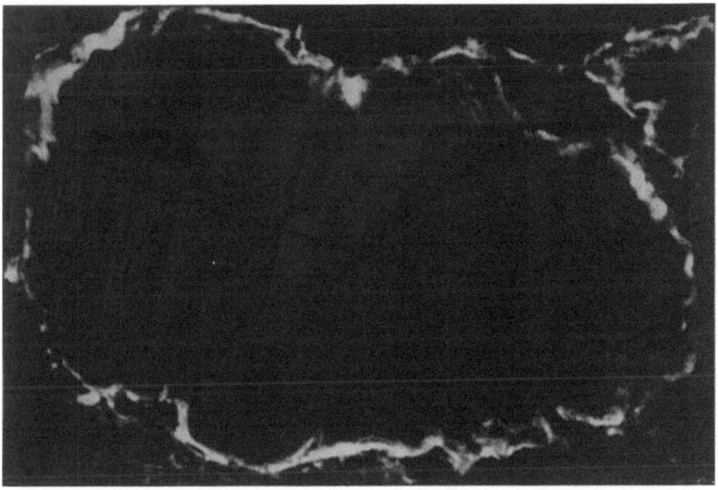

Fig. 14. Cryostat section from a chorionic villus: specific fluorescence of trophoblastic cells after incubation of the tissue section with a FITC-conjugated crude globulin fraction of a rabbit-anti-placenta serum. 400 ×

Hence we concluded, that the organ-specific antigen of the chorionic villus is located exclusively in the trophoblastic cells.

The antigen was further studied by fractionating the crude extract of perfused fetal placental tissue by column-chromatography and agarose-electrophoresis. The chromatographic separation produced 4 maxima, the first peak containing the major part of the material. The small fraction preceding peak I was neglected (Fig. 15). Fraction I comprises 65.25 % of the total material applied (Fig. 16).

After electrophoretic separation of the crude extract of perfused fetal placental tissue, only one anodal fraction + R_3 gave a precipitation line in the double diffusion test of OUCHTERLONY, with immune sera from rabbits immunised with extracts of perfused fetal placental tissue. We therefore, assume that this fraction contains the organ-specific antigen from the trophoblastic cells. This fraction, is relatively pure by electrophoretic criteria (Fig. 17), it exhibits heterogenicity when chromatographed on DEAE-Sephadex (Fig. 18).

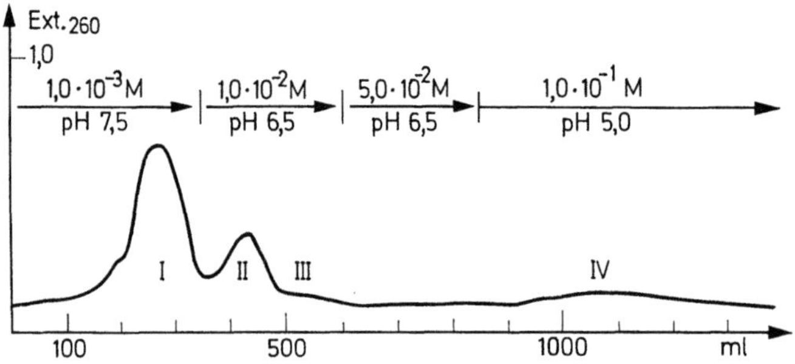

Fig. 15. Fractionating of 2 mg crude extract of perfused fetal placental tissue on DEAE-Sephadex A-50 (20 × 2 cm); continuous registration of the absorbancy of the effluent

Fraction	%
I	65.25
II	24.15
III	3.71
IV	6.89

Fig. 16. Percentage of the fractions obtained after chromatography of the crude extract of perfused fetal placental tissue

The precipitating fraction $+ R_3$ will only stain with amido Schwarz when at least 40 mg of the lyophilised material was applied to the reservoir. We therefore assume, that its protein content is relatively low. Neither the crude extract nor fraction $+ R_3$ showed a positive reaction when stained for lipids with Sudan black. The low protein content of the fraction is also supported by the results of the amino acid analysis: (Fig. 19); when 800 µg of the lyophilised placental extract was applied, only 9.46 % protein was recovered.

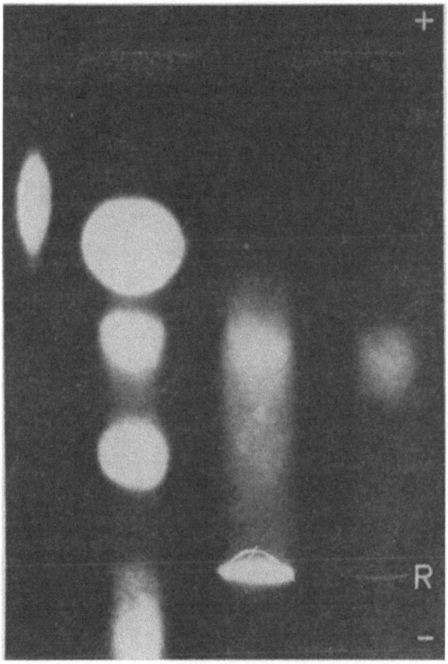

Fig. 17. Agarose electrophoresis. *R* = reservoir. Starting from left: bromphenolblue labelled albumin, normal human serum, crude extract of perfused fetal placental tissue, precipitating fraction + R_3. Stain: amido Schwarz

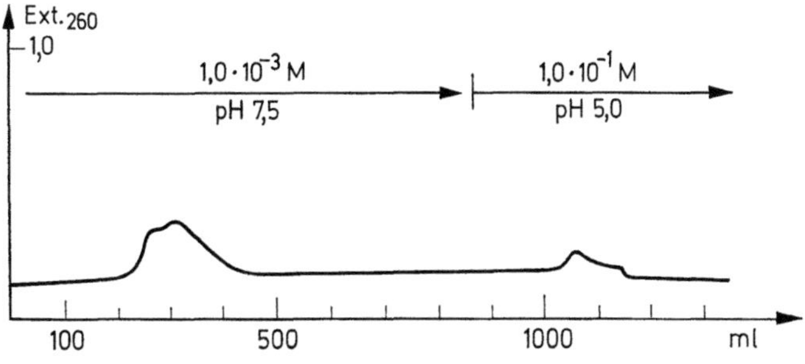

Fig. 18. Fractionating of 2 mg precipitating fraction + R_3 purified by preparative electrophoresis on DEAE-Sephadex A-50 (20 × 2 cm); continuous registration of the absorbancy of the effluent

Amino acid	μMole · 10^{-2}	mg · 10^{-3}
Lys	3.78	5.53
His	2.93	4.56
Arg	4.81	8.38
Asp	5.59	7.44
Thr	2.94	3.49
Ser	3.93	4.13
Glu	8.18	12.03
Pro	4.91	5.65
Gly	4.70	3.53
Ala	4.42	3.94
Ile	2.69	3.53
Len	6.74	8.84
Tyr	1.10	1.99
Phe	1.60	2.64

Fig. 19. Amino acid composition of the crude extract of perfused fetal placental tissue

It is assumed that the organ-specific antigen of the human trophoblastic cell is a conjugated protein. The contaminants are either polysaccharide and/or lipid components in a concentration too low to be detected by colorimetric methods or by the absorbancy at 280 mμ.

For the exclusion of an immunologic identity of the organ-specific antigen with hormones we tested commercial chorionic-gonadotropin, estrogen and progesterone by the double diffusion test of Ouchterlony with the rabbit-anti-placental sera: no precipitation lines were obtained.

Since the deportation of trophoblastic cells into the maternal circulation is considered to be a natural process during pregnancy [19, 95] the mother is continuously exposed to a fetal antigen stimulus. According to Iklé [37, 38] the daily desquamation rate of trophoblastic cells into the maternal circulation is about 100,000. These cells were found in maternal blood as early as the fourth month of pregnancy [4].

We examined the sera of 1226 women for the presence of circulating anti-trophoblast antibodies with the passive hemagglutination technique of Boyden [11]. The sera were obtained during pregnancy, during delivery, 6 days post partum, 6 weeks post partum and 6 days after spontaneous abortions.

The passive hemagglutination reaction was positive in 18.8 % of the total number of patients whose sera were tested. However, the antibody titer was relatively low and ranged between 1 : 10 and 1 : 160. Only 12 women had higher antibody titers. The number of pregnancies, Rh-factor, the presence of Rh-antibodies, toxemia, diseases of the thyroid, the liver, and rheumatic disease had no influence on the incidence of positive reactions in the Boyden technique.

The following reasons may be responsible for the rather low antibody titers by the passive hemagglutination method:

1. the weakened immunologic responsiveness of the mother
2. a weak immunogenicity of the organ-specific antigen of the trophoblastic cell [52]
3. a partial immune tolerance induced by the continuous deportation of antigenic trophoblast cells into the maternal circulation. The weaker the antigen the more readily is tolerance induced [88].

It must be emphasized that only term placentas were examined, there is no information regarding the antigenicity of human trophoblastic cells of the earlier stages of pregnancy. The possibility of an antigen-antibody reaction at the human trophoblastic cell with possible wastage of the embryo or the fetus exists at least theoretically, if the trophoblast is antigenic in early pregnancy too. Though the presence of immunologically competent cells seems to be necessary [91], circulating antibodies also may play some role in graft rejection [88]. Disruption of the integrity of the protective fibrinoid layer might expose the trophoblast antigens to maternal antibodies. To date there is, however, only preliminary and fragmentary information on the possibility of immunological rejection as a cause of spontaneous abortion [44].

References

1. Andresen, R. H., Monroe, C. W.: Experimental study of the behaviour of adult human skin homografts during pregnancy. Amer. J. Obstet. Gynec. **84**, 1096 (1962).
2. Bandhauer, K., Marberger, E., Marberger, H.: Immunologische Einflüsse auf männliche Fertilitätsstörungen. Urologe **3**, 222 (1964).
3. — Immunreaktionen bei Fertilitätsstörungen des Mannes. Urol. Int. **21**, 247 (1966).

4. Bardawil, W. A., Toy, B. L.: The natural history of choriocarcinoma: problems of immunity and spontaneous regression. Ann. N. Y. Acad. Sci. **80**, 197 (1959).
5. — Mitchell, G. W., McKeogh, R. P., Marchant, O. J.: Behaviour of skin homografts in human pregnancy I. Habitual abortion. Amer. J. Obstet. Gynec. **84**, 1283 (1962).
6. Behrman, S. J., Amano, Y.: Immunochemical studies on human seminal plasma. Int. J. Fert. **12**, 291 (1967).
7. Billingham, R. E., Lampkin, G. H.: Further studies in tissue homotransplantation in cattle. J. Embryol. exp. Morph. **5**, 351 (1957).
8. — Silvers, W. K.: Skin transplants and the hamster. Sci. Amer. **208**, 118 (1963).
9. — Transplantation immunity and the maternal-fetal relation. New Engl. J. Med. **270**, 667 (1964).
10. Boss, J. H.: The antigen distribution pattern of the human placenta. Lab. Invest. **12**, 332 (1963).
11. Boyden, S.: The adsorption of proteins on erythrocytes treated with tannic acid and subsequent hemagglutination by antiprotein sera. J. exp. Med. **93**, 107 (1951).
12. Bradbury, S., Billington, W. D., Kirby, D. R. S.: A histochemical and electron microscopical study of the fibroid of the mouse placenta. J. roy. micr. Soc. **84**, 199 (1965).
13. Cohen, B. H., Sayre, J. E.: Further observations on the relationship of maternal AB0 and Rh types to fetal death. Amer. J. hum. Genet. **20**, 310 (1968).
14. Coombs, R. R. A., Edwards, R. G., Gurner, B. W.: Cited in: Brit. med. Bull. **26**, 72 (1970) by Edwards, R. G.: Immunology of conception and pregnancy.
15. Cronquist, S.: Spermatic invasion of the epididymis. Acta path. microbiol. scand. **26**, 786 (1949).
16. Currie, G. A., Doorninck, W. V., Bagshawe, K. D.: Effect of neuraminidase on the immunogenicity of early mouse trophoblast. Nature (Lond.) **219**, 191 (1968).
17. Curźen, P.: The antigenicity of the human placenta. Proc. roy. Soc. Med. **63**, 65 (1970).
18. Dougherty, T. E., Berliner, M. L., Schnecbeli, G. L., Berliner, D. L.: Hormonal control of lymphatic structure and function. Ann. N. Y. Acad. Sci. **113**, 825 (1964).
19. Douglas, G. W., Thomas, L., Carr, M., Cullen, N. M., Morris, M.: Trophoblast in the circulating blood during pregnancy. Amer. J. Obstet. Gynec. **78**, 960 (1959).
20. — The immunologic role of the placenta. Obstet. Gynec. Survey **20**, 442 (1965).
21. Dubrauszky, V.: Personal communication.
22. Edwards, R. G.: Transmission of antibodies across membranes of the reproductive tract. In: Edwards, R. G., Immunology and reproduction, Hertford: Stephan Austin and Sons, Ltd. 1969.

23. EDWARDS, R. G.: Immunology of conception and pregnancy. Brit. med. Bull. **26**, 72 (1970).
24. EYQUEM, A., JOUANNE, C., LANDREIN, S.: Contribution a l'étude des constituants antigéniques du plasma seminal humain. Ann. Inst. Pasteur **110**, 89 (1966).
25. FELTKAMP, F. E. W., KRUYFF, K., LADIGES, N. C. J. J., RÜMKE, P.: Auto-spermagglutinins. II. Immunofluorescent studies. Ann. N. Y. Acad. Sci. **124**, 702 (1965).
26. FJÄLLBRANT, B.: Sperm antibodies and sterility in men. Acta obstet. gynec. scand. **47**, Suppl. 4, 1 (1968).
27. FRANKLIN, R. R., DUKES, C. D.: Further studies on sperm-agglutinating antibody and unexplained infertility. J. Amer. med. Ass. **190**, 682 (1964).
28. — — Antispermatozoal antibody and unexplained infertility. Amer. J. Obstet. Gynec. **89**, 6 (1964).
29. FRIEDMAN, N. B., GARSKE, G. L.: Inflammatory reactions involving sperm and the seminiferous tubules: extravasation spermatic granulomas and granulomatous orchitis. J. Urol. (Baltimore) **62**, 363 (1949).
30. GERMUTH, F. G.: The role of adrenocortical steroids in infection immunity and hypersensitivity. Pharmacol. Rev. **8**, 1 (1956).
31. GROSS, S. J.: The current dilemma of placental antigenicity. Obstet. Gynec. Survey **25**, 105 (1970).
32. HALPERN, B. N., KYAND, T., ROBERT, B.: Clinical and immunological study of an exceptional case of reaginic type sensitization to human seminal fluid. Immunology **12**, 247 (1967).
33. HEKMAN, A., RÜMKE, P.: The antigens of human seminal plasma. Fertil. Steril. **20**, 312 (1969).
34. HESLOP, R. W., KROHN, P. L., SPARROW, E. M.: The effect of pregnancy on the survival of skin homografts in rabbits. J. Endocr. **10**, 325 (1954).
35. HULKA, J. F., HSU, K. C., BEISER, J. M.: Antibodies to trophoblast during the post-partum period. Nature (Lond.) **191**, 510 (1961).
36. — BRINTON, V., SCHAAF, J., BANEY, C.: Appearance of antibodies to trophoblast during the post-partum period in normal human pregnancies. Nature (Lond.) **198**, 501 (1963).
37. IKLÉ, A.: Trophoblastzellen im strömenden Blut. Schweiz. med. Wschr. **91**, 943 (1961).
38. — Dissemination von Syncytiotrophoblastzellen im mütterlichen Blut während der Gravidität. Bull. schweiz. Akad. med. Wiss. **20**, 62 (1964).
39. ISOJIMA, S., LI, T. S., ASHITAKA, Y.: Immunologic analysis of sperm-immobilizing factor found in sera of women with unexplained sterility. Amer. J. Obstet. Gynec. **101**, 677 (1968).
40. JANOVSKI, N.: Personal communication.
41. KAKU, M.: Placental polysaccharide and the aetiology of the toxaemia of pregnancy. J. Obstet. Gynec. Brit. Emp. **60**, 148 (1953).
42. KATSH, S.: In vitro demonstration of uterine anaphylaxis in guinea pigs sensitized with homologous testis or sperm. Nature (Lond.) **180**, 1047 (1957).

43. Katsh, S.: Demonstration in vitro and anaphylactoid response of the uterus and ileum of guinea pigs injected with testis or sperm. J. exp. Med. **101**, 95 (1958).
44. Kerr, M. G.: Immunological rejection as a cause of abortion. J. Rep. Fert. Suppl. **3**, 49 (1968).
45. Kibrick, S., Belding, D. L., Merrill, B.: Methods for the detection of antibodies against mammalian spermatozoa. Fert. Steril. **3**, 430 (1952).
46. King, E. S. J.: Spermatozoa invasion of the epididymis. J. Path. Bact. **70**, 459 (1955).
47. Kirby, D. R. S., Billington, W. D., Bradbury, S., Goldstein, D.: Antigen barrier of the mouse placenta. Nature (Lond.) **204**, 548 (1964).
48. — — James, D. A.: Transplantation of eggs to the kidney and uterus of immunized mise. Transplantation **4**, 713 (1966).
49. — Transplantation and pregnancy. In: Rapaport, F. T. and J. Dausset: Human transplantation, p. 565. New York: Grune and Stratton 1968.
50. Krieg, H., Kasper, K.: Zur Frage der AB0-Inkompatibilität als Abortursache. Geburtsh. Frauenheilk. **27**, 679 (1967).
51. — Blood group substances and infertility. In: Batanov, K.: Immun. Spermatozoa and Fertilization, p. 46. Varna: Bulgarian Acad. Sci. 1967.
52. — Immunologisch bedingte Fertilitätsstörungen. Med. Klin. **64**, 1223 (1969).
53. — Effect of blood group differences between parents on fertility and embryonic development. In: Edwards, R. G.: Immunology and reproduction, p. 234. Hertford: Stephen Austin and Sons, Ltd. 1969.
54. Landsteiner, K.: Zur Kenntnis der spezifisch auf Blutkörperchen wirkenden Sera. Zbl. Bakt. **25**, 546 (1899).
55. Lanman, J. T., Dinerstein, J., Fikrig, S.: Homograft immunity in pregnancy: lack of harm to the fetus from sensitization of the mother. Ann. N. Y. Acad. Sci. **99**, 706 (1962).
56. — Transplantation immunity in mammalian pregnancy, mechanisms of fetal protection against immunologic rejection. J. Pediat. **66**, 525 (1965).
57. Leithoff, H., Leithoff, J.: Neue Ergebnisse immunoelektrophoretischer Untersuchungen des menschlichen Samenplasmas. Med. Welt **4**, 181 (1962).
58. Little, C. C.: Genetics of tissue-transplantation in mammals. J. Cancer. Res. **8**, 75 (1924).
59. Marine, D., Manley, O. T., Baumann, E. J.: Influence of thyroidectomy, gonadectomy, suprarenalectomy and splenectomy on thymus gland of rabbit. J. exp. Med. **40**, 429 (1924).
60. Matsunaga, E., Itoh, S.: Blood groups and fertility in a Japanese population, with special reference to maternal-fetal incompatibility. Ann. hum. Genet. **22**, 111 (1958).
61. Medawar, P. B.: Immunity to homologous grafted skin. III. Fate of skin homografts transplanted to brain, to subcutaneous tissue and to anterior chamber of eye. Brit. J. exp. Path. **29**, 58 (1948).

62. MEDAWAR, P. B.: Some immunological and endocrinological problems raised by evolution of viviparity in vertebrates. In: Soc. for exp. Biology Evolution Symposia No. 11, pp. 320. New York: Academic Press, Inc. 1953.
63. — SPARROW, E. M.: Effects of adrenocortical hormones, adrenocorticotropic hormone and pregnancy on skin transplantation immunity in mice. J. Endocr. **14**, 240 (1956).
64. METCHNIKOFF, S.: Etudes sur la résorption des cellules. Ann. Inst. Pasteur **13**, 737 (1899).
65. NELSON, J. H.: Alterations in immune mechanisms in pregnancy. Clin. Obstet. Gynec. **8**, 263 (1965).
66. OUCHTERLONY, O.: Antigen-antibody reactions in gels. Acta path. microbiol. scand. **26**, 507 (1949).
67. PARISH, W. E., WARD, A.: Studies of cervical mucus and serum from infertile women. J. Obstet. Gynec. Brit. Cwth **75**, 1089 (1968).
68. PHADKE, A. M.: Fate of spermatozoa in cases of obstructive azoospermia and after ligation of vas deferens in man. J. Rep. Fert. **7**, 1 (1964).
69. PRESSMAN, D., KORNGOLD, L.: Localizing of anti-placenta serum. J. Immunol. **78**, 75 (1957).
70. QUINLIVAN, W. L. G.: The specific antigens of human seminal plasma. Fert. Steril. **20**, 58 (1969).
71. RAO, S. S.: Cited in: EDWARDS, R. G.: Immunology of reproduction, p. 170. Hertford: Stephan Austin and Sons Ltd. 1969.
72. REID, B. L.: The behaviour of human sperm towards cultured fragments of human cervix uteri. Lancet **21**, 7323 (1964).
73. ROIG DE VARGAS-LINARES, C. E.: Plasma cells in the hamster vagina: cyclical and experimental variations. J. Rep. Fert. **15**, 389 (1968).
74. RÜMKE, P.: The presence of sperm antibodies in the serum of two patients with oligozoospermia. Vox. Sang. **4**, 135 (1954).
75. — HELLINGA, G.: Auto-antibodies against spermatozoa in sterile men. Amer. J. clin. Path. **32**, 357 (1959).
76. — Sperm-agglutinating autoantibodies in relation to male infertility. Proc. roy. Soc. Med. **61**, 275 (1968).
77. — Clinical aspects of autoimmunity to spermatozoa in men. In: EDWARDS, R. G.: Immunology and reproduction, p. 251. Hertford: Stephen Austin and Sons Ltd. 1969.
78. SCHLESINGER, M.: Uterus of rodents as site for manifestation of transplantation immunity against transplantable tumors. J. nat. Cancer Inst. **28**, 927 (1962).
79. — Serologic studies of embryonic and trophoblastic tissues of the mouse. J. Immunology **93**, 255 (1964).
80. SCHOYSMAN, R.: Treatment of male infertility due to auto-agglutination of spermatozoa. Proc. 6th World Congr. Fertility and Sterility, Tel Aviv, 112 (1968).
81. SCHWIMMER, W. B., USTAY, K. A., BEHRMAN, S. J.: An evaluation of immunologic factors of infertility. Fert. Steril. **18**, 167 (1967).

82. Simmons, R. L., Russell, P. S.: The antigenicity of mouse trophoblast. Ann. N. Y. Acad. Sci. **99**, 717 (1962).
83. — — Failure to demonstrate immunological competence in term mouse placenta cells. Transplantation **2**, **431** (1964).
84. — — Histocompatibility antigens in transplanted mouse eggs. Nature (Lond.) **208**, 698 (1965).
85. — Mechanisms of trophoblast non-antigenicity. Fed. Proc. **25**, 356 (1966).
86. — Russell, P. S.: The histocompatibility antigens of fertilized mouse eggs and trophoblast. Ann. N. Y. Acad. Sci. **129**, 35 (1966).
87. — — Immunological interactions between mother and fetus. Adv. Obstet. Gynec. **1**, 38 (1967).
88. — Histoincompatibility on the survival of the fetus: current controversies. Transplantation Proc. **1**, 47 (1969).
89. Sobrero, A., McLeod, J.: An immediate post-coital test. Fert. Steril. **13**, 184 (1962).
90. Speiser, P.: Über die bisher jüngste menschliche Frucht (27 mm/2. 2 g) an der bereits die Erbmerkmale A_1, M, N, s, Fy (a+), C, c, D, E, e, Jk (a+ ?) im Blut festgestellt werden konnten. Wien. klin. Wschr. **71**, 549 (1959).
91. Steffen, C.: Allgemeine und experimentelle Immunologie und Immunpathologie. Stuttgart: Georg Thieme 1968.
92. Strauss, F.: Personal communication.
93. Szulman, A. E.: The histological distribution of blood group substances in man as disclosed by immunofluorescence. The A, B, and H antigens in embryos and fetuses from 18 mm in length. J. exp. Med. **119**, 503 (1964).
94. — The ABH antigens in human tissues and secretions during embryonal development. J. Histochem. Cytochem. **13**, 752 (1965).
95. Veit, J.: Über Deportation von Chorionzotten. (Verschleppung von Zotten in mütterliche Blutbahnen.) Z. Geburtsh. Gynäk. **44**, 466 (1901).
96. Volkova, L. S., Maysky, I. N.: Immunological interaction between mother and embryo. In: Edwards, R. G.: Immunology and reproduction, p. 211. Hertford: Stephen Austin and Sons Ltd. 1969.
97. Wilken, H.: Der Antigencharakter der Plazenta. Z. ärztl. Fortb. **23**, 1290 (1962).
98. — Über den Nachweis von Gewebs-Antikörpern gegen Plazenta, Leber und Niere in der Schwangerschaft. Z. Geburtsh. Gynäk. **161**, 113 (1963).
99. Wilson, L.: Spermagglutinins in human semen and blood. Proc. Soc. exp. Biol. (N. Y.) **85**, 652 (1954).
100. Woodruff, M. F. A.: Transplantation immunity and immunological problem of pregnancy. Proc. roy. Soc. London **148**, 68 (1958).
101. Zimmer, F., Parks, J.: Die Schwangerschaft als immunologisches Problem. Fortschr. Med. **85**, 1013 (1967).

Genetic Aspects of Early Mammalian Development

R. G. Edwards

Physiological Laboratory, Cambridge, Great Britain

Knowledge of the genetics of pre-implantation development in mammals and man is increasing rapidly with our improving control over these stages. It would be impossible to do full justice to the entire subject in one lecture. Fortunately, Dr. Ralph Brinster has covered major areas of this topic in his description of the synthesis of macromolecules during preimplantation development. I will concentrate largely, although not entirely, on our own work. Most of my remarks will also be devoted to the oocyte and the embryo, but first a few comments are called for on the genetics of spermatozoa, especially with reference to attempts at the identification of different types of spermatozoa from their haploid phenotype.

Is there Haploid Expression in Mammalian Spermatozoa?

This question has been asked repeatedly, with divergent answers. Differences have been reported to exist between X- and Y-bearing spermatozoa for example, in response to changes in pH or time of insemination in relation to ovulation. This contention implies that phenotypic differences exist between these two types of spermatozoa, i.e. in their charge or metabolic pathways. Yet when we examine the biochemistry of spermatogenesis the evidence tends to suggest that such differences do not exist. RNA synthesis by the XY bivalent ends before meiosis and does not resume in the spermatid (Henderson, 1963; Monesi, 1966). This bivalent also has the typical heterochromatic appearance of chromosomal regions inactive in RNA synthesis. Phenotypic characteristics determined by sex-linked genes would thus display 'diploid' inheritance from the spermatogonium rather than haploid inheritance from the spermatid. Moreover, large areas of the cell wall disappear or are not formed between neighbouring spermatids. This situation could result in considerable mixing of the products of adjacent spermatids; 'metabolic compensation' occurs between some adjacent cells in culture when cell contacts are present (SUBAK-SHARPE, BÜRK and PITTS, 1969). It therefore appears doubtful that only gene products typical of X or Y-linked genes are present in spermatozoa.

The situation with autosomal genes could be different, because there is evidently weak synthesis of RNA by autosomes in spermatids. Again, however, the dominant contribution to the differentiating spermatozoa would appear to be provided from the earlier diploid stages.

When we actually look for firm evidence of expression in mammalian spermatozoa, the only clear cut example is that in Tt^{12} heterozygous male mice. Differences between spermatozoa carrying T and t^{12} can be illustrated merely by altering the time of mating with respect to the time of ovulation (BRADEN, 1958). If the mating and ovulation coincide, each class of spermatozoa fertilises one-half of the eggs; however, if the spermatozoa have to wait during a period of delayed ovulation then the great majority of eggs are fertilised by t^{12} spermatozoa. This is clear evidence of a 'haploid' effect in spermatozoa. Other examples quoted of haploid expression are very doubtful. Blood-group antigens provide a classic example: after many claims that these antigens could be detected on spermatozoa and showed "haploid" expression, opinion has now changed and there is a major probability that the earlier workers merely detected "secretor" antigens adhering to spermatozoa (EDWARDS, FERGUSON, and COOMBS, 1964; see the review by POPIVANOV and VULCHANOV, 1969). The suggestion that anomalies in the inheritance of a human translocation were due to distortion in fertilization ratios has been withdrawn (HAMERTON, 1969). Another example may be the melanising activity of rabbit spermatozoa, heterozygotes for pigment genes producing two types of spermatozoa – one with and the other without melanising activity (BEATTY 1956).

Recently fresh data has been brought forward revealing the existence of transplantation antigens in human and mouse spermatozoa, those on human spermatozoa showing haploid expression (VOLTISKOVA, 1969; FELLOUS and DAUSSET, 1970). It will be of interest to follow subsequent developments in this field.

The foregoing examples show that haploid expression might exist, but the evidence is largely unconvincing except in one case.

There is clear evidence that the diploid soma of the male can exert profound influences on the inheritance of characteristics in spermatozoa. These data have been extensively analysed by BEATTY and his collaborators (reviewed by BEATTY, 1970). Before leaving this topic, I must stress that techniques designed to detect the X or Y chromosome in the sperm-head (BARLOW and VOSA, 1970), methods to separate X or Y-bearing spermatozoa by sedimentation, or other methods relying on physical differences between sperm due to the X-Y difference, could be of potential value. But repeated claims to have used these methods of separation are consistently denied; a recent paper provides details (BEATTY, 1969).

Gametogenesis and the Origin of Chromosomal Anomalies in Animal and Human Embryos

A considerable number of human abortuses are non-diploid. Of the total, 20 % are triploid, and the origin of these is probably through delayed fertilization resulting in polygyny or, more likely, polyandry. But the majority of non-diploids are trisomics. How do these arise?

The possession or lack of one chromosome could obviously be due to non-disjunction, and this could occur during meiosis or mitosis. Errors in mitosis after the early cleavage divisions could result in mosaicism; mosaics occur but most trisomics are uniform. If the origin of trisomy is due to the formation of univalents, it can best be detected in diakinesis; if failure of separation of chromosomes is the reason, then anaphase is the optimal stage for examination. Studies in meiosis in human spermatocytes has shown that univalents are exceedingly rare among the autosomes, but that the XY often shows end-to-end association (KJESSLER, 1970). In contrast, examination of mouse oocytes revealed that with increasing maternal age, the number of univalents increased markedly (HENDERSON and EDWARDS, 1968). Similar anomalies were seen in human oocytes but much more data on frequency is needed (EDWARDS and HENDERSON, unpublished; see EDWARDS and FOWLER, 1970). Univalents were found among all sizes of chromosomes, perhaps more frequently among the smaller sizes.

Based on these observations, autosomal trisomics would arise through errors in oocytes, but sex-chromosome trisomics could arise in either spermatocyte or oocyte. There is strong supporting evidence for this suggestion. Using the Xg^a marker on the sex chromosomes, RACE and SANGER (1969) found that sex chromosome trisomy could arise in spermatogenesis or oogenesis. Different methods are needed for the analysis of autosomal trisomy, and one important piece of evidence is that the incidence of such human trisomies is correlated with maternal age and not with paternal age. Clearly these data support the suggestion that autosomal trisomies arise in oocytes, and that sex-chromosome trisomies arise in both spermatocyte or oocyte.

Other suggestions have been made for the origin of trisomy in oocytes, e.g. that decreased frequency of intercourse in older couples leads to delayed fertilization. The experimental data from animals supports neither this contention, nor various others (reviewed by EDWARDS and FOWLER, 1970).

Genetic Aspects of Pre-Implantation Development

Studies on invertebrates have shown that the early period of development after fertilization is largely controlled by maternal templates inherited from the oocyte. The situation in mammalian embryos appears to be different. Transcription evidently begins in the 4-celled stage or earlier (WOODLAND and GRAHAM, 1968); Dr. BRINSTER has reviewed other pertinent evidence in this symposium. Various other data can be added in support of this suggestion, e.g. the death of YO mouse embryos during early cleavage (MORRIS, 1968).

Moreover, genes are also active in the late morula, e.g. those causing the death of homozygous $t^{12}\,t^{12}$ mouse embryos. By the blastocyst stage there is considerable evidence of genetic activity, as shown by the death of some species hybrids at this stage, and by the presence of sex chromatin in rabbits and other blastocysts. Some claims have been made that the transplantation antigens of the foetus are expressed in these stages (SIMMONDS and RUSSELL, 1966; OLDS, 1968). If true, these reports could obviously be important in our understanding of maternal/foetal interactions. In our own work we could not detect these antigens using mixed cell anti-globulin tests (SELL, COOMBS, and EDWARDS, unpubl.).

Despite the initiation of genetic activity in these early stages, it is clear that major alterations in chromosome number are compatible with early development. Triploids can survive to post-implantation stages, but die shortly after implantation (BEATTY, 1957; BOMSEL-HELREICH, 1970). A curious fact about triploids is that the number of sex chromatin bodies in the nuclei is determined by different rules to those effective in the sex chromosome polysomies.

Recently, haploid and diploid parthenogenetic mouse embryos have been shown to survive to implantation or beyond (TARKOWSKI et al., 1970; GRAHAM, 1970). Haploidy was induced by electrical stimulation of oocytes or by exposure of the oocytes to enzymes. It is surprising that earlier attempts to induce gynogenesis (i.e. the initiation of the development of the egg by a spermatozoon with its chromosomes previously inactivated by irradiation, dyes, etc.) merely resulted in abnormal early cleavage by the haploid embryos (EDWARDS, 1957a, b).

Manipulations on Embryos

Earlier work has shown that the blastomeres resulting from the early cleavage divisions of mammalian embryos are very labile. Some of the blastomeres can be destroyed without preventing embryonic development (TARKOWSKI, 1959), and fusion of cleaving eggs and morulae can result in chimaeric embryos (TARKOWSKI, 1961; MINTZ, 1962).

The blastocyst can now be manipulated in various ways, and still give rise to full-term foetuses. Pieces of trophoblast can be dissected from sheep or rabbit blastocysts (ROWSON and MOOR, 1967; GARDNER and EDWARDS, 1968). The trophoblast excised from rabbit foetuses served to sex the blastocyst by means of the sex chromatin body in females (GARDNER and EDWARDS, 1968). A recent advance promises to extend this work, for the long arm of the Y chromosome distal to the centromere can now be stained selectively in interphase or mitosis by means of the dyes quinacrine mustard or quinacrine hydrochloride (PEARSON, BOBROW, and VOSA, 1970; GEORGE, 1970). These dyes should serve as excellent markers for male embryos at all stages of development.

Equally challenging is the recent work showing that entire inner cell masses or individual cells can be inserted into blastocysts, and can give rise to cells colonising large areas of the resulting foetus

(Gardner, 1968, 1970). The inner cell mass can be withdrawn entire from one blastocyst, and placed into a recipient blastocyst. The inner cell mass of the recipient and donor can intermingle, and so lead to the formation of chimaeric foetuses. Alternatively, the inner cell mass of the host can be excised before the donor inner cell mass is inserted. Even more astonishing is the insertion of a single cell into a recipient blastocyst, the donated cell producing cell lines that colonise vast areas of the developing foetus (Gardner, 1968, 1970).

Sex-linked genes are proving of great value in plotting events during cellular differentiation in the early mammalian embryo. According to the Lyon hypothesis, one X chromosome is inactivated during early development. Inactivation is usually random, so that approximately one half of the cells in the embryo express genes on one X chromosome, and the remaining half express genes on the other. Deviations from equality could obviously occur in particular tissues where there are only one or a few stem cell precursors in the embryo. Gandini et al. (1968) analysed tissues of humans heterozygous for mutants at the glucose-6-phosphate dehydrogenase locus for the ratio of the two types of cell, and indicated that a limited number of stem cells (8 or less) are precursors of the haemopoietic tissue in the human foetus (see also Gartler, 1970). This form of "clonal" development could obviously apply to other tissues. The conclusions of Gandini et al. are negated if non-random inactivation of the X's, or if cell selection for one cell type occurs extensively. These phenomena were evidently absent from their material and did not invalidate their conclusions. But non-random inactivation, or cell selection apparently does occur in the erythropoietic line of cells in the Lesch-Nyhan syndrome (Nyhan et al., 1970).

One consequence of clonal differentiation from a few precursor cells could be to provide an opportunity for selectively colonising particular tissues of a host embryo. Placing two or three precursor cells of a tissue in a blastocyst could lead to the extensive colonisation of that tissue in the foetus. Cells specific for particular tissues might be obtained by using the embryonic inducers (Tiedemann, 1970) to control the differentiation of cells in the donor blastocyst. Outgrowths of cells from rabbit blastocysts show considerable differentiation in culture in the presence or absence of inducers (Cole, Edwards, and Paul, 1965, 1966).

Can we Obtain Human Embryos for Study?

I would like to make a brief mention of our work on human fertilization and cleavage. Details have been given elsewhere (EDWARDS, 1970). Many of the obstacles to obtaining cleaving human embryos have now been overcome. To ensure successful embryonic development, the oocytes will have to be taken from patients just before ovulation. The patients are therefore treated with gonadotrophins in order to impose some control over their menstrual cycle, menopausal gonadotrophins being used to stimulate the development of many follicles, and chorionic gonadotrophin to induce ovulation (STEPTOE and EDWARDS, 1970). Laparoscopic methods serve to recover the oocytes which are removed just before ovulation. Studies on fertilization in vitro have advanced to the stage where the conditions necessary for fertilization *in vitro* to occur have been identified (BAVISTER, 1969; EDWARDS, BAVISTER, and STEPTOE, 1969; BAVISTER, EDWARDS, and STEPTOE, 1969). We are now studying the cleavage of the embryos, and have regularly observed 8-celled embryos in culture (EDWARDS and STEPTOE, in preparation).

The medium used for cleavage is based on those developed by BRINSTER (1963), WHITTEN and BIGGERS (1968) and WHITTINGHAM (1968) for mouse embryos, but with the addition of glucose and with and osmolarity below 290 m. osmols./Kg. Human blastocysts might soon be available using these techniques. The culture of mouse embryos to the blastocyst now presents few problems, largely due to the recognition that a reduced oxygen tension in the gas phase and a low osmolarity in the medium are of primary importance (WHITTEN, 1970). Yet the culture of these embryos from the 1-celled stage appeared to be difficult only 2—3 years ago. Present difficulties with human embryos should also be rapidly overcome, and a plentiful supply soon be available for clinical and scientific purposes

References

BARLOW, P., VOSA, C. G.: Nature (Lond.) **226**, 961 (1970).
BAVISTER, B. D.: J. Reprod. Fertil. **18**, 544 (1969).
— EDWARDS, R. G., STEPTOE, P. C.: J. Reprod. Fertil. **20**, 159 (1969).
BEATTY, R. A.: Proc. roy. Soc. Edinb. **25**, 39 (1956).
— Parthenogenesis and polyploidy in mammalian development. Cambridge University Press 1957.

Beatty, R. A.: J. Reprod. Fertil. **19**, 379 (1969).
— Biol. Rev. **45**, 73 (1970).
Bomsel-Helmreich, O.: In: Intrinsic and Extrinsic Factors in Early Mammalian Development. Ed. G. Raspé. Advances in the Biosciences, Vol. 6. New York: Pergamon; Braunschweig: Vieweg 1970.
Braden, A. W. H.: Nature (Lond.) **181**, 786 (1958).
Brinster, R. L.: Exp. Cell Res. **32**, 205 (1963).
Cole, R. J., Edwards, R. G., Paul, J.: Exp. Cell Res. **37**, 501 (1965).
— — — Devlop. Biol. **13**, 385 (1966).
Edwards, R. G.: Proc. roy. Soc. B. **146**, 469 (1957a).
— Proc. roy. Soc. B. **146**, 488 (1957b).
— In: Symposium on Actual Problems in Infertility, Stockholm 1970. Int. Fertility Assocn. 1970.
— Bavister, B. D., Steptoe, P. C.: Nature (Lond.) **221**, 632 (1969).
— Ferguson, L. C., Coombs, R. R. A.: J. Reprod. Fertil. **7**, 153 (1964).
— Fowler, R. E.: Modern Trends in Human Genetics Chap. 6. Ed. A. Emery. London: Butterworths 1970.
Fellous, M., Dausset, J.: Nature (Lond.) **225**, 191 (1970).
Gandini, E., Gartler, S. M., Angioni, G., Argioles, N., Dell'Acqua, G.: Proc. nat. Acad. Sci. (Wash.) **61**, 945 (1968).
Gardner, R. L.: Nature (Lond.) **220**, 596 (1968).
— In: Intrinsic and Extrinsic Factors in Early Mammalian Development. Ed. G. Raspé. Advances in the Biosciences, Vol. 6. New York: Pergamon; Braunschweig: Vieweg 1970.
— Edwards, R. G.: Nature (Lond.) **218**, 346 (1968).
Gartler, S. M.: In: Intrinsic and Extrinsic Factors in Early Mammalian Development. Ed. G. Raspé. Advances in the Biosciences 6. New York: Pergamon; Braunschweig: Vieweg 1970.
George, K. P.: Nature (Lond.) **226**, 80 (1970).
Graham, C. F.: Nature (Lond.) **226**, 165 (1970).
Hamerton, J. L.: In: Human Population Cytogenetics. Ed. P. A. Jacobs, W. H. Price, P. Law. Edinburgh: Edinburgh University Press 1970.
Henderson, S. A.: Chromosoma **15**, 345 (1964).
— Edwards, R. G.: Nature (Lond.) **218**, 22 (1968).
Kjessler, B.: Modern Trends in Human Genetics. Chap. 7. Ed. A. Emery. London: Butterworths 1970.
Mintz, B.: Science **138**, 594 (1962).
Monesi, V.: Exp. Cell Res. **39**, 197 (1966).
Morris, T.: Genet. Res. **12**, 125 (1968).
Nyhan, W. L., Bakay, B., Connor, J. D., Marks, J. F., Keele, D. K.: Proc. nat. Acad. Sci. (Wash.) **65**, 214 (1970).
Olds, P. J.: Transplantation **6**, 478 (1968).
Pearson, P. L., Bobrow, M., Vosa, C. G.: Nature (Lond.) **226**, 78 (1970).
Popivanov, R., Vulchanov, V. H.: In: Immunology and Reproduction. Ed. R. G. Edwards. Int. Planned Parenthood Fedn. 1969.
Race, R. R., Sanger, R.: Brit. med. Bull. **25**, 99 (1969).
Rowson, L. E. A., Moore, R. M.: J. Anat. **100**, 77 (1966).

SIMMONDS, R. L., RUSSELL, P. S.: Ann. N. Y. Acad. Sci. **129**, 35 (1966).
STEPTOE, P. C., EDWARDS, R. G.: Lancet **1970 I**, 683.
SUBAK-SHARPE, H., BÜRK, R. R., PITTS, J. D.: J. Cell Sci. **4**, 353 (1969).
TARKOWSKI, A. K.: Acta Theriol. **3**, 191 (1959).
— Nature (Lond.) **190**, 857 (1961).
— WITKOWSKA, A., NOWICKS, J.: Nature (Lond.) **226**, 162 (1970).
TIEDEMANN, H.: In: Intrinsic and Extrinsic Factors in Early Mammalian Development. Ed. G. RASPÉ. Advances in the Biosciences 6. New York: Pergamon; Braunschweig: Vieweg 1970.
VOJTÍŠKOVÁ, M.: Nature (Lond.) **22**, 1293 (1969).
WHITTEN, W. K.: In: Intrinsic and Extransic Factors in Early Mammalian Development. Ed. G. RASPÉ. Advances in the Biosciences, Vol. 6. New York: Pergamon; Braunschweig: Vieweg 1970.
— BIGGERS, R. L.: J. Reprod. Fertil. **17**, 399 (1968).
WHITTINGHAM, D. G.: Nature (Lond.) **220**, 592 (1968).
WOODLAND, H. R., GRAHAM, C. F.: Nature (Lond.) **221**, 327 (1969).

Maintenance of Pregnancy under the Aspects of Uterus-Inhibiting Substances

H. Jung

Vorstand der Frauenklinik der Techn. Hochschule Aachen, D-5100 Aachen, Goethestr. 27/29

With 20 Figures

To understand the pharmacodynamic and therapeutic relaxation of the uterus during pregnancy, it is expedient to first describe the physiological properties of the myometrium and the dependence of its function on control processes of the autonomic nervous system and hormones. Particular emphasis must be laid on the fact that the uterus

1) exhibits an autonomy that is rare for a muscle tissue,
2) depends on control mechanisms of the central nervous system, and
3) is subject to humorally activating and inhibiting factors.

Fundamentally, the excitatory processes at the uterine muscle are comparable with those of other smooth muscle organs and excitable tissues, such as at the nerve and striated muscle. However, as compared with the more differentiated functional responses of the nerve and striated muscle, it can now be considered as established that among its elementary functional properties the uterine muscle has acquired myogenic, fiber-bound excitation conduction. In our own investigations, during which in 1955 we first referred to myogenic excitation conduction of the myometrium, we found a mean velocity of excitation conduction of about 6.7 cm/sec. The velocity of excitation conduction depends on the hormonal situation of the myometrium.

Contraction is maintained by tetanic series of peak potentials. The intensity of contraction is controlled by the frequency of tetanic excitations and by spatial summation (Fig.1). The membrane potential of the uterine muscle fiber is slightly lower than at the striated muscle. The level of mean potential depends on the endocrine situation of the myometrium. This process is mainly caused by a shift of cellular electrolyte concentration and membrane changes.

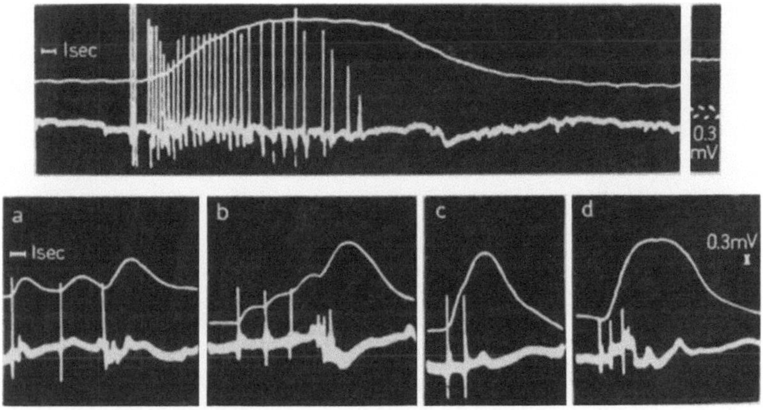

Fig. 1. Top: Mechanogram and action potentials of in situ contraction of a rat uterus. Below: a) Individual mechanical contractions at a fairly long interval of excitation. If the interval becomes shorter, they change via a superposition (b) to incomplete (c) and complete tetanus (d)

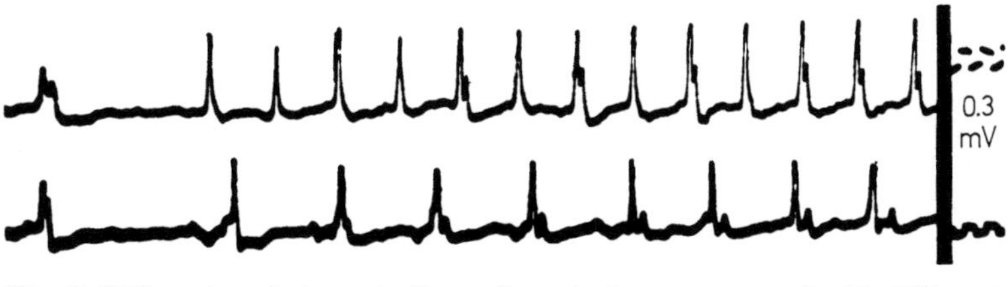

Fig. 2. Different excitatory rhythms of a cat uterus, measured with different electrodes at only a few millimeters' distance from each other

We first reported in 1955 that excitation conduction also takes place myogenically in the uterine muscle tissue. This assumption was the result of our observation of different excitatory rhythms on areas of muscle a few millimeters apart and adjacent to each other

(Fig. 2). Later on, we observed in investigations of single fiber that within a tetanic excitatory volley between two peak potentials some of the muscle fibres exhibit slow depolarizations, such as were described by TRAUTWEIN and ZINK as being typical of the sino-atrial node of the heart and interpreted as signs of local origin of excitation conduction (Fig. 3).

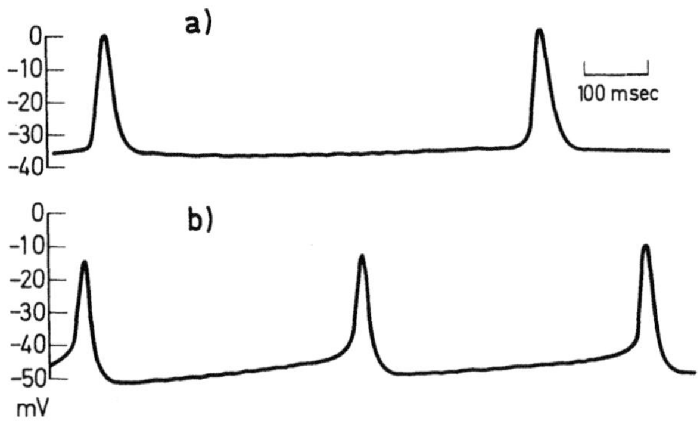

Fig. 3. a Action potentials with single fiber conductions from rat uterus and horizontal course of membrane potential between two excitations. b Between two excitations the rest potential does not have a horizontal course but shows slow depolarization up to the next peak potential as the expression of local origin of excitation

Under certain conditions, the membrane of the myometrial cell is relatively labile and has a tendency toward spontaneous depolarizations. The state of lability depends on the hormonal situation. Thus, in a castrate uterus we find particular tendencies toward spontaneous depolarizations. Estrogen can stabilize the membrane at the same time as the rest potential increases. Simultaneously, a wide safety zone is produced up to the critical potential at which conducted excitations can be induced. As a consequence, in conjunction with the induced metabolic activity of the muscle fiber there is an increase in contraction amplitude (Fig. 4) and a decrease in spontaneous contraction frequency under the influence of estrogen. In this estradiol is much more effective than estrone and estriol (Fig. 5). Accordingly, there is also a sharper increase in membrane potential under the influence of the three different classic estrogens (Fig. 6).

Progesterone has the additional capacity of blocking sodium conductivity. MARSHALL and CSAPO describe inter alia an increase in membrane potential with progesterone. Therefore, at high concentration both hormones have a synergistic action in the sense

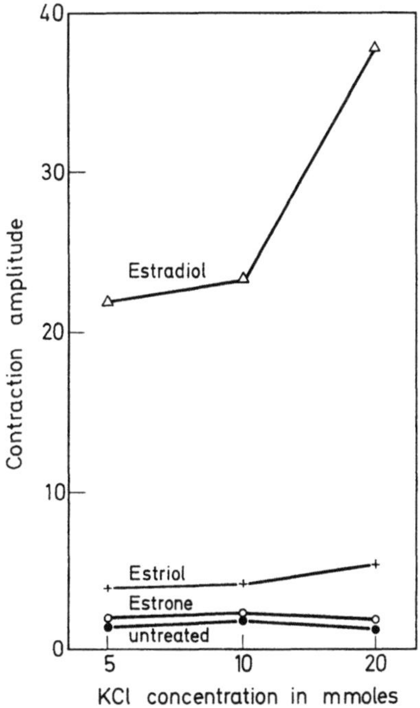

Fig. 4

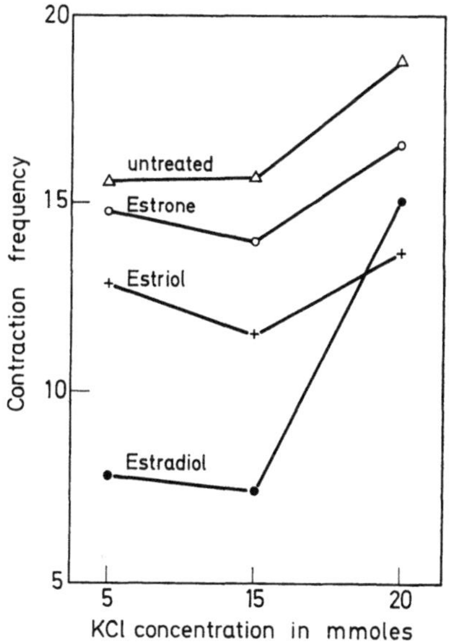

Fig. 5

Fig. 4. Contraction amplitude of isolated uterus of castrated rat, untreated, treated with estrone, estriol, and estradiol

Fig. 5. Contraction frequency of isolated uterus of castrated rat after treatment with estrone, estriol, and estradiol

that they increase membrane potential, stabilize the membrane, and thus inhibit excitability and motility of the myometrium. It follows from this that the activity balance of the uterine musculature is mainly influenced humorally by estrogens and progesterone. According to our present knowledge, other steroids of the organism have no great significance for functional processes at the uterine muscle cell.

On the other hand, concrete indications can be found of the dependence of labor and onset of delivery on factors pertaining to the central nervous system. In recent investigations, we ascertained that immediately before the onset of delivery there is a statistically significant drop of general neuromuscular excitability, measured on the rheobase and the stimulation time-current intensity curve (Fig. 7). For 30 parturients in the stage of the dilatation of the

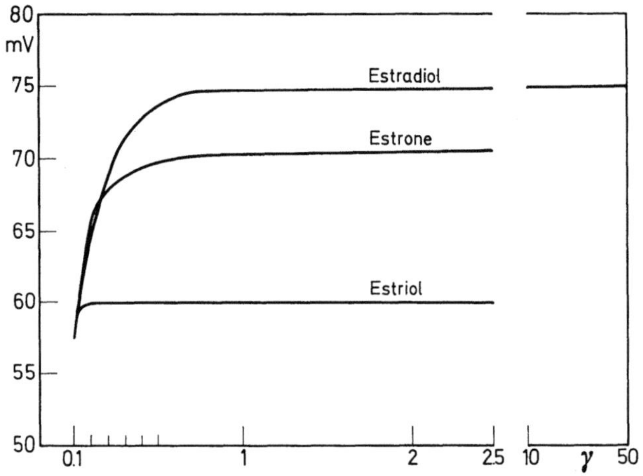

Fig. 6. Course of mean membrane potential at the castrated rat uterus after treatment with estriol, estrone, and estradiol

cervix at the time of onset of labor, the rheobase is already 2 mA. lower than for 30 pregnant and 30 nonpregnant controls. The same process can be induced by Oxytocin, depending on the dose (Fig. 8). However, comparative investigations in women with spontaneous onset of delivery at term and in controls with real overcarriage of the fetus showed that the found rheobase drop cannot be caused by Oxytocin alone. This is also in agreement with the unaminous results of experience gained in the meantime, which show that Oxytocin undoubtedly is a decisive factor in labor but that ultimately it cannot alone initiate delivery. Thus, even today, it is not possible to give a satisfactory answer to the question as to the causes of onset of delivery. In recent years, various hypotheses concerning the cause of induction of labor have been discussed:

1) the Oxytocin theory,
2) the estrogen withdrawal theory,
3) the theory of the estrogen-progesterone quotient,
4) the progesterone withdrawal theory,
5) the theory of a hormonal metabolic change,
6) the control of endocrine and functional metabolism of the uterus and of the placenta by the fetus.

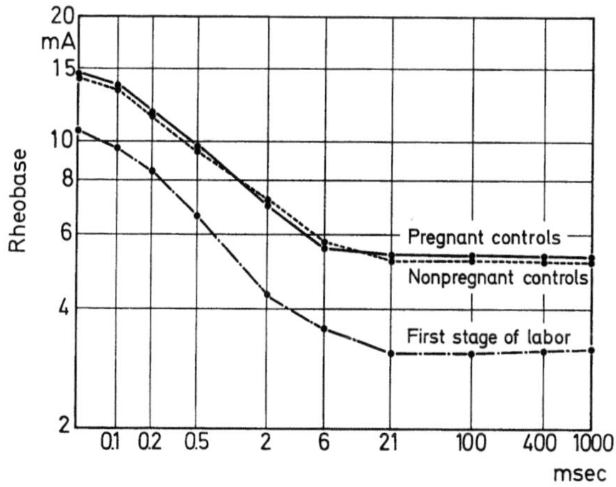

Fig. 7. Course of mean stimulation time-current intensity curve and rheobase for 30 normal pregnant women and 30 nonpregnant women compared with values of the first two hours of labor during dilatation of the cervix

None of these theories has been able to explain in detail the changes which occur at the uterine muscle in connection with the onset of delivery. Since there has recently been a particularly lively discussion about the importance of membrane processes for excitation and their control by hormones and nerval influences. CSAPO thought a drop in membrane potential could be made decisively responsible for induction of delivery. In our own tests on a large number of pregnant rats, in which we measured membrane potential throughout pregnancy and at term (Fig. 9), we found however that membrane potential is constant up to the time of parturition. It is not until immediately after delivery that membrane potential drops within 4 days from an average of —67 mV to —59 mV. Today, we may assume that the safety mechanism for maintenance of pregnancy is a very complex process, and that at the time of delivery this safety system has to be outmaneuvered by setting in motion a "feedback mechanism". At the moment, we can only say that the onset of delivery triggers a process which overcomes the hormonal and autonomic blockage of uterine musculature excitability. It cannot be decided whether a drop in concentration or

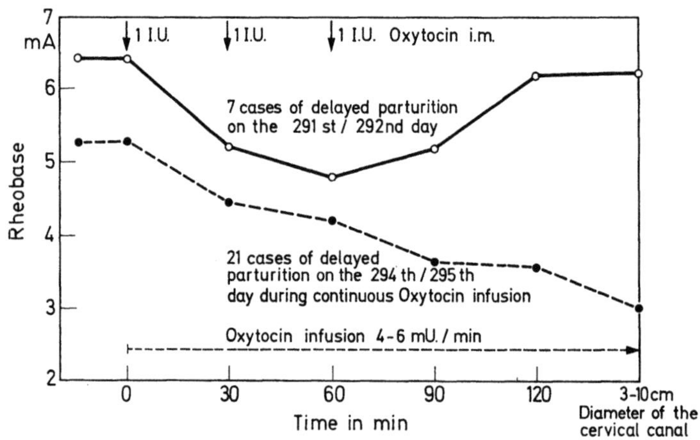

Fig. 8. Mean rheobase values for pregnant women on the 291st and 292nd day of gestation after single injection of 1 I.U. of oxytocin. Mean values of 21 women on the 294th to 295th day of gestation during continuous infusion of oxytocin

metabolic change of a hormone is responsible, or whether the increase in activity of an, as yet, unknown substance abolishes motility blockage at the myometrium, for whose existence the placental hormones have an important partial significance. According to investigations into the significance of α- and β-receptors under the influence of pregnancy and sex hormones, the control changes to be expected at the myometrium in connection with delivery may be even more complex. The complexity of the safety contrivances of the uterus to maintain pregnancy makes it obvious that due to the failure of *one* of the safety factors or due to early disturbances in the fetoplacental unit, there may be the risk of before-term expulsion of the embryo, as is the case with abortions and premature labor.

In recent times, greater importance has also been attached to immunological processes in connection with the protective mechanism of pregnancy and induction of delivery. Pregnancy is looked upon as a heterologous, immunologically incompatible implant, so that the fetus could be expected to trigger an immunological reaction in the mother and be expelled as an incompatible foreign protein.

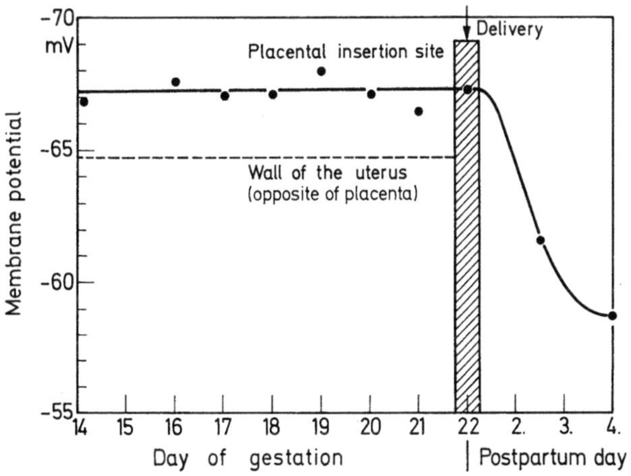

Fig. 9. Mean membrane potential at the pregnant rat uterus during gestation, during delivery and within 4 postpartum days at placental insertion site and opposite the placenta

Today, various authors (LANMAN, WOODRUFF, HULKA) explain the absence of this primarily to be expected reaction by two theories:

1) the trophoblast tissue is not antigenic,

2) the placenta could represent an immunological protective barrier against immunizing antigens of the fetus, or against induction of an immune reaction leading to expulsion of the fetus.

Although, to disprove the first point, HULKA thinks he has demonstrated localization with specificity of antibodies to the trophoblast and on the other hand, contrary to the theory of a placental barrier, the passage of fetal lymphocytes, erythrocytes and whole villi into the maternal blood is exactly established, the immunological aspect of pregnancy, abortion and labor is now the subject of particularly lively discussions (Hellmann, Haines, Johnson, and Petering), especially in connection with the deformities caused by thalidomide. In the meantime, the immunodepressive action of thalidomide seems established. Hellmann considers it possible that thalidomide-deformed babies were preserved from abortion and carried to full term because of the drug's immunodepressive action, and were then born in this unfortunate condition. Indeed, in our own fairly large-scale statistical studies of women who were given treatment for threatened abortion and whose pregnancy could be maintained up to full term by conservative therapy, we found deformities in 6.7 % of the offspring as compared with a normal deformity rate of 1.0 %. However, it must be emphasized that we never observed malformation of extremities. Although at present an immunological process in the form of a "transplant rejection" is not established for normal parturition, in pathologic cases of pregnancy an immunological process is thoroughly capable of determining the fate of the fetus prematurely. This may explain cases of habitual abortion, whose causes were hitherto unknown.

In this context recent results form Pelner and Rhoades, Haines, Johnson and Petering are interesting, according to which progestational substances, such as hydroxyprogesterone caproate (Delalutin-Squibb) and 6-alpha-methyl-17-hydroxyprogesterone acetate (Provera-Upjohn), have an antiimmunizing effect in experiments with skin graft rejections. Simmons, Price, and Ozerkins observed an anti-immunizing effect against skin grafts in mice due to high doses of estrogens, not so much due to progestational compounds. They deduced from their tests that the immunity status of pregnancy is not so much tied up with the fibrinoid of the trophoblast as with the endocrine balance. The significance of immunological processes for the fate of the embryo and their relationship to estrogens and progestagens shows the influence of endocrine control for maintenance of pregnancy and parturition from a new aspect.

According to present clinical experience, utilization of the knowledge I have just described concerning control processes for maintenance of pregnancy and expulsion of the conceptus must take two different gestational periods into account:

1) the "threatened early weeks of pregnancy" and the "threatened abortion" up to the end of the 4th month of pregnancy, and

2) the "threatened abortion" from the 5th month of pregnancy onwards up to and including the "threatened premature labor" and continuing up to full-term delivery of a viable child.

The "threatened early weeks of pregnancy" and "threatened abortion" up to the end of the first half of pregnancy require a

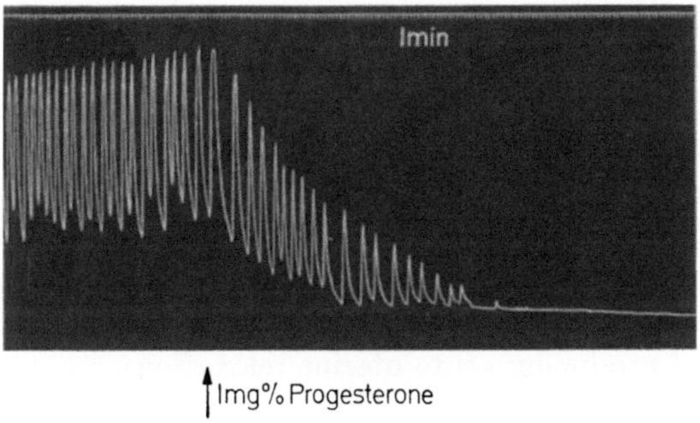

Fig. 10. Inhibition of spontaneous motility of isolated human unterus strip by progesterone

fundamentally different treatment from "threatened premature labor" or threatened late abortions between the 5th and 7th month of pregnancy, because the causes are fundamentally different. According to present knowledge, the principal causes of "threatened abortion" during the first months of pregnancy are malformations of the ovum, organic disturbances of the uterus, such as deformities and hypoplasias, retroflexio uteri, embryopathies and fetopathies. Among the causes of expulsion of the embryo between the 5th and 7th month of pregnancy, I must mention disturbances of the placenta, cervical insufficiency, myomas and lesser-degree uterine deformities.

Despite the steadily increasing number of publications dealing with this subject, the problem of treating "threatened abortion" is still unsolved. Not only the appropriateness and kind of conservative therapeutic measures but also the hitherto imperfect diagnostic methods for detecting the degree of disturbance and making a prognosis are still a controversial topic. However, successful therapy of a threatened abortion is directly linked to the need to obtain as quickly as possible a reliable statement about the state of pregnancy.

Whereas in past decades obstetricians have mainly treated threatened abortion by giving progesterone, on the assumption that this hormone has a reliable muscle-relaxing effect, we have now become more critical in assessing the mechanism of action of progesterone. It is indeed established that progesterone inhibits the isolated human and animal uterine muscle (Fig. 10), and in individual tests we did demonstrate that very large doses of progesterone, administered intravenously, have an inhibitory action at the pregnant uterus in vivo (JUNG, 1965) and at the whole isolated, perfused human uterus in vitro (MOSLER, 1968). However, on the other hand, it is a fact that as a result of the extraordinarily short progesterone turnover of 1—2 minutes the concentration required for a convincing relaxation of the uterus in vivo cannot be attained by therapeutic intravenous administration of progesterone with the intention of producing acute uterine relaxation.

Mean progesterone concentration at the end of pregnancy (ZANDER et al.)

Placental tissue	2×10^{-6} g/g
Uterine musculature	2×10^{-7} g/g
Uterine venous blood	5×10^{-7} g/ml
Peripheral venous blood	1.5×10^{-7} g/ml

Pharmacological action of progesterone at the intact isolated human uterus during pregnancy

Increase of excitability	10^{-7} to 3×10^{-7} g/ml
Coordination of uterine activity and increase of efficiency:	5×10^{-7} to 3×10^{-6} g/ml
Inhibition of uterine activity and decrease of efficiency:	10^{-5} to 3×10^{-5} g/ml

Fig. 11. Comparison of mean uterine and placental progesterone concentration at the end of pregnancy with the in vitro progesterone concentration necessary to inhibit the uterus (according to MOSLER)

MOSLER has compared the progesterone concentrations determined by ZANDER et al. in the uterine muscle, in uterine blood, and in placental tissue with the progesterone concentrations required for reliable muscular inhibition of the uterus (Fig. 11). On the basis of the progesterone concentrations required for inhibition, which he found in his own investigations with the isolated, perfused human uterus, he came to the conclusion that the progesterone concentrations normally circulating in the uterine muscle and uterine blood during pregnancy could not be sufficient for physiological inhibition and relaxation of the uterus during pregnancy. Indeed, he thinks that an increase in uterine activity in vitro could be achieved with only slightly lower progesterone concentrations.

Today, natural progesterone is not considered to be a suitable substance for therapy of threatened abortion, although high-dose depot progestagens do have their therapeutic justification for once via a pharmacological long-term effect in the sense of a relaxation of the musculature, especially in cases of hypoplasia or insufficient muscular blood flow. Besides, we must think of improving decidualization already at the time of implantation of the ovum. Among the group of women with habitual abortions without recognizable cause, we find a very large number whose endometrium responds insufficiently to the body's own production of estrogen and progesterone. Therefore, on the basis of our own experience of the physiological action of estrogens at the myometrium and in the interest of better proliferation and decidualization, we recommend combined high-dosage estrogen-progestogen therapy. The results of our comparative clinical trials involving two different treatment groups and a relatively large number of patients, in contrast to the small numbers of patients mentioned in most publications, encourage us to retain this therapy for the conservative treatment of threatened early abortion.

Among 4,785 abortions in various stages, we divided 511 cases of threatened abortion into two different treatment collectives (Fig. 12). Group I comprised patients who in our opinion received optimal treatment, namely 250 mg of 17-alpha-hydroxyprogesterone caproate together with 10 mg of estradiol benzoate every day up to at least every third day. In addition, on the first day of therapy, the patients were given an injection of 250 mg of 17-alpha-hydroxyprogesterone caproate.

In the comparison group, there were 180 cases who were merely confined to bed without hormonal treatment or who in an early phase of hormonal therapy had been given a single estrogen dose of

5—10 mg of stilbene or an intravenous dose of 20 mg of Lutocyclin. Today, the latter can be described as having no effect and, from modern viewpoints, the patients of Group II can be considered to have been untreated or to have been given completely inadequate treatment. As a criterion of therapeutic success, we evaluated all cases of threatened abortion in both groups who could be discharged

Discharged with pregnancy maintained

	total	without known causes
Group I 111 cases	77.5%	83.0%
Group II 180 cases	56.2%	65.5%

Fig. 12. Treatment result of 2 groups of women with "threatened abortion" as a percentage of patients discharged with maintained pregnancies. Group I was treated with 250 mg of 17-alpha-hydroxyprogesterone caproate + 10 mg of estradiol benzoate daily. In addition, 250 mg of 17-alpha-hydroxyprogesterone caproate were given on the first day of therapy. Group II was confined to bed without treatment or given an intravenous injection of 20 mg of Lutocyclin or Proluton i.v. and up to 10 mg of estradiol. By comparison with Group I, it can be regarded as having received no or insufficient treatment

with clinically intact pregnancies. In Group I we arrived at a success rate of 77.5 %, which is statistically significant by comparison with the maintenance rate of only 56.2 % for Group II, whose patients received no or inadequate hormonal therapy. It stands to reason that we cannot expect this percentage success rate to correspond with the real number of pregnancies which went to full term. It is probable that a certain percentage of these women did after all have abortions later on, but this proportion is doubtless relatively evenly distributed over the two comparison groups.

In spite of our critical attitude to a purely inhibitory action of progesterone at the uterus, on the basis of our results we continue to recommend combined, high-dose estrogen-progestogen depot therapy for the at-risk early weeks of pregnancy and for threatened abortion up to the end of the 4th month. The main basis of our assumption is a pharmacological long-term effect of both hormones to relax the musculature and to improve blood flow in the uterus and endometrium. According to SCHREINER's investigations. both hormones also improve utero-placental blood flow as well as trans-

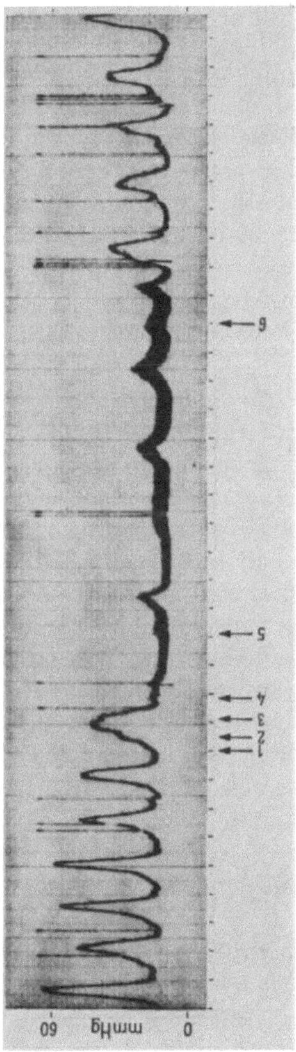

Fig. 13. Labor inhibition with Fluothane during intrauterine pressure recording according to SAAMELI. International togography. Fluothane anesthesia and spontaneous labor. Same patient as in Table IV. For Fluothane 1 denotes 1%, 2 2%, 3 3%, 4 4%, 5 reduced to 3%, 6 Fluothane discontinued

port capacity and metabolic potency of the placenta. According to investigations by AHLQUIST, CIECIOROWSKA, POL, and TELKO, the reactivity of the α- and β-receptors of the myometrium's smooth musculature is also essentially influenced by an adjusted estrogen-progestogen balance. With a unilateral drop in progesterone and a relatively stable estrogen level, there is increased response of the α-receptors to noradrenaline so that the uterus tends to increase motility and expel the embryo. Let me remind you in this context of the discussions on improving immunotolerance in the maternal organism to the homograft of an embryo by means of high estrogen-progesterone doses. Because of completely different causes of embryo expulsion, threatened pregnancy beyond the 5th month and threatened premature birth necessitate other therapeutic consequences. Here the sole preoccupation is the risk of expulsion of an intact embryo due to a purely motor-uterine dysregulation. In recent years pharmacologists, pharmacists, and obstetricians have jointly been on the lookout for substances with a motor uterus-relaxing action.

Today, the group of "*classic spasmolytics*" based on papaverine and similar substances are regarded as unsuitable for relaxation of the uterus during pregnancy due to their side effects or poor therapeutic efficiency. *Psychopharmacological drugs* with a spasmolytic action from the series of benzodiazepin compounds (Librium, Valium) also only have a sufficiently relaxing effect at the isolated human and animal muscle strip. In conformity with other authors, we did not find an inhibitory action of Valium in our own in vivo tests on the uterus during premature labor and full-term deliveries with tocographic control and recording of labor. *Inhalation anesthetics*, such as ether and halothane (Fluothane), guarantee reliable relaxation of the uterus. Fig. 13 shows the effect of Fluothane inhalation in a patient during parturition from a figure by SAAMELI, obtained by means of in vivo intra-uterine pressure recording. However, inhibition of uterine activity by halothane (Fluothane) inhalation is only recommended for a very limited time in cases of acute uterine hypermotility or threatened hysterorrhexis. For effective labor inhibition, a Fluothane concentration of 3—5 percent by volume is necessary at the start of inhalation. If there is sufficient influx of oxygen, the concentration must be reduced to 1—2 percent by volume after a few minutes.

Gaseous anesthetics are unsuitable for long-term therapy of threatened premature labor.

Just recently, particular attention has been paid to the therapeutic use of *ethyl alcohol* for threatened premature labor. FUCHS and FUCHS were the first to refer to the inhibitory action of ethyl alcohol on uterine activity in in vitro and in vivo tests, and they have already achieved satisfactory results in several hundred cases of using intravenous ethyl alcohol therapy for threatened premature labor. The mechanism of action has not yet been uniformly elucidated. Whereas FUCHS and FUCHS pinpoint a diencephalic inhibition of pituitary Oxytocin secretion for clinical success of ethanol treatment, HÜTER on this basis of investigations with the isolated mammalian uterus believes in a direct muscle-sedative action of ethanol. However, in a more recent study FUCHS, FUCHS et al. have ascertained that ethanol concentrations which are equivalent to the blood alcohol value for clinical inhibition of labor (1.0 to 1.6 mg/ml) do not inactivate the isolated human uterus. Much higher alcohol concentrations (3.2 mg/ml) are required to inactivate the isolated human myometrium in vitro, so that the authors attribute clinically effective labor inhibition at low alcohol concentration in cases of threatened premature labor to action on the central nervous system.

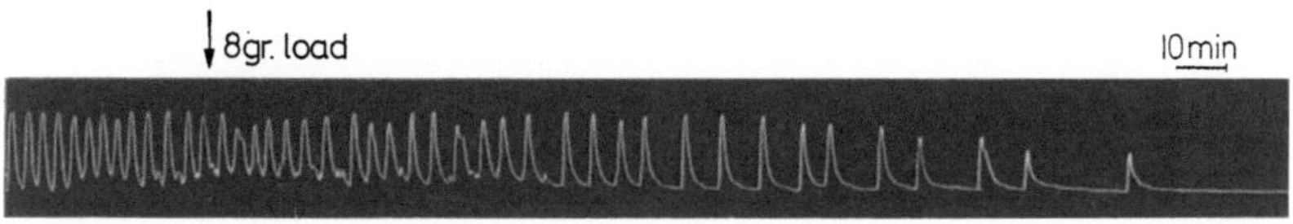

Fig. 14. Inhibition of spontaneous activity of isolated human uterus strip by the "uterus-inhibiting substance", which is liberated when a second muscle strip suspended in the same muscle bath is subjected to a load of 8 grams

We ourselves have been on the lookout for efficient labor inhibitors for treatment of threatened abortions and premature births. In 1967 we first described a hitherto unknown "*inhibitory substance*" which is liberated by the human and animal myometrium at a certain degree of extension (Fig. 14).

As Fig. 14 shows, if a strip of human myometrium is subject to a load of 8 grams, an active strip of muscle suspended in the same fluid is inhibited in its spontaneous motility. Further investigations have shown that the passively loaded myometrium strip liberates a substance which can be collected in an aqueous solution and is capable of completely inhibiting an active myometrium strip in a second bath solution. The "inhibitory substance", which is also liberated by the extended rat uterine cornu, has a converse, i.e. an activating effect, at the rat uterus. The "inhibitory substance", whose chemical nature has not so far been elucidated, has inter alia an Oxytocin-antagonistic action. The substance is stable when heated to 70° C for 20 minutes, can be dialyzed, and according to our investigations hitherto has a low molecular weight. Since in cases of passive overextension with an acute hydramnion or with multiple pregnancies the human uterus may also liberate such a substance, we tested blood serum extracts from such patients for an inhibitory action at the isolated human myometrium against control sera from patients with normal pregnancies.

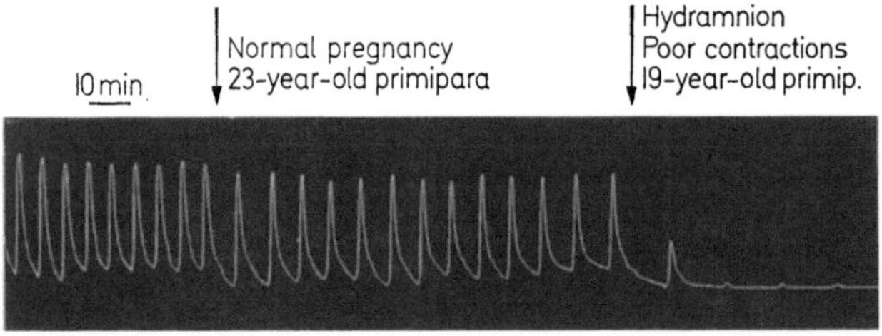

Fig. 15. Inhibition of spontaneous motility of isolated human uterus strip by a serum extract from the blood of a patient with considerable passive overextension of the uterine wall with hydramnion and poor contractions, compared with non-inhibitory action from the serum of a normal parturient

As Fig. 15 shows, in the blood serum of 8 out of 10 cases we demonstrated a distinct inhibitory action, similar to that of the "inhibitory substance" we discovered, by comparison with control sera.

Because even small human myometrial strips produce large quantities of this extraordinarily active "inhibitory substance", we have made further efforts in recent years to elucidate its chemical nature. Inter alia we tried to exclude the possibility that fractions of the prostaglandin group were involved. According to more recent studies of my coworker, Klöck, prostaglandin is indeed detectable inter alia in the effective extracts which we obtained from passive

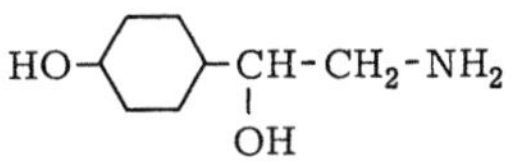

Noradrenaline

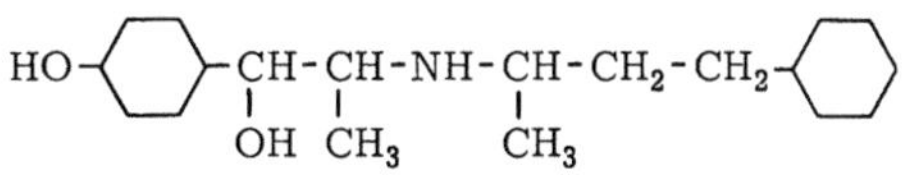

Buphenine

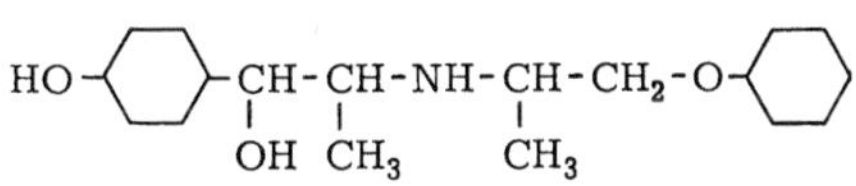

Isoxsuprine

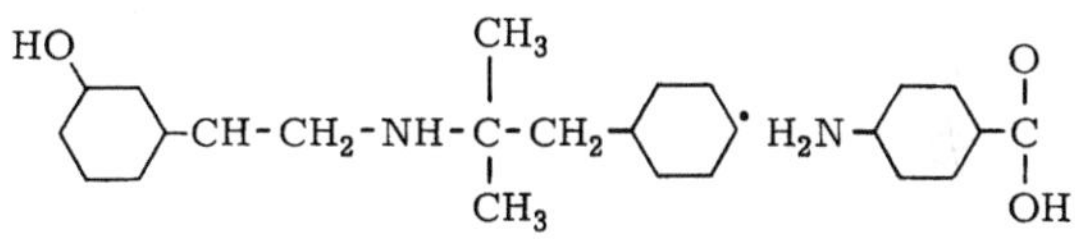

TV 399

Fig. 16. Structural formula of noradrenaline, buphenin, isoxsuprine, TV 399 (Tropon-Werke, Cologne)

extension of the human uterine strip. However, it is unlikely that these are identical with the "inhibitory substance" because prostaglandins do not withstand heating to 70° C for 20 minutes without loss of biological activity, and furthermore upon access to air they very quickly become inactive due to oxidation. The "uterus-inhibiting substance" does not lose its activity upon access to air and retains its property for days at a refrigerator temperature of about +5° C.

The very intensive investigations during recent years into the mode of action of prostaglandins and possible clinical applicability for uterine relaxation deserve particular attention. The substance with a uterus-inhibiting or -activating property, which was first detected in fresh human semen by KURZROCK and LIEB in1930 and by EULER in 1934, has meantime proved to be a group of substances with relatively nonuniform pharmacological activities. According to more recent studies by EMBREY, when injected intravenously E- and F-prostaglandins appear to have an activating effect at the human uterus so that it is unlikely that they will be clinically applicable for inhibition of labor in the case of threatened abortion and premature birth.

Other substances, such as nonestrogenic and nonprogestational steroids, HCG, relaxin, phenothiazides and vegetative hormones, like adrenaline and histamine, have no practical significance for clinical treatment of abortions and premature labor.

On the other hand, recently the therapeutic optimism of the obstetrician has been backed particularly by rapid developments in the field of sympathomimetic drugs with β-adrenergic action. Already at the beginning of the 1960's there were first reports by U.S. obstetricians (HENDRICKS et al.) on successful inhibition of labor using paraoxyephedrine derivatives, such as isoxsuprine (Vasodilan, manufacturer Mead Johnson and Co., Duvadilan, manufacturer: Philips Duplar GmbH). Another preparation from the series of β-adrenergic substances with good uterus-inhibiting efficiency is Buphenin (Nylidrin), (Dilatol, manufacturer: Tropon-Werke, Cologne) — see Fig. 16. Although uniform reports on a clinically effective inhibition of labor through Buphenin and isoxsuprine are available (HENDRICKS et al., 1961; BAYER, 1966; WOLFF, 1966 and 1967; HÜTER, 1968; MOSLER, 1968), all authors are just as unanimous in their reports of severe side effects at a fairly high dosage, such as decrease in blood pressure, increase in pulse rate, subjective sensations. feeling of unrest, palpitations,

---

Fig. 17. External tocography and above the cardiac sound curve in the so-called "cardiotocogram" of a parturient during the first stage of labor. Complete extinction of labor upon intravenous administration of the β-adrenergic labor inhibitor TV 399 (Tropon-Werke, Cologne). The mean heart rate curve of the fetus only shows a slight increase in frequency in the physiological range. In lower figure, resumption of labor after TV 399 has been discontinued

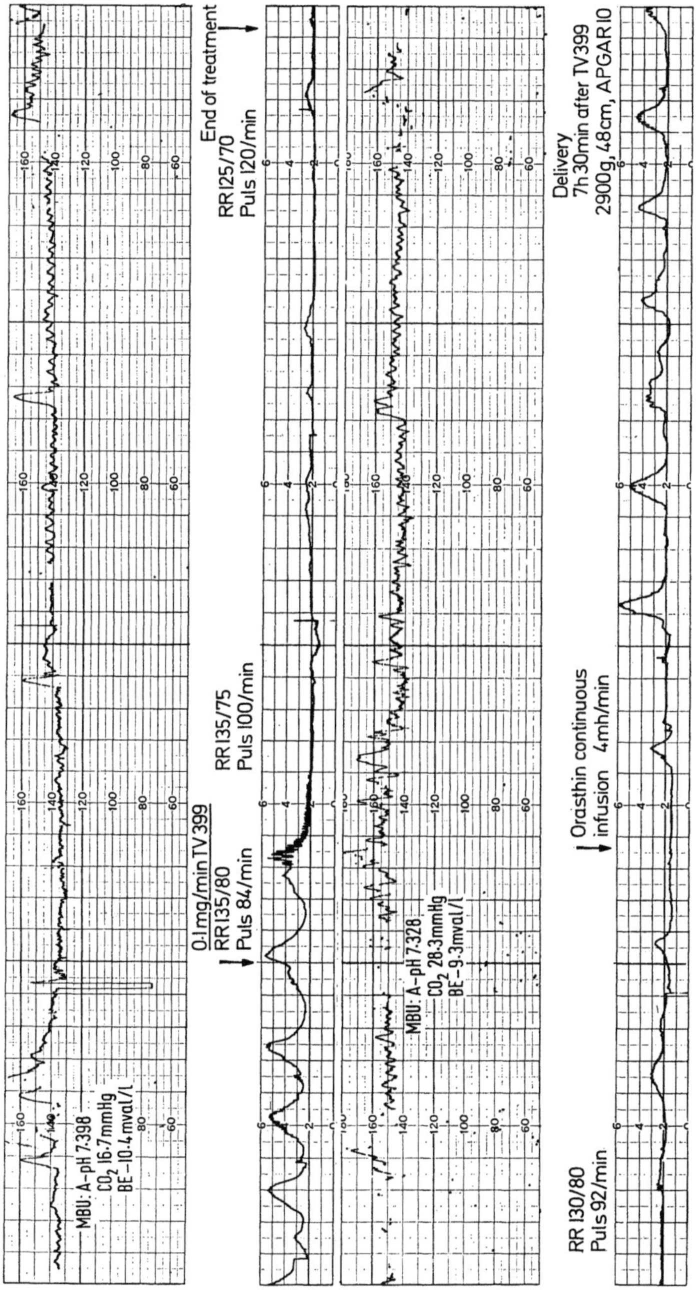
MBU: A-pH 7.398
CO_2 16.7 mmHg
BE -10.4 mval/l
0.1 mg/min TV 399
RR 135/80
Puls 84/min
RR 135/75
Puls 100/min
End of treatment
RR 125/70
Puls 120/min
MBU: A-pH 7.328
CO_2 28.3 mmHg
BE -9.3 mval/l
RR 130/80
Puls 92/min
Orasthin continuous
infusion 4mh/min
Delivery
7h 30min after TV399
2900g, 48cm, APGAR 10

Fig. 17

nausea, vomiting, and trembling hands. The long side chains of Buphenin and isoxsuprine at the nitrogen atom are responsible for the pronounced affinity to the β-receptors of the smooth musculature (Schuh, 1969). There is a recognizable relationship to the structural formula of noradrenaline and adrenaline, which differ only by a methyl substituent at the amino group. Besides, Buphenin and Isoxsuprine have an additional methyl group at the ethanol side chain and only one phenolic hydroxyl group, which results in inhibition of the inactivating enzymes, monoamine oxidase and benzcatechol-0-methyl transferase. They can therefore be administrated orally as well and have a prolonged therapeutical activity.

In our attempt to find similar substances with fewer side effects and the same or a better uterus-inhibiting effect, we have recently obtained even better clinical results with 1-hydroxyphenol-ethanol-p-aminobezoate, which was derived by Tropon-Werke, Cologne, from the substance group of Buphenin (Dilatol). Fig. 17 shows the very prompt and total inhibition of labor during the first stage in a 23-year-old primipara in response to intravenous infusion of the test substance, which was designated TV 399. The figure also shows that depsite high concentrations of the β-adrenergic substance the cardiac action of the child was not adversely affected. Simultaneous examination of the child showed that the actual pH remained practically unchanged before and after TV 399. The base excess improved slightly from —10.4 to —9.3 and the other blood gas values were also within the normal range. A spontaneous, live birth occurred a few hours later.

By comparison with Buphenin (Dilatol), TV 399 at the same concentration also has more therapeutical potency at the uterus in vivo in women with premature and first stage labor. However, in comparative tests with Dilatol and TV 399 at the isolated human uterus strand we surprisingly arrived at different results, which also illustrates the dubiousness of making deductions from in vitro tests on the isolated human uterus strand or on the animal uterus for β-adrenergic substances.

Thus, in contradistinction to the greater potency of TV 399 as against Dilatol at the uterus in vivo, Fig. 18 shows a much stronger inhibitory action of Dilatol at the same concentration at the human uterus strip in vitro. These results are statistically significant in the case of fairly large series of comparative tests.

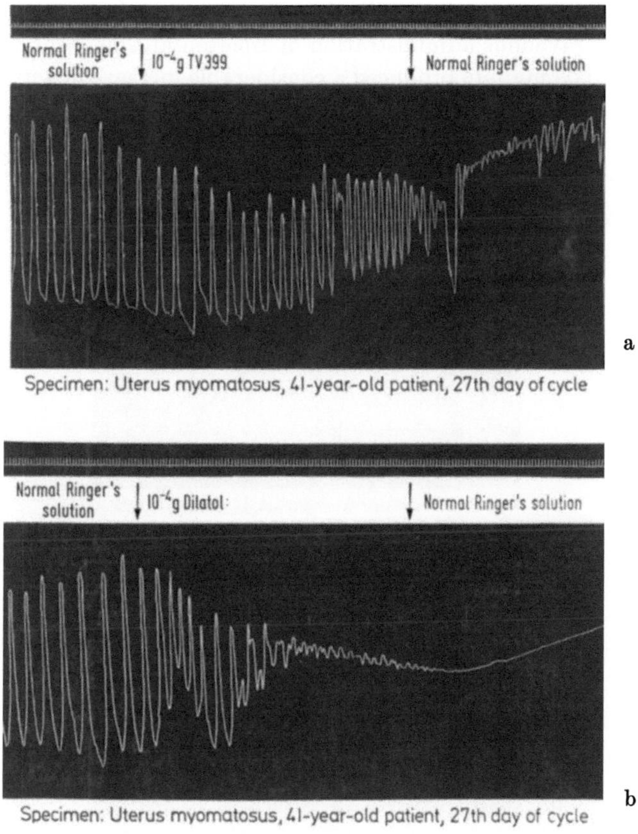

Fig. 18. Comparison of effect of Dilatol and TV 399 at the isolated human uterus strip. In (a) Dilatol at a concentration of 10^{-4} g/ml produces complete labor inhibition. The same concentration of TV 399 (b) has less influence on the isolated human uterus strip, although in the clinical comparison TV 399 is more effective in situ

We therefore carried out tests of the potency of other β-adrenergic substances on the isolated human uterus strip parallel to clinical testing. Theoretically, propranolol (Dociton) being a β-receptor blocking agent can be expected to activate the smooth musculature, but Dociton causes distinct inhibition of the spontaneously active, isolated human uterus strip. By way of contrast, in clinical investigations with the uterus in vivo during premature labor intravenous administration of Dociton at a concentration of 0.07 to 0.08 mg/min produced a considerable increase in activity up to a partial contracture.

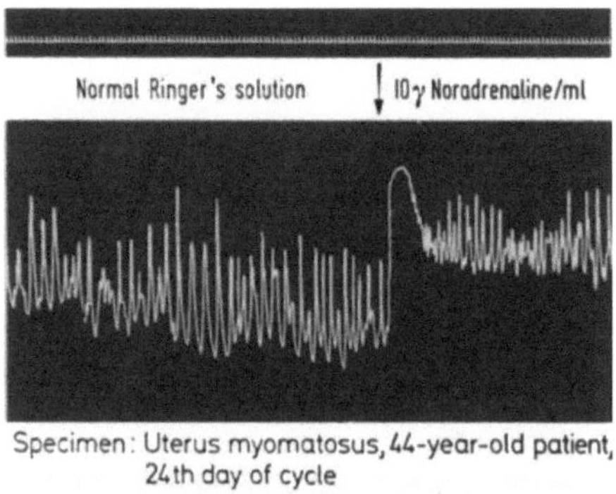

a

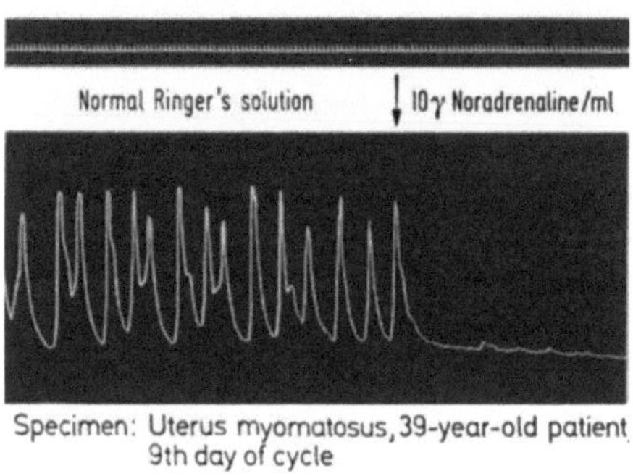

b

Fig. 19. Converse effect of noradrenaline at the human uterus in vitro a in the corpus luteum phase on the 24th day of the cycle, b in the estrogen phase on the 9th day of the cycle

We also included in our tests (Fig. 19) the fundamental question as to the action of noradrenaline, which is considered to be a neurotransmitter with a high affinity to α-receptors, after the results of our in vitro and in vivo tests with β-adrenergic substances at the human uterus were so different from theoretical expectation.

We know from experience that noradrenaline has a stimulating action at the human uterus during parturition in vivo. It has the same effect at the isolated, nonpregnant human uterus strand on the 24th day of the cycle, namely after ovulation, under the influence of predominantly progesterone stimulation. On the 9th day of the cycle, namely during the estrogen phase, noradrenaline produces distinct inactivation at the nonpregnant uterus strip. However, *it is possible* that these observations are only basically valid for the isolated human uterus, because in in vivo tests with β-adrenergic substances in some cases we obtained opposite results. We can endorse the opinion of other authors (AHLQUIST, CIECEOROWSKA, TELKO) that the sensitivity of receptors of the sympathetic system depends decisively on the actual estrogen-progesterone ratio. Although further detailed studies are necessary to elucidate the dissimilar actions of β-adrenergic substances as a function of the hormonal situation, the species and the method of testing, we are today already unanimous in our opinion that at the end of pregnancy and during parturition noradrenaline has a contraction releasing effect on the uterus in vivo in women. In addition, we already have a lot of β-stimulating sympathomimetic drugs which guarantee clinically unequivocal and reliable inhibition of labor in the case of treatened premature births and abortions. Among the more recent preparations we have tested, such as TV 399 (Tropon-Werke, Cologne) or Th 1165a (Boehringer-Ingelheim), the side effects are less pronounced than with buphenin and isoxsuprine although the uterus is reliably inactivated. Hypotension was not observed at the recommended therapeutic doses with the two preparations mentioned. At higher dosages slight hypertension was more likely. According to our experience hitherto, subjective cardiac sensations at a very high dosage were slightly less perceptible with Th 1165a than with TV 399. In the upper range of the therapeutic dose both preparations increase the pulse rate to values of 100 to 140/min. In a patient with insulin-stabilized diabetes, who was given high doses of TV 399, we observed temporarily a rather sharp *increase in blood glucose* corresponding to the known glycogenolysis in skeletal muscle (SCHUH, 1969) (Fig. 20). Of course, the diabetogenic effect of β-adrenergic substances must be considered for therapeutic use in premature labor. During simultaneous investigations with β-adrenergic substances, we have

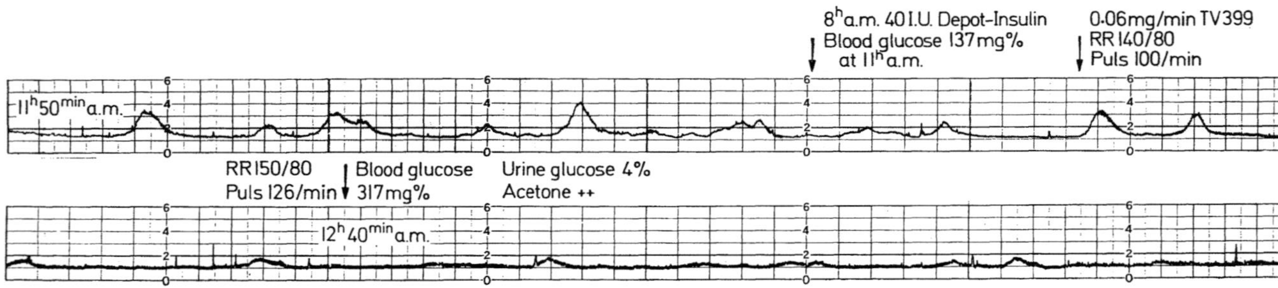

Fig. 20. External tocography for threatened abortion mens VII/3 with severe juvenile diabetes and total placenta praevia. Intravenous administration of TV 399 produces complete inhibition of labor with a rapid increase in blood glucose from 137 mg-% to 317mg-%, accompanied by glycosuria and acetonuria

so far found no evidence of adverse effects on the fetus. A slight increase in the heart rate by 10—15 beats/min during cardiotocographic control was harmless in all cases and reversible after reducing the dosage. Tests by means of blood gas analyses showed an improvement of actual pH and base excess for normal, uninjured babies. All babies for whom the preparation was used temporarily to test for side effects during normal delivery were born alive and delivery was uneventful.

Thus, we can already say today that the sympathomimetic β-adrenergic substances are the most potent therapeutics we have ever had for treatment of premature labor in obstetrics. At present, developments in the field of this group of drugs are so extensive and rapid that in the near future the obstetrician can expect a final and clinically satisfactory solution to this problem.

References

Ahlquist, R. P.: A Study of the Adrenotropic Receptors. Amer. J. Physiol. **153**, 586 (1948).

Bayer, R.: Attempts to Suppress Contractions with Duvadilan in Women in Labour. Uterine Contractility, p. 217. Basel-New York: Karger 1966.

Cieciorowska, A., Telko, M.: Die Wirkung des Adrenalins und des Noradrenalins auf die Kontraktionen des menschlichen Uterus und ein Erklärungsversuch dieser Wirkung. Gynaecologia (Basel) **152**, 39 (1961).

Csapo, A.: Progesterone-"Block". Amer. J. Anat. **98**, 273 (1956).

— Progesterone and the defence mechanism of pregnancy. CIBA Found. study group 9. London: J. & A. Churchill Ltd. 1961.

Embrey, M. P.: The Effect of Prostaglandins on the Human Pregnant Uterus. J. Obstet. Gynaec. Brit. Cwth. **76**, 738—789 (1969).

Euler, U. S. v.: Zur Kenntnis der pharmakologischen Wirkungen von Nativsekreten und Extrakten männlicher accessorischer Geschlechtsdrüsen. Arch. exp. Path. Pharmak. **175**, 78 (1934).

Fuchs, F., Fuchs, A. R., Poblete, V. F., Risk, A.: Effect of Alcohol on threatened premature labour. Amer. J. Obstet. Gynec. **99**, 627 (1967).

Haines, R. F., Johnson, A. G., Petering, H. G.: Variable influences of antitumor drugs progestional agens on immune responses in rodents. Fed. Proc. **26**, 3 (1967).

Hellmann, K. 1968: persönliche Mitteilung.

Hendricks, Ch. H., Cibils, L. A., Pose, S. V., Eskes, Th. K.: The Pharmacologic Control of Excessive Uterine Activity with Isoxsuprine. Amer. J. Obstet. Gynec. **82**, 1064 (1961).

Hüter, J.: Situationsgerechte Toko- und Tonolyse in der Geburtshilfe. Das ärztliche Gespräch, Troponwerke Köln Nr. 10, 1968, S. 52.

Hulka, J. F. et al.: zit. N. Haines et al.

Jung, H.: Ein myogener Hemmstoff des menschlichen Uterus. Arch. Gynäk. **203**, 279 (1966).

— Zur Physiologie und Klinik der hormonalen Uterusregulation. Basel-New York: Karger 1965.

Klöck, F. K., Jung, H.: Extraktion und dünnschichtchromatographische Untersuchungen des „Uterus-Hemmstoff". Symposion über pharmakologische Geburtserleichterung und Wehenhemmung, Aachen 12.—14.6.70.

Kurzrok, R., Lieb, C. C.: Biochemical studies of human semen. Proc. Soc. exp. Biol. (N. Y.) **28**, 268 (1930).

Lanman, J. T.: The role of immune phenomena in labour — Initiation of labour — Interdisciplinary Conference on the Initiation of labour. Departement of Health, Education and Wellfare Bethesda Maryland No. 1390.

Marshall, J. M.: Effects of oestrogen and progesterone on single uterine muscle fibres in the rat. Amer. J. Physiol. **197**, 935 (1959).

Mosler, K. H.: Zur Dynamik des Uterus in der Schwangerschaft. — Das ärztliche Gespräch. Troponwerke Köln 10, 26 (1968).

— The Dynamics of Uterine Muscle. Basel-New York: Karger 1968.

Pol, M. C. van der: The effect of some sympathicomimetics in relation to the two receptor-theory. Acta Physiol. Pharmacol. Neerl. **4**, 524 (1956).

Pelner, L., Rhoades, M. E.: zit. n. Haines et al.

Saameli, K.: Pharmakologische Beeinflussung der Uterusaktivität, Gynäkologie u. Geburtshilfe, S. 571. Bd. 2. Stuttgart: Thieme 1967.

Schreiner, W. E.: Die plazentare Funktion und ihre Störungen. Gynaecologica (Basel) **161**, 372 (1966).

Schuh, F.: Wehenhemmung durch Stimulierung adrenerger β-Rezeptoren des Uterus. Geburtsh. u. Frauenheilk. **30**, 143 (1970).

Telko, M.: Die Rolle des Noradrenalin bei der Regelung der gestörten Wehentätigkeit. Zbl. Gynäk. **82**, 817 (1960).

Trautwein, W., Zink, K.: Über Membran- und Aktionspotentiale einzelner Myocardfasern des Kalt- und Warmblüterherzens. Pflügers Arch. ges. Physiol. **256**, 68 (1952).

Wolf, C. H.: Die Sedation des irritierten Myometrium. Das ärztliche Gespräch. Troponwerke Köln Nr. 10, 1968, S. 5.

The Role of the Conceptus in the Control and Maintenance of Pregnancy

R. M. Moor

Agricultural Research Council, Unit of Reproductive Physiology and Biochemistry, University of Cambridge

The profound endocrine change that characterises pregnancy in most mammals is dependent upon an early maternal response to the presence of a conceptus in the uterus. It is with the mechanisms underlying the relationship between the mother and conceptus, and in particular with the way in which the conceptus acts to convert the corpus luteum of the cycle to one of pregnancy, that this paper will be concerned. The subject will be approached in a comparative manner using three species, namely the human and rhesus monkey on the one hand and sheep on the other as examples, since these represent extremes in the endocrine spectrum when considered in terms of the way in which the conceptus acts on the maternal organism. Reviews by Psychoyos (1967), Greenwald and Rothchild (1968), Moor (1968a, b) and Short (1969) provide recent detailed information on the relationship between the conceptus and ovaries in a variety of Eutherian mammals.

In women, the most obvious early sign of pregnancy is the absence of the menses. However, many critical events have occurred in the endocrine balance of the mother before this time. By the sixth day after ovulation, the conceptus has implanted (Boyd and Hamilton, 1952) and four days later is secreting chorionic gonadotrophin (HCG) in detectable quantities (Hammerstein, 1962; Goldstein, Aono, Taymor, Jochelson, Todd, and Hines, 1968). The secretion of HCG by the trophoblast corresponds closely with the stage at which the corpus luteum of the cycle is transformed into a corpus luteum of pregnancy. This apparent relationship between HCG and the maintenance of the corpus luteum has been interpreted as indicating that HCG is the luteotrophic hormone in women (Hisaw and Astwood, 1942). The absence of HCG, it is suggested, results in the termination of luteal function at the end of the menstrual cycle. Experimental evidence in support of the above hypothesis has been obtained by a number of independent investigators (Browne and Venning, 1938; De Watteville, 1949; Bradbury, Brown, and Gray, 1950; Segaloff, Sternberg, and Gaskill, 1951). It has been demonstrated by these investigators that menstruation can be significantly delayed in non-pregnant women by the administration of physiological levels of HCG. Menstruation is not, however, delayed by giving daily doses of prolactin or follicle-stimulating hormone (FSH) to patients during the menstrual cycle (Bradbury et al., 1950). More recently it has been reported by Savard, Marsh, and Rice (1965) that the presence of LH or HCG greatly stimulates progesterone synthesis by slices of human corpora lutea *in vitro:* progesterone secretion is unaffected by the addition of prolactin or FSH.

The life-span of the corpus luteum in rhesus monkeys is, as in humans, significantly extended by the daily injection of HCG during the luteal phase of the cycle (Hisaw, 1944; Bryant, 1951). Furthermore, the daily administration of very low levels of HCG (25 i.u.) induces a pattern of progesterone secretion that closely resembles the secretory pattern observed in monkeys during early pregnancy (Neill and Knobil, 1969). Thus, immediately after implantation, or after the injection of HCG, a significant but transient increase in the systemic blood progesterone level is recorded; this increase occurs at a time when the circulating progesterone levels are very low in non-pregnant monkeys. The elevated levels of progesterone that are characteristic of females during early pregnancy do not persist, probably because the ovaries become refractory to the continual presence of chorionic gonadotrophin in the maternal circulation. The administration of prolactin rather than HCG does not extend the life-span or stimulate the function of the corpus luteum in the rhesus monkey.

There is thus little or no disagreement that, in primates, a hormone secreted by the conceptus is essential for the maintenance of the corpus luteum of pregnancy. However, some workers (Diczfalusy and Troen, 1961; Short, 1967) advocate extreme caution in accepting that chorionic gonadotrophin is the placental luteotrophin in primates. Short (1969) details major weaknesses in the evidence suggesting that HCG and MCG are the principal luteotrophic hormones of pregnancy; he points to the possibility that placental lactogen might be the more important luteotrophin.

In the sheep, the corpus luteum of the cycle undergoes regression and stops secreting progesterone abruptly on day 15 but is maintained in a fully functional state in both pregnant and hysterectomised ewes (EDGAR and RONALDSON, 1958; MOOR, ROWSON, HAY, and SHORT, 1970). The sheep, therefore, stands in marked contrast to the primates, since the results of the hysterectomy experiments indicate that placental luteotrophic support is not required for the maintenance of the ovine corpus luteum beyond the end of the cycle. It is indeed probable that the inherent life-span of the corpus luteum of the cycle in the sheep and in many other Eutherian mammals is at least equivalent to the length of the gestation interval. The presence of a lytic stimulus of uterine origin (referred to in generic terms as a uterine luteolysin) is necessary for the rapid involution of the corpus luteum at the end of each oestrous cycle in these species (see CALDWELL, 1970, for references).

How then does the ovine conceptus overcome the lytic action of the uterus and so ensure the survival of the corpus luteum of pregnancy? That the corpus luteum is not influenced by the conceptus during the first 12 days after oestrus has been established by a series of experiments involving embryo transfer and embryo removal techniques (MOOR and ROWSON, 1966a, b). It is between the 12th and 13th day that a close relationship forms between the conceptus and the ovary; this vital relationship underlies the mechanism responsible for the maintenance of the corpus luteum during the first trimester of pregnancy (MOOR, ROWSON, HAY, and CALDWELL, 1969).

Recent evidence strongly suggests that the corpus luteum is influenced in a local manner by the early conceptus. For example, only those corpora lutea on the ovary adjacent to the conceptus are maintained if luminal continuity between the two uterine horns is surgically disrupted (MOOR and ROWSON, 1966c). The corpora lutea in the ovary next to the non-pregnant horn will regress in such sheep unless the non-pregnant horn is removed by partial hysterectomy.

Mechanical stimuli do not appear to play a significant role in the maintenance of the corpus luteum of pregnancy (ROWSON and MOOR, 1968). On the other hand, the intra-uterine infusion of embryonic homogenates maintains the corpora lutea of non-pregnant sheep (Rowson and Moor, 1967). Luteal function can also be prolonged by infusing monolyer cultures of disaggregated sheep conceptuses into the uterine lumen (Hay, Moor, Rowson, and Lawson, unpublished observations), but not by similar infusions of either homogenised pig conceptus or various control solutions The active principle in the sheep conceptus is thus a chemical substance, highly specific in nature and able to act only when in direct contact with the uterine lumen.

The lytic effect of the non-pregnant uterus, the apparent absence of a placental luteotrophin and the local specific action of the conceptus on the uterus in sheep has led to the development of a new hypothesis for the sheep (Moor, 1968a, b). It is suggested that the influence of the conceptus in this species is "anti-luteolytic" rather than directly luteotrophic in action; the principal role of the conceptus being to prevent the uterus from acquiring its lytic properties. In pregnant rodents and hystricomorphs on the other hand, the maintenance of the corpus luteum appears to depend both upon an anti-luteolytic and a directly luteotrophic effect of the conceptus. The relationship between the conceptus and ovaries in these animals could, therefore, be considered to occupy an intermediate position between the directly placental luteotrophic effect of the human conceptus and the anti-luteolytic action of the conceptus in the sheep.

References

Boyd, J. D., Hamilton, W. J.: In Marshall's Physiology of Reproduction, 3rd Ed., Vol. 2, p. 1. Ed. A. S. Parkes. London: Longmans, Green, 1952.

Bradbury, J. T., Brown, W. E., Gray, L. A.: Rec. Prog. Hormone Res. **5**, 151 (1950).

Browne, J. S. L., Venning, E. H.: Amer. J. Physiol. **123**, 26 (1938).

Bryant, F. E.: Endocrinology **48**, 733 (1951).

Caldwell, B. V.: Mammalian Reproduction. Proc. 21st Mosbach Colloquium. Berlin-Heidelberg-New York: Springer 1970.

Diczfalusy, E., Troen, P.: Vitam. and Horm. **19**, 230 (1961).

Edgar, D. G., Ronaldson, J. W.: J. Endocr. **16**, 378 (1958).

Goldstein, D. P., Aono, T., Taymor, M. L., Jochelson, K., Todd, R., Hines, E.: Amer. J. Obstet. Gynec. **102**, 110 (1968).

Greenwald, G. S., Rothchild, I.: J. Anim. Sci. **27** Suppl. 1, 139 (1968).

Hammerstein, J.: 9th Symposion der Deutschen Gesellschaft für Endokrinologie, p. 181 (1962).

Hisaw, F. L.: Yale J. biol. Med. **17**, 119 (1944).
— Astwood, E. B.: Ann. Rev. Physiol. **4**, 542 (1942).
Moor, R. M.: J. Anim. Sci. **27**, Suppl. 1, 97 (1968a).
— Proc. roy. Soc. Med. **61**, 1217 (1968b).
— Rowson, L. E. A.: J. Endocr. **34**, 233 (1966a).
— — J. Endocr. **34**, 497 (1966b).
— — J. Reprod. Fertil. **12**, 539 (1966c).
— — Hay, M. F., Caldwell, B. V.: J. Endocr. **43**, 301 (1969).
— — — Short, R. V.: J. Reprod. Fertil. **21**, 319 (1970).
Neill, J. D., Knobil, E.: Fed. Proc. **28**, 772 (abs,) (1969).
Psychoyos, A.: In: Advances in Reproductive Physiology, Vol. 2, p. 257. Ed. A. McLaren. London: Logos Press 1967.
Rowson, L. E. A., Moor, R. M.: J. Reprod. Fertil. **13**, 511 (1967).
— — Proc. VIth Int. Congr. Anim. Reprod. and A. I. (Paris) Vol. 1, p. 727 (1968).
Savard, K., Marsh, J. M., Rice, B. F.: Rec. Prog. Hormone Res. **21**, 285 (1965).
Segaloff, A., Sternberg, W. H., Gaskill, C. J.: J. clin. Endocr. **11**, 936 (1951).
Short, R. V.: A. Rev. Physiol. **29**, 373 (1967).
— Ciba Fdn. Symp. on Foetal Autonomy, p. 2. de Watteville, H. (1949). In: Livre jubilaire Memoires originaux dédiés au Professeur Gaston Cotte. Lyons, p. 519 (1969).

The Role of the Uterus in the Regulation of Ovarian Periodicity

Burton V. Caldwell

Worcester Foundation for Experimental Biology Shrewsbury, Massachusetts

With 3 Figures

Introduction

In order to adequately assess the possible role of the uterus in the regulation of ovarian periodicity it is necessary to consider in some detail the overall concept of ovarian cyclicity. Following the development and wide use in recent years of highly sensitive and reliable radioimmunoassay methods for the estimation of pituitary and gonadal hormones, considerable research effort has been expended which has increased our basic understanding of the mechanisms which are involved in the control of ovarian periodicity. Principally these advances have come in the area of the possible means by which the pituitary exerts a regulatory action on the growth and maturation of the follicle, its ovulation, and the subsequent morphological and functional development of the corpus luteum. Although much additional information remains to be gathered before a definitive statement can be made on the precise sequence of endocrinological events which characterize normal ovarian periodicity, there has emerged a scheme which will be presented as a working model from which this paper will attempt to postulate the role of the uterus in the control of corpus luteum function. Fig. 1 is a graphic illustration of the interrelationships between the hypothalamus, pituitary and ovary and includes the possible regulatory influence of the uterus and embryo.

Since it is fairly arbitary where one begins in the description of a true cycle, for purposes of emphasizing the possible eventual involvement of the uterus, the initiating event of the model cycle will be the elaboration of follicle stimulating hormone releasing

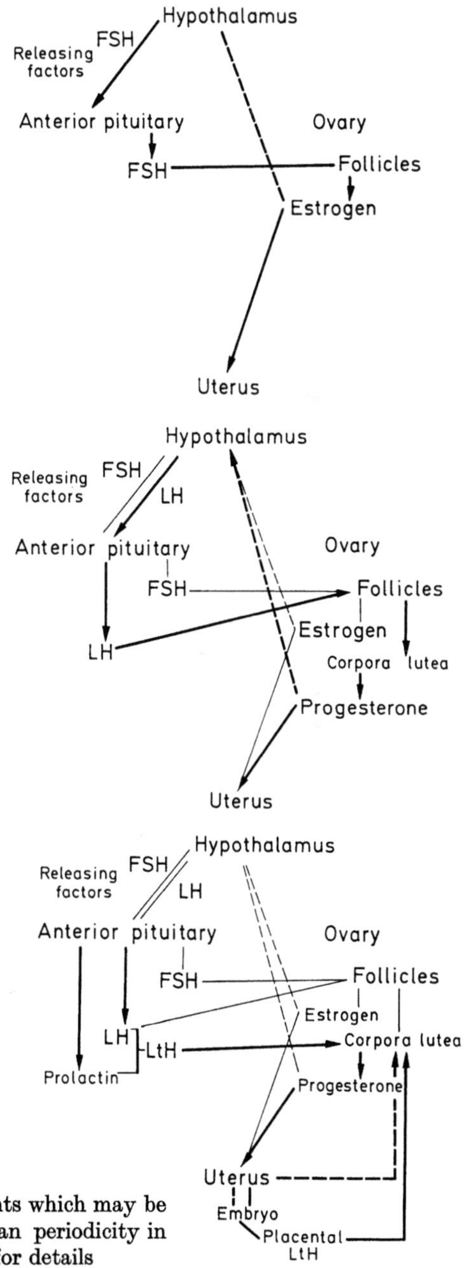

Fig. 1. The sequence of events which may be operating to regulate ovarian periodicity in mammals. See text for details

factor from the hypothalamus (FSH RF) which acts to promote the secretion of FSH from the pituitary. Although there remains considerable speculation on the exact role of FSH in the regulation of ovarian cycles, there seems to be a consistent pattern which, as has been suggested by others [1—4], indicates that the general rising of FSH levels leads to a growth and maturation of the developing follicle. The maturing follicles begin to secrete more estrogens which act on their target cells in the uterus and also on the hypothalamo-pituitary axis. As the estrogen levels pass a threshold amount it seems quite clear that the hypothalamus responds by elaborating luteinizing hormone releasing factor (LH RF) which promotes the secretion of LH from the anterior pituitary.

There is well documented evidence that suggests a direct relationship between the increase in estradiol concentrations and the release of the LH "surge" [1, 5]. The estrogen may also act as a feedback mechanism to inhibit the further release of FSH from the pituitary. The effect of LH on the developing follicle is presumably to promote its further growth and eventual ovulation with the formation of a corpus luteum. LH may also act at this time to stimulate the secretion of progesterone from the corpus luterum in some species [3, 6]. The pre-ovulatory rise in estrogen which acted to stimulate cell growth and proliferation of the endometrium is followed by a post-ovulatory rise in progesterone which results in the proper physiological development of the uterine milieu. Thus the uterus becomes prepared to accept an embryo for implantation.

It is felt by many that the continued secretion of progesterone depends upon a continuous pituitary stimulation. It is known that in some species a second luteotrophin is essential to initiate and maintain normal corpus luteum function. Prolactin, an anterior pituitary hormone, is also apparently controlled by the hypothalamus but in a negative manner. The hypothalamus seems to elaborate a prolactin inhibiting factor (PIF) which, under the proper conditions is itself inhibited, resulting in the release of prolactin which probably acts in concert with other gonadotrophins to maintain the secretory function of the corpus luteum.

The net effect of all of the above events is the formation of a corpus luteum which secretes large amounts of progestins and which is essential for the successful maintenance of pregnancy. If conception has not taken place during the sequence of events described above, the corpus luteum of most mammalian species regresses at a fixed time, an event which initiates the next ovarian cycle. However, if fertilization of the ovum has occurred, the corpus luteum is maintained and continues to secrete progesterone for an elongated period to ensure the proper implantation and maintenance of the developing conceptus. It is this critical point in the cycle which has come under close scrutiny by many workers in the last few years, that is, what actually controls the functional life span of the mammalian corpus luteum ? Why does it regress in the absence of a developing conceptus, and conversly why is it maintained when the egg has been fertilized ? It will be the aim of this paper to explore the possibility that the uterus may be directly involved in this process by controlling the regression of the corpus luteum.

As a part of the concept which will be presented, it will be shown that the embryo in some mammalian species may have an effect which is to reverse or inhibit the "luteolytic" effect of the uterus, allowing the corpus luteum and pregnancy to be maintained.

Prevalent Theories to Account for the Cyclic Regression of Corpora Lutea

Over the last several years a number of reviews have been written on the broad subject of utero-ovarian control mechanisms to which the reader is referred for a more detailed presentation of the specific data in support of the different concepts [7—12]. Of the many suggestions which have been made to account for the cyclic regression of the corpus luteum, four have emerged which have been most often cited. Each of these possibilities have certain drawbacks which, at the moment cannot be reconciled with some of the experimental evidence. However, each has a particular attribute which its supporters contend makes it the most likely explanation. 1) In 1923 Loeb observed that following hysterectomy in the cycling guinea pig the functional life span of the corpora lutea was extended for a much elongated period [13]. He proposed at that time that the uterus may be producing a substance which acts to regulate the functional life span of the corpora lutea, a substance which has since been called by different names but now generally known as the hypothetical "luteolysin" or "luteolytic factor". This view holds that under the proper sequence of endocrinological events, the endometrium produces a "luteolysin" which causes the regression of corpora lutea. 2) Some workers have proposed that the pituitary may play a role in the process of luteal regression as it now is known to play a major role in luteal formation. This theory proposed that the pituitary may withdraw the support of a luteotrophin at the proper point in the cycle, or may even release a luteolysin. 3) The involvement of neural pathways has also been invoked to explain the early regression of the corpus luteum based on findings following the insertion of foreign bodies or beads into the lumen of the uterus. Surgical denervation of the uterus has been shown to reverse this luteolytic effect of the beads giving rise to the view that neural pathways, acting perhaps via the pituitary, control luteal function. 4) To partially explain the effects of hysterectomy on luteal function some workers have proposed that the uterus may metabolize, or in some way alter, a substance which is essential for the maintenance of the corpus luteum, such as estrogen or LH. Following the removal of the uterus this luteotrophin would then remain in the circulation at a high level and forstall regression of the corpus luteum.

As mentioned previously each of these views can be only partially supported, and it will be the purpose of this paper to examine the experimental evidence, especially that relating to the role of the uterus in the regulation of ovarian periodicity, and to speculate from available information of the mechanism by which the ovary undergoes cyclic variation in function.

The Utero-Ovarian Relationship

Comparative Aspects

The most decisive evidence that the uterus may exert a regulatory action on the ovary comes from the studies which have followed the effects of hysterectomy on the functional life span of the corpus luteum in many mammalian species. This subject has been exhaustively reviewed in the past few years and Table 1 shows the effects of removing the uterus on ovarian periodicity in those mammalian species that have been thus far studied. It is still questionable as to what role the uterus may have in primates, since several investigations have produced conflicting results. Andreoli [14] had shown that urinary pregnadiol secretion was significantly prolonged after hysterectomy during the luteal phase in woman (Days 17 or 18 after menstruation); however, this work has not been confirmed in other laboratories. Recent evidence in the monkey

Table 1. *The effects of hysterectomy on ovarian periodicity in various mammalian species*

Species	Parameter studied	Effect	Reference
Human	urinary progestins	prolonged	ANDREOLI [14]
Monkey	peripheral progesterone	none	NEILL et al. [15]
Cow	length of estrous cycle	prolonged	WITBANK and CASIDA [16]
Sheep	length of estrous cycle	prolonged	MOOR and ROWSON [18]
Pig	length of estrous cycle	prolonged	DU MESNIL and DAUZIER [25]
Guinea pig	length of estrous cycle	prolonged	LOEB [13]
Rat	length of estrous cycle	none	BRADBURY [84]
	length of pseudopregnancy	prolonged	
Hamster	length of estrous cycle	none	CALDWELL et al. [27]
	length of pseudopregnancy	prolonged	
Mouse	length of estrous cycle	none	BARTKE [107]
	length of pseudopregnancy	prolonged	
Rabbit	length of pseudopregnancy	prolonged	ASDELL and HAMMOND [106]
Dog Cat Ferret Badger Opposum 13-lined ground squirell	ovarian cycles	none	MELAMPY and ANDERSON [10]

[15] would seem to suggest that uterine removal in this species was without effect on corpus luteum life span. Other work is currently in progress to further examine the possibility that the primate corpus luteum may have a fixed life span which is independent of the uterine influence. However, until more definitive evidence is available it must be accepted that there has been no demonstrable "luteolytic" relationship detected between the primate ovary and uterus.

Another group of animals in which no definite utero-ovarian regulatory relationship has been shown includes the ferret, 13 lined ground squirrel, opposum, cat and dog. Each of these species has unique features about its reproductive cycles and they will not be discussed in this paper. For further details see ANDERSON et al. [10].

In turning to the domestic species, and to the rodents and guinea pigs, however, a clear relationship has been unquestionably established between the presence of the uterus and the functional life span of the corpus luteum. As shown in Table 1 removing the uterus from the sheep, cow, pig, guinea pig, rat, hamster, mouse and rabbit results in luteal maintenance under the proper set of circumstances. In all cases, the extension of the corpus luteum life span following hysterectomy elongates the cycle to that approximating gestation. We will examine closely two representative species, the sheep and hamster, about which most is known concerning the possible existence of a uterine "luteolytic" hormone.

Experimental Alteration of Luteal Life Span

Many workers have now confirmed the original observation of WITBANK and CASIDA [16, 17] that hysterectomy in the cyclic ewe results in the prolonged function of the corpus luteum for a period which closely approximates the normal gestation period. This means an extension in life span from 16 days to about 145 days. A most striking feature of the utero-ovarian relationship in many species is the well documented local nature of the regulatory influence [18, 19, 10]. As shown in Fig. 2, Groups 1—3, when one uterine horn is removed from a sheep in which corpora lutea are present on both ovaries, only the corpus luteum on the ovary adjacent to the remaining uterine horn regresses at the normal time. The luteal tissue on the ovary which no longer has the normal tissue continuity with a uterine horn remains functional for a much elongated period (usually greater than 30 days). This local effect is further demonstrated by the fact that merely surgically separating the ovary and uterus through the broad ligament prevents the corpus luteum from undergoing regression at the normal time [19]. This type of work was also extended to the guinea pig and rat showing that the normal tissue continuity between the ovary and uterus is of utmost importance [20]. The guinea pig cycle was extended by placing ligatures around the ends of both uterine horns [21] or by separating the ovaries from the uterine horns [22]. When the fallopian tube was interrupted, however, there was no alteration in cycle activity. In the rat [23] pseudopregnancy was extended by the cutting of

	Treatment	Reproductive tract (Right - Left)	Life span of Corpus luteum: Right	Life span of Corpus luteum: Left
1	Control		16 days	16 days
2	Total hysterectomy		145 days	145 days
3	Hemi - hysterectomy		<30 days	16 days
4	Left ovary to neck		Removed	Prolonged and variable
5	Left uterine horn to neck		Prolonged and variable	Removed
6	Left ovary and uterine horn to neck		Removed	16-20 days
7	Endometrium to flank		30-50 days	None present
8	Uterus to omentum		16-20 days	None present

Fig. 2. Graphic illustration of the local relationship between the ovary and uterus of sheep showing various experimental modifications of the reproductive tract

the oviduct and the uterine ovarian vessels. Severing of the oviducts by itself was not sufficient to induce a longer luteal phase. Merely cutting the blood vascular connections between the ovary and uterus was effective in prolonging the luteal phase of rats [24]. The local effect of the uterus on the adjacent ovary has also been clearly established by a series of experiments in the pig [25, 26] which showed that in this species as well, the primary action of the uterus is mediated through a local mechanism. When all of the work is viewed together there would appear to be a blood-vascular or tissue transmission of a substance emanating from the uterus to the ovary.

The relationship between the uterus and ovary in the rodent species deserves some special consideration since it is not as clear as that seen in the domestic species. As typical of most of the rodent species thus far studied, the hamster uterus would also seem to have a regulatory effect on luteal function, however, total hysterectomy has no prolonging effect on the corpora lutea which are formed during the normal estrous cycle. It should be pointed out that this luteal tissue does not reach full function in the hamster rat or mouse, without the further stimulation by what most investigators feel is a luteotrophic complex originating in the pituitary. This luteotrophin may be prolactin, or a combination of the other gonadotrophins, but this is a matter separate from the reverse process, that is, luteal regression. It is sufficient to report that without the proper stimulation, the corpora lutea of the estrous cycle regresses rapidly to give a normal cycle length of 4 days. If, however, the animals are mated with a sterile male, pseudopregnancy results and the functional life span of the luteal tissue is extended with a cycle length of about 9 days [27]. In contradistinction to the effects of hysterectomy on the cyclic ovary in the hamster, after induction of pseudopregnancy, total hysterectomy causes a significant prolongation of luteal function to about 18 days, a period which closely approximates gestation in this species [27]. Just as there was a clear local effect of the adjacent uterine horn in the sheep and other domestic species described previously, so too is there a marked initial regression of the corpora lutea in the ovary which remains attached to the remaining uterine horn of hemihysterectomized animals [10].

Another aspect of the studies involving the removal of the uterus is the apparent quantal relationship. In several species it has been reported that as an increasing amount of uterine tissue is removed from a cycling animal the functional life span of the corpus luteum is proportionately increased. In the sheep [28], pig [29], cow [30], rat [31], and guinea pig [32], and rabbit [33] the degree of prolongation was directly related to the amount of uterine tissue removed. This finding of course further supports the concept that the uterus is actively producing a luteolytic substance.

Corpus Luteum Function in Animals with Congential Uterine Malformations

There has been apprehension expressed by some workers who cannot accept the results of hysterectomy as being definitive in establishing the regulatory relationship between the ovary and uterus. Such "mullerian duct mutilations" they feel do not provide evidence from a normal animal and the effects of surgery are difficult to assess [41]. In the pig [42, 43], guinea pig [44], cow [45] and sheep [46] evidence from congenitally deformed animals has

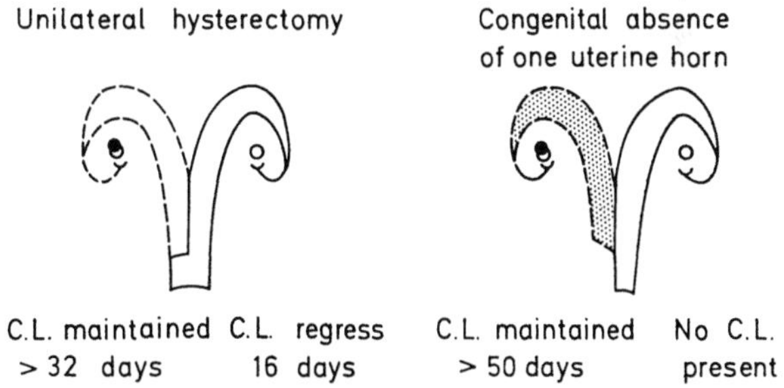

Fig. 3. A comparison of the effects of unilateral hysterectomy with observations on an ewe with congenital absence of one uterine horn

added additional support to the results noted following uterine removal. Fig. 3 shows the results of hemihysterectomy experiments compared with the findings reported from observations of a sheep with the congenital absence of one uterine horn. Unilateral hysterectomy in sheep in which corpora lutea are present on both ovaries results in the regression of only the corpus luteum on the ovary adjacent to the remaining uterine horn at the normal time of 16 days. In the congenitally deformed ewe a similar result was obtained since fifty days after the last ovulation there was viable luteal tissue on the ovary which had no direct connection with any uterine tissue, while the ovary on the side with the intact uterine horn had no luteal tissue present. Although evidence of this type is not, in itself, definitive, it does tend to support the experimental findings.

In the human, various workers have reported similar congenital deformaties which are not as rare as those found in the domestic species. However, no extension of the luteal phase of the menstrual cycle has been noted in women who do not have the normal uterine tissue connection to the ovaries [47].

The Timing of the Luteolytic Influence

A series of studies have been conducted in several species which have provided an invaluable insight into the timing over which the "luteolytic" process may be operating. One procedure which has been used in this type of study has been particularly effective. Animals are allowed to ovulate and, following the formation of the corpora lutea (original CL), a new set of corpora lutea are induced (induced CL) by the exogenous administration of gonadotrophins (usually PMS and HCG). This results in their being two sets of corpora lutea on the ovaries, which differ in age according to when the induced set were formed. In sheep it was found that corpora lutea which were induced even as late as day 13 of a 16-day cycle all regressed with the original CL 2—4 days later [34]. This indicated that there was a factor extrinsic to the luteal tissue which was responsible for the regulation of the functional life span and that luteal tissue in sheep does not have a predetermined life span. This finding was in direct contrast to another study using the same experimental procedure in pigs [35]. In this work NEILL and DAY reported that new corpora lutea formed between day 9 and 13 of the estrous cycle did not regress with the original corpora lutea but remained functional and extended the estrous cycle. This study suggested that the two sets of corpora lutea had approximately the same functional life span regardless of the day on which new ovulations were induced. It was possible therefore that pig luteal tissue had a fixed life span independent of extrinsic factors. Since other work was being conducted which indicated that the mechanism operating to regulate ovarian periodcity in pigs seemed to be initiated earlier than studied by the above workers, this laboratory repeated the experiment of NEILL and DAY but induced a second set of ovulations and corpora lutea earlier in the estrous cycle. We concluded that corpora lutea which were present in the cycling pig by day 6 of the estrous cycle regressed at the same time as the original corpora lutea whereas those formed by day eight resulted in an extension of the estrous interval [36] (Table 2). The last finding confirmed the previous work but when a calculation of the functional life span of the induced corpora lutea was made for each set of new luteal tissue there were significant differences found between those corpora lutea formed on day 6, 8, or 10, when compared to those formed on day 16. In the pig therefore we asserted that the corpora lutea do not have a predetermined life span and that the period between day 6 and day 8 may be critical in establishing the factors which influence the regression of the corpora lutea. This data is in keeping with morphological studies performed recently [37].

Table 2. *Estimated functional life span of original and induced corpora lutea in pigs and sheep*

Species	Estimated day of ovulation original	induced	Estimated functional life span original	induced	Cycle length days
Pig	2	—	16	—	21.0
	2	6	16	12.5	21.5
	2	8	16	14.5	25.5
	2	10	16	15.0	28.0
	2	16	16	19.3	38.3
Sheep	1	—	13	—	15—18
	1	5	13	10.0	15—18
	1	9	13	6.0	15—18
	1	13	13	2.0	15—18

The disparity between the results obtained in these two species, ovine and porcine, is not surprising since the involvement of the pituitary in this process would appear, from other studies, to be different. This latter aspect will be discussed in another section.

A recent experiment by Moor et al. [38] has provided additional valuable information on the timing of the luteolytic influence in sheep while also demonstrating that the lytic effect of the uterus can be halted by hysterectomy. Briefly, these workers hysterectomized a large number of sheep on day 15 of the estrous cycle and withdrew blood from the ovarian vein of the ovary containing the now regressing corpus luteum. Normally, these animals would all be expected to return to estrus within the next 24—48 hours if the uterus were left *in situ*. As a result of the hysterectomy procedure only fifty percent of the animals returned to behavioral estrus and when the progesterone concentration from the ovarian venous effluent was determined, all of these were below 40 μg/100 ml. If the level of progesterone in the blood was above this value the corpora lutea were maintained for a greatly elongated period and, most interestingly, at approximately the same functional state as was assessed at the time of hysterectomy. When they sacrified these animals at varying periods following hysterectomy, and took ovarian venous samples again, they found that 66 % of the animals had plasma progesterone levels that were approximately the same as when hysterectomy was performed. They quite naturally concluded that "the uterus is essential both for the initiation and for the continuation of the degenerative processes that occur during the regression of the corpus luteum."

In the hamster the uterus seems to exert its "luteolytic influence' mainly between the afternoon of day 7 and the morning of day 8, in a 9-day pseudopregnancy [39]. The guinea pig estrous cycle of 16 days is not prolonged if the uterus remains after days 13—15, indicating that the "luteolytic" activity of the uterus in this species operates over this time period [40]. In summary, the duration over which the "luteolytic" influence is operating varies considerably between the species thus far studied. However, the role of the uterus in this process is clear since it is possible to interrupt the regression process by removing the uterus as long as the corpora lutea have not regressed too far.

Hysterectomy and Uterine Transplantation

Since the effects of hysterectomy on the prolongation of luteal function would appear to be well documented in several species it was only natural to attempt to re-establish normal ovarian periodicity by transplanting the uterus to ectopic sites. This procedure has been followed by most workers who have, in the past, tried to establish the possible existence of new hormones emanating from a suspected tissue. The subject of uterine transplantation in hysterectomized animals has been recently reviewed by this author [17] but certain works deserved full attention in this paper as well.

In the pig the uterus has been transplanted to the abdominal oblique musculature, a procedure which resulted in the re-establishment of near normal cycles in hysterectomized animals [48]. Similar findings were reported when the uterus was placed in the peritoneum [49]. However the transplantation of uterine tissue to the small intestine or to the abdominal wall in other studies were without effect in reducing the functional life span of corpora lutea in these hysterectomized pigs [50].

In the rat, several workers have reported the successful reversal of prolonged luteal function evident in hysterectomized pseudopregnant animals when the uterus was placed in the peritoneal cavity, abdominal wall or kidney capsule [10, 31, 51]. In the guinea pig, both success [32] and failure [52] to re-establish near normal ovarian cycles has been reported following uterine transplantation procedures. The transplantation of the uterus or endometrium to hysterectomized hamsters has produced consistent shortening of the induced pseudopregnancy when these grafts were made to the cheek pouch [27]. Although human endometrium transplanted to the hamster cheek pouch was ineffective in reducing the pseudopregnancy length, rat and sheep endometrium [27, 53] did cause a significant reduction in pseudopregnancy. These findings lave been recently confirmed by other workers [54].

The transplantation studies that have been performed in sheep have yielded the most meaningful results. Fig. 2 (Groups 4—8) is a summary of the effects of various surgical procedures involving the uterus with respect to corpus luteum function, which will relate these studies to the local effects of hemihysterectomy previously described. Central to these studies is an experimental surgical procedure which allows the investigator to place various organs into a specially prepared skin loop in the neck. The blood supply to the organ, in this case the ovary, uterus, or both, is re-established by the anastomosis of the ovarian artery (and the uterine artery where the uterus is transplanted as well) to the carotid artery. The common utero-ovarian vein is likewise anastomosed to the jugular vein, which results in a flow of blood from the carotid artery, through the ovarian artery and ovary, and then through the utero-ovarian vein into the jugular. By occluding the various vessels it is possible to infuse substances directly into the ovary and to withdraw blood from the jugular vein which is essentially ovarian venous blood (see Godding et al. [55]).

When the left ovary alone is placed in the neck and the other ovary and left uterine horn are removed while the right uterine horn remains *in situ* there is an extension of the luteal phase of the estrous cycle (Fig. 2, Group 4). The same results were found when one uterine horn was placed in the neck with the ovary remaining in the abdomen [55] (Group 5). However these workers have recently reported the very significant finding that when the ovary and uterus were transplanted together to the neck (Group 6) with no remaining reproductive organs in the abdomen, the estrous cycles were of near normal duration [56]. We have been able to confirm these findings [57] and have studied the cyclic variation in ovarian steroids and LH levels in blood. The cycle lengths and secretion of these hormones indicated that the ovary was functioning in a normal manner.

Other less sophisticated approaches produced similar results. When endometrium was transplanted to the flank of hysterectomized sheep the resulting cycle length was approximately 35 days [7] which was a marked shortening from the expected 145 days found in hysterectomized animals (Group 7). Another site for the uterus was the ommentum which, again, resulted in near normal ovarian periodicity [58].

In summary of the effects of hysterectomy and uterine transplantation it may be stated with some conviction that the increased functional life span of corpora lutea resulting from uterine removal can be at least partially reversed by the presence of the uterus in ectopic sites, but for normal ovarian function it is essential that the ovary and uterus be transplanted as close to one another as possible. The systemic effect of transplanted uterine tissue indicates that the "luteolytic" principle may be blood-borne. However, this systemic influence would probably not be a normal physiological process since the local effects which are so obvious in the species discussed here undoubtedly acts to control luteal function in the intact animal. The fact that corpora lutea regress 30—50 days after placing endometrium in the flank may indicate that the endometrium is able to elaborate a "luteolytic principle," but its effects may be diluted in the general circulation, whereas under normal circumstances the "luteolysin" may be more directly transported to the ovary. This, of course, brings up the possible means by which any secretion of the uterus may get to the adjacent without traveling in the general circulation. Many workers in the field have attempted to demonstrate a blood vascular, lymph, or tissue pathway to account for the local effects of the uterus. However, no clear evidence has been established for linking the uterine secretions with the ovary in a local manner, and until this aspect is settled, there will continue to be justifiable hesitance to accept the theory of a local acting "uterine luteolysin."

The Possible Role of the Pituitary in Corpus Lueum Maintenance

A series of rather complex experiments have been performed in the sheep and pig which point out some interesting comparisons between these two species. It is clear that the pituitary plays a role in the maintenance of the corpora lutea in both species even when luteal function has been extended by hysterectomy. The most comprehensive treatment of this subject can be found in a review by DENAMUR [26] and only the most pertinent points will be detailed here, and summarized in Table 3.

In sheep, removing the pituitary early in the estrous cycle did not cause regression of the corpora lutea until after day 9 in one study [59]. However, other workers, performing the same experiment concluded that the corpora lutea regressed very quickly after hypophysectomy [60]. The disparity in these two reports has yet to be reconciled, but it may be said that the ovine corpus luteum of the estrous cycle probably needs some pituitary support if the cycle is to be of normal duration. When hypophysectomy is performed on previously hysterectomized sheep, once again the corpora lutea regress within 4 days of pituitary removal [59]. In fact, the progesterone concentration in the ovarian vein blood began to show a significant decline after only 48 hours and fell to undetectable levels 96 hours after hypophysectomy. This hypophysectomized-hysterectomized test animal was then used for studying the effects of various pituitary and ovarian hormones on corpus luteum function to determine what substances might be able to maintain luteal function after pituitary removal. It should also be pointed out that stalk transection in the hysterectomized sheep did not cause the rapid regression seen after the hypophysectomy procedures

Table 3. *The effects of hypophysectomy, hysterectomy and gonadotrophin replacement on corpus luteum function*

Species	Surgical procedure	Exogenous gonadotrophin	Effects on corpus[a] luteum
Sheep	hypophysectomy	—	regress within 4 days ? regress after 10 days
	hypophysectomy and hysterectomy	—	regress within 4 days
	hypophysectomy and hysterectomy	infusion of LH (crude)	maintained
		injections of LH	regressed
		injections of FSH	regressed
		injections of prolactin	maintained
		injections of LH and prolactin	maintained
Pig	hypophysectomy	—	maintained 10 days
	hypophysectomy and hysterectomy	—	maintained 10 days
	hypophysectomy and hysterectomy	injections of LH	maintained 20 days
		injections of HCG	maintained 20 days
		injections of prolactin	regressed
Guinea pig	hypophysectomy	—	maintained
	hypophysectomy and hysterectomy	—	maintained

[a] See text for references

described above. The authors reasoned, therefore that prolactin, the major pituitary product following stalk transection, was enough to maintain the functional life span of corpora lutea. However, these corpora lutea regressed after a while and were not maintained for the period of time expected of hysterectomized animals. From the infusion of various pituitary substances it was concluded that a combination of gonadotrophins is probably necessary to maintain luteal function since pure LH or FSH were not sufficient on their own but a combination of these two was able to maintain the secretory activity of the corpus luteum, and the addition of prolactin was not of any further advantage. Prolactin with LH also acted to maintain the luteal tissue [26]. Another study reported that the constant infusion of a crude LH mixture into sheep was able to curtail luteal regression as long as the infusion was not interupted for as little as a few hours [6]. There is a high probability that in this latter work the LH was contaminated with some other gonadotrophins. In the sheep, therefore, it would seem that the corpus luteum is completely dependent upon pituitary secretions for its normal functional life span.

The relationship between the pituitary and ovary in the pig is a bit different. In this species hypophysectomy soon after the onset of estrus had no effect on the development of the corpora lutea but regression became evident about day 14 of the cycle [61]. These results supported a previous view in the pig that the pituitary was only needed for a brief period early in the cycle but then was of little importance to the functional life span of the corpora lutea. In the hysterectomized-hypophysectomized animal the corpora lutea regressed 5—12 days after pituitary removal [61]. In this respect the results are similar to those found in the sheep, the pituitary is responsible for producing a luteotrophic effect which is responsible for mantaining the life span of the luteal tissue which had been extended following hysterectomy. In the pig, LH, or HCG infused into the hysterectomized-hypophysectomized animals could maintain the secretory function of the corpora lutea. Prolactin alone could not [62]. None of these pituitary products could maintain corpora lutea in the presence of the uterus, however.

The guinea pig represents a third class of animals in the effects of hypophysectomy on luteal function. When the pituitary is removed after the 3—4th day following estrus the luteal tissue continues to develop as normal and is maintained in previously hysterectomized animals for the same duration as was found in hysterectomized animals with the pituitary intact [63]. In all of the animals discussed above the pituitary is essential for the maintenance of pregnancy.

The most obvious suggestion in reviewing these results from the pig, sheep and guinea pig, is that there may be a dual mechanism for regulating the functional life span of the corpora lutea. The pituitary is essential for the development of the luteal tissue and in the sheep and pig a low level of support is probably needed to maintain secretory function. In the guinea pig this continued support is probably not needed. At a specific time, and under the proper sequence of ovarian steroid secretion, the uterus probably produces a "luteolytic" influence which is able to overcome the trophic stimulus from the pituitary, resulting in the regression of the corpora lutea. In the hysterectomized animals the trophic support remains sufficient to maintain the luteal tissue. The view that the uterus may be responsible for inhibiting this low level trophic support from the pituitary as a means of inducing luteal regression is attractive, but not in agreement with the other experimental findings. Exogenous gonadotrophins cannot prolong the function of the corpus luteum in pigs with the uterus intact. Also the very clear local effects of the adjacent uterine horn in hemihysterectomized animals could not be adequately explained using this theory. We must conclude this aspect by admitting that the exact inter-relationship between the ovary, uterus and pituitary remains to be elucidated.

The Effects of Exogenous Steroids and Oxytocin on Corpus Luteum Function

The effects of exogenously administered hormones on the utero-ovarian relationship have received much attention in the past, but have not provided a clear picture of the complex inter-relationships. In the rat [65] estrogens were administered and resulted in the initiation and maintenance of pseudopregnancy. A series of experiments in the intact and hypophysectomized rabbit clearly show that estrogens are luteotrophic since corpus luteum maintenance has been reported even after pituitary removal [66]. Further evidence for this view has been obtained with the use of X-irradiated rabbit ovaries which results in the complete loss of endogenous estrogen [67]. These workers reported that progesterone secretion from the corpora lutea would not occur if estrogen was not administered. In sheep, estrogen has also been shown to be luteotrophic [68]; however, the effects of this steroid are probably mediated through the pituitary since hypophysectomy inhibits the luteotrophic influence of the estrogen. In the pig, estrogens were luteotrophic as well [69], whereas no demonstration of this effect has been found in the cow [70]. In the guinea pig [71], like the cow [70], and hamster [72], estrogen has been shown to have a luteolytic influence on the corpora lutea of the estrous cycle.

One of the most interesting and significant findings using exogenous estrogen was reported by ANDERSON et al. [62]. As mentioned previously these workers were not able to show maintenance of procine corpora lutea following LH injections into intact pigs. However, when estradiol and LH were given together, the luteal tissue was maintained.

Progesterone administration to the cow and sheep early in the estrous cycle has a luteolytic effect as determined by the reduction in estrous cycle length [73]. These workers extended their original findings in intact animals and noted that following hemihysterectomy and progesterone administration, the luteolytic effect was local in nature [74, 75]. Absence of the uterine horn on the side of the ovulating ovary prevented largely the ability of the exogenous progesterone to reduce corpus luteum life span. However, if the uterine horn was removed opposite the ovary with the functional corpus luteum, leaving a direct connection between the ovary with the luteal tissue and the uterine tissue, the progesterone exerted a luteolytic influence. Thus it would appear that the luteolytic effect of progesterone in these two species is probably mediated through the uterus.

In the pig [76] the lytic influence of exogenous progesterone was evident in both intact and hysterectomized gilts, however, local injections directly into the corpora lutea had no noticeable effect. In the guinea pig no evidence has been reported that progesterone has a luteolytic effect [71].

From the work described above, following the administration of exogenous steroid hormones, two possibilities have been most often expressed.

1. Estrogen may exert a luteotrophic influence by preventing the uterus from acquiring the ability to produce a luteolytic substance.

2. Progesterone may exert a luteolytic influence by hastening the formation and release of a luteolytic substance from the uterus, a substance which these studies show acts in a local manner.

Both of these concepts fit into the view that the uterus only develops the potential to curtail luteal function after a normal sequence of endocrinological events have "programmed" the endometrium to secrete the "luteolysin." Therefore, administration of estrogen or progesterone would probably alter the normal sequence of events sufficiently to disrupt the normal control mechanisms. With respect to this view it is perhaps pertinent to mention briefly the experiments which have been designed to show that the uterus may be involved in the metabolism of either estrogens or progestins. The supporters of this concept reason that the uterus may actively change these hormones under normal conditions and that removal of the uterus results in luteal maintenance because the level of these steroids present in the general circulation would then remain high and act as a luteotrophin. There have been papers written on this subject which have thus far produced no conclusive evidence; therefore the reader is merely referred to these works for the details of the procedure [77, 11, 33].

Oxytocin has been reported to have a definite luteolytic effect only in cows and only when administered for the first seven days of the cycle [78]. Bovine LH, the principle luteotrophin in cows, was able to reverse this luteolytic effect [79] as was HCG [80]. Several other interesting points that have been brought up concerning the interrelationship between the ovary, pituitary and uterus with special reference to the use of exogenous oxytocin make a full conclusion difficult. It would appear that the uterus is essential for the luteolytic effect of the oxytocin since hysterectomy prevents the early demise of the bovine CL following oxytocin; however, partial hysterectomy did not prevent the premature luteal regression [78]. Also, oxytocin has a local effect as demonstrated by the fact that its primary action is on the ovary which is adjacent to the uterine horn that remains following hemi-hysterectomy [81]. This has led to the suggestion that in the cow oxytocin may act to stimulate the uterus to produce the luteolytic influence which acts in a local manner.

The Use of Assay Systems for Attempting to Characterize a Uterine Luteolytic Factor from the Endometrium

In Vivo Techniques

Several attempts have been made to demonstrate the existence of a substance that can be extracted from the endometrium which will induce the regression of corpora lutea. These studies naturally follow the hysterectomy and uterine transplantation studies since classicaly this has been the procedure used for establishing the existence of a new hormone within a specific tissue. Although there has been no conclusive evidence presented, certain studies appear promising and worthy of consideration in this review. For a complete treatment of all the studies using endometrial extracts in test animals see CALDWELL [17].

In the hysterectomized hamster the length of copulation induced pseudopregnancy is approximately 18 days as opposed to the duration of pseudopregnancy in the intact animal which is normally about 9 days. CALDWELL et al. [27] described this animal as a possible bioassay model with the knowledge that uterine transplantation to the cheek pouch of hysterectomized animals significantly reduced the functional life span of the corpora lutea of pseudopregnancy from 18 to about 13 days. It was attempted, therefore, to re-establish normal ovarian periodicity in these hysterectomized hamsters by administering extracts of the hamster uterus from different days of the normal pseudopregnancy. MAZER and WRIGHT [82] reported success in this approach only when the extracted uterine tissue was taken from hamsters on days 6 or 7 of pseudopregnancy.

Other workers have used the hysterectomized hamster to assay the effects of heterologous endometrial tissue extracts from the cow and have recently reported these to cause the premature regression of hamster corpora lutea [83]. The most effective extracts were those prepared from endometrium taken between days 10—13 of the bovine estrous cycle. Control tissue from skeletal muscle, myometrium and bovine serum albumin were not effective in this respect. Because of its behavior on Sephadex gel filtration columns these workers have estimated their active endometrial fraction to be a large molecular weight substance or some smaller molecule bound to a protein. Perhaps one of the most significant findings reported by the above work was the lack of behavioral oestrus in a high percentage of the treated animals. This might be important since other workers have had marginal success in reversing the elongated functional life span of corpora lutea when they used the return to oestrus as the sole means for assessing the life span of the hamster luteal tissue. Perhaps it means that the more specific parameter used by LUKASZWASKA and HANSEL is necessary in the hysterectomized hamster bioassay. They used progesterone concentration and size of the corpora lutea instead of the behavioral estrus to obtain these significant findings.

This laboratory has used the hysterectomized hamster and reported that endometrial extracts taken from days 13—15 of the sheep estrous cycle were effective in reducing the length of hysterectomized pseudopregnancy, while extracts from days earlier in the cycle were not effective in this respect [52]. However, the specificity and reliability of the hysterectomized hamster as a bioassay animal is still open to question since some workers have not been able to use this model with success [17].

The hysterectomized rat [84, 85, 86], rabbit [87], sheep [88], guinea pig [21, 10] and pig [89, 30] have all been used to study the effects of various endometrial extracts with very equivocable results [17]. However, the specific procedures often employed in these studies leave much to be desired and the negative findings should not be taken as definitive evidence against the existence of a luteolytic hormone in endometrial extracts. It appears from all of the work using the above *in vivo* experiments that more and better approaches must be sought if we are to obtain unequivocable evidence of the presence of a uterine luteolytic hormone in endometrial extracts.

One such possible method has been employed in our laboratory with some initial success. Sheep, with ovaries autotransplanted to a skin loop in the neck, may provide a suitable assay system for assessing the possible effects of various substances on luteal function. Investigators have used this experimental preparation to show the effects of a variety of substances on the ovary, most notably the work of the Cambrisge workers whose papers are suggested for further reading on this subject [55, 56]. As previously mentioned this experimental preparation allows an investigator to infuse a substance directly into the carotid artery to which the ovarian artery has been anastomosed. By occluding the area above the transplanted ovarian artery one is assured that the substance to be infused gains direct access to the ovarian tissue. Also, since the utero-ovarian vein has also been anastomosed, in this case to the jugular vein, and using the same principle, the investigator can withdraw blood from the jugular which is essentially ovarian venous effluent. Therefore, the direct effects of any infused substance can be estimated without the problems associated with studies of this kind using other previous approaches.

In two experiments, an extract of sheep endometrium which had been previously tested on other bioassay systems, including the hysterectomized pseudopregnant hamster described above, was infused into the transplanted ovary and secretion of progesterone from the corpus luteum was measured. An infusion of six hours duration resulted in a 50 % reduction in progesterone concentration, however, the animal did not return to estrus. Perhaps the progesterone level had not fallen far enough to allow the appearance of behavioral estrus. These results were promising, however, and considerable effort along these lines is being undertaken. It would appear that a longer infusion would be necessary to insure the total regression of the luteal tissue which would allow the appearance of estrous behavior (Table 4).

Table 4. *The effects of sheep endometrial extracts on the secretion of progesterone in a sheep with an ovary transplanted to a loop in the neck*

Time period	n[a]	Mean concentration μg/25 ml	Mean secretion rate μg/hr	% Control
Before infusion (control)	4	11.57	657.0	100
During infusion	8	9.08	455	69
After infusion[b]	4	6.60	333	51

[a] Separate collections of 25 ml blood.
[b] 12—24 hours after stopping the infusion.

The autotransplanted ovary has been used to study the effects of prostaglandins, specifically $F_{2\alpha}$, and confirmed the earlier findings of other workers that this prostaglandin is able to induce the regression of luteal tissue [90, 91]. Whenever small concentrations of prostaglandin $F_{2\alpha}$ were infused into a transplanted ovary the concentration of progesterone fell rapidly and the ewe returned to behavioral estrus shortly thereafter. This well-documented effect of $F_{2\alpha}$ has brought about a greatly deal of effort and speculation as to the possibility that it may be the "luteolytic factor." Some feel that it meets all of the necessary requirements for such a substance and it has been shown to cause regression of corpora lutea in monkeys [93], rats [91], guinea pigs [92], hamsters [94] and sheep [90]. However, most investigators view its action as primarily pharmacological and its relationship to the substance which may control normal uterine regression remains to be demonstrated. The possibility is intriguing, however, and the proposed mechanisms by which this substance has been thought to exert its effect has some basis. Pharriss has proposed that the potent vasoconstricting properties of this prostaglandin may allow it to regulate the blood flow through the common utero-ovarian vein. It is thought that this may give rise to a constriction of blood supply and allow a build of products which could conceivably result in a feedback "product" inhibition, if the secretory products of the luteal tissue cannot be quickly removed. Another explanation is that the reduction in blood flow through the ovary could result in a loss of the necessary steroid precursor substances, cofactors and metabolic oxygen for progesterone synthesis which, in turn, could result in the demise of the luteal tissue. This is a possible explanation, perhaps as plausible as any proposed by the "luteolysin" supporters, however, the following points must be considered:

1. If prostaglandin $F_{2\alpha}$ is the luteolysin then it must be secreted from the uterus at a specific time during the reproductive process, evidence for this is not yet available.

2. The pathway for which the prostaglandin can get to the ovary is not clear, but then neither is the pathway for the luteolysin, however, it should be shown to have a local effect to be in keeping with the other experimental evidence.

3. It is difficult to reconcile the *in vivo* effects of prostaglandin $F_{2\alpha}$ with the *in vitro* effects. Most workers have found this substance to be luteolytic *in vivo*, but there is no report of it being luteolytic *in vitro*. In fact, in many studies there was a marked stimulation of progesterone synthesis following the addition of $F_{2\alpha}$ to corpora lutea tissue incubations. It should also be noted here that the same disparity has also been reported following the use of endometrial extracts in *in vitro* incubations. None of the considerations should eliminate the possibility that $F_{2\alpha}$ may be the "luteolysin" since there is no clear reason to suspect that the mechanism operating *in vivo* should be the same as that *in vitro*. In fact, if the theory

proposed by PHARRISS is correct then the lack of an *in vitro* inhibition is partly a positive result. Clearly more work is being done, and is needed before it can be established what normal physiological role might be ascribed to the prostaglandins.

In Vitro Assay Systems

Of primary importance to attempts to demonstrate and characterize a "luteolysin" would be a rapid, sensitive and reliable means for detecting substances which may be effective in causing corpus luteum regression. The use of corpora lutea tissue slices or homogenates have not met with much success using changes in progesterone concentration as the sole measurable parameter. In fact, the results of studies using luteal tissue have been paradoxical when the effects of uterine extracts have been tested. One report showed a decrease in progesterone synthesis by porcine luteal tissue using filtrates from pig endometrium taken from days 16—18 of the estrous cycle [95]. In the same study they reported that filtrates from days 12 and 13 of the estrous cycle actually stimulated progesterone synthesis. A similar increase in the synthesis of progesterone was noted in sheep *in vitro* incubations when sheep endometrial slices were incubated with luteal tissue [96]. However, other workers have reported no effects of endometrial slices incubated *in vitro* with pig luteal tissue.

The one *in vitro* system which has enjoyed the widest use and some measure of success is the granulosa cell monolayer sulture technique developed by CHANNING [97] and employed in the assay of endometrial flushings by SCHOMBERG [98]. The basis of the assay comes from studies which have reported that granulosa cells, when harvested from the developing follicle, are able to differentiate in a monolayer culture system and secrete progestins in the proper ratio when compared with mature luteal tissue. The luteinization of these granulosa cells in culture makes it theoretically possible to test the endometrial extracts on a very specific assay system.

SCHOMBERG has been very successful in employing this technique for the assay of pig uterine flushings and has reported that a lytic substance can be isolated from the flushings of the pig uterus from days 12—18. These flushings were able to cause the granulosa cells to drop off the coverslip cultures and to stop the production of progesterone. Flushings from uteri that were from an earlier period of the pig estrous cycle were ineffective in killing the cultures or diminishing progesterone secretion [99]. Unfortunately considerable restraint has had to be placed on any wide interpretation of these studies and it is difficult to reconcile the fact that the same substances that were effective in destroying the granulosa cell monolayers also had a deleterious effect on other monolayer cultures [99].

We have used this culture system, as modified by Dr. Hay, to test the effects of sheep endometrial extracts on sheep granulosa cell monolayer cultures. We reported that 66 % of the sheep granulosa cells were destroyed when endometrial extracts were taken from days 14 or 15 of the sheep estrous cycle, a period when the appearance of a luteolytic substance has been postulated. However, 25 % of the extracts prepared from other times in the cycles were also judged lytic to the cultures [7]. Our results also were difficult to interpret since we found that the extracts that were lytic to the granulosa cells were also lytic to other monolayer cultures. This, of course, suggests a lack of specificity in the assay and although it may be of some value as a screening procedure for testing various potentially luteolytic substances, its value is probably limited in this respect. Therefore some restraint must be placed on the interpretation of the results using the above *in vitro* methods until the specificity and reliability of these methods can be assured.

The Initial Role of the Embryo in the Maintenance of the Corpora Lutea

For pregnancy to be successful in most mammalian species the corpus luteum must be functional for a period generally longer than the normal estrous cycle. There is a variable period, therefore, for which time the luteolytic process, whatever it may be, must be overcome to allow the ovary to produce enough progesterone to insure the maintenance of pregnancy. To many investigators this phase of the life span of a corpus luteum has become of considerable interest since it marks a normal physiological event resulting in the extension of luteal function. Moor and Rowson have studied this interval between the time when the corpora lutea of sheep would have regressed if a conceptus was not present, and the time when the conceptus becomes sufficient to maintain itself without the help of the ovary. They have reported a series of superb findings [100, 101, 102, 103, 104], which Dr. MOOR will detail in the present colloquim. However, for the sake of providing a complete picture of the events which involve the uterus in ovarian periodicity a brief discussion of their work will be presented here.

It was found that the sheep embryo must be present in the uterus before day 12 in order to have a luteotrophic effect on the corpora lutea. If embryos were removed from the uterus before day 12, or, if they were transferred into the uterus after day 12, the chances of the corpus luteum persisting was greatly diminished. Also when homogenates of properly timed embryonic material were infused into the uterus, luteal function was maintained for an extended period of time [102].

Of considerable interest were their findings that the embryos protective effect on the corpora lutea was manifested by a local mechanism in the same sense that the uterus was luteolytic in a local manner. When an embryo was isolated in one uterine horn, surgically separated from the non-gravid horn, corpora lutea present on both ovaries behaved independently. Only the corpora lutea on the ovary which was directly adjacent to the gravid horn was maintained, while the luteal tissue on the ovary that was adjacent to the sterile horn regressed at the normal time. This led to the concept that the embryo was perhaps acting as an "antiluteolytic" influence, rather than a luteotrophic influence. This debate can not be resolved at the present time. In continuation of the finding that the embryo had a local protective influence over the adjacent ovary, MOOR et al. [104] then established the duration of time over which this influence was acting. The experimental procedure was rather complex but the results were most clear. The embryo acts to protect the corpora lutea of the adjacent ovary for about the first fifty days of pregnancy before it is able to act to protect luteal tissue on both ovaries. This finding was in good agreement with other work which had suggested that the ovary was essential throughout this same period but could be removed thereafter without the loss of pregnancy [105].

Summary

A summary of the complex inter-relationships between the ovary and uterus is difficult to formulate. The concept of a "luteolysin" produced by the endometrium, is based primarily on experimental evidence without any real definitive studies on the isolation and purification of this substance. However, it is possible to develop a reasonable hypothesis taking into account all of the results reported in the text of this review. Since the bulk of the experimental evidence has come from studies with sheep, the following scheme is proposed to account for normal ovarian periodicity in this species. To a large extent the concept can also be applied to the cow, pig and guinea pig, and with some slight modifications may also include the rodent species.

The pituitary is obviously the primary control center for the initiation of follicle growth, ovulation, and the formation of the corpus luteum, under the direction of hypothalamic releasing factors. The proper sequence of ovarian steroid secretion is probably essential for the "programming" of the endometrium. This sequence includes an estrogen "surge" to promote the growth and proliferation of the endometrium, and a subsequent increase in progesterone to aid in the differentiation and prepare the uterus for the possible arrival of a fertilized egg. If fertilization has not occurred the endometrium may produce a "luteolysin" the effect of which is to curtail any further secretion of progesterone from the corpus luteum and initiate the beginning of the next ovulatory cycle. When an embryo is present in the uterus there may be an inhibition of the endometrial "luteolysin" which allows the luteal tissue to remain functional for the necessary period prior to the time when the conceptus is able to produce its own steroids. This effect of the early embryo may be a luteotrophic one, although the clear local relationship between the embryo and the corpus luteum makes this interpretation more difficult to justify.

It should be stressed that the uterus probably only develops the ability to regulate ovarian function after the proper sequence of endocrinological events. The many effects of exogenously administered steroid and protein hormones may be at least partially explained by their possible interruption of this normal series of events. It is also possible that the pituitary must continue to secrete a weak luteotrophic influence to maintain the luteal tissue, an influence which is overcome by the endometrial "luteolysin."

There are at least two points which concern investigators in this field. What controls the functional life span of the corpora lutea in the primate? and why have attempts to isolate the hypothetical "luteolysin" from endometrial tissue so far been without complete success? Obviously the principle need for further progress in these studies is a sensitive and reliable assay for a "luteolysin." Without such an assay little progress can be expected since the use of the hysterectomized hamster and the granulosa cell monolayer cultures have not provided the specificity needed to definitively establish the existence of this suspected new hormone.

With the great impetus of new workers in this field many investigators have reason to hope that results will be forth coming which will support or destroy the concept of a uterine luteolytic hormone.

Acknowledgments

The author wishes to acknowledge the active participation of the following individuals in the research and preparation of the manuscript. Dr. R. M. Moor, Dr. Mary F. Hay, Mr. L. E. A. Rowson, Dr. J. A. McCracken, Dr. S. A. Tillson, and Dr. R. J. Scaramuzzi. The work presented here was supported in part by a grant from The Agency for International Development (CSO-2169).

References

1. Vande Wiele, R. L., Bogumil, J., Dyrenfurth, I., Ferin, M., Jewelewicz, R., Warren, M., Rizkallah, T., Mikhail, C.: Recent Progress in Hormone Research. New York-London: Academic Press 1970.
2. Swerdloff, R. S., Odell, W. D.: Calif. Med. **109**, 467—485 (1968).
3. Strott, C. A., Yoshimi, T., Ross, G. T., Lipsett, M. B.: J. clin. Endocr. **29**, 1157 (1969).
4. Goebelsmann, U., Midgley, A. R., Jaffe, R. B.: J. clin. Endocr. **29**, 1222 (1969).
5. Caldwell, B. V., Scaramuzzi, R. J., Thorneycroft, I. H., Tillson, S. A.: In Immunological Methods in Steroid Determinations. Eds. Peron and Caldwell. New York: Appleton Century Crofts 1970.
6. McCracken, J. A., Baird, D. T.: In: The Gonads. Ed. K. W. McKerns. New York: Appleton Century Crofts 1970.
7. Caldwell, B. V., Rowson, L. E. A., Moor, R. M., Hay, M. F.: J. Reprod. Fertil. Suppl. 8 (1969).
8. Bland, K. P., Donovan, B. T.: J. Endocr. **34**, **3** (1966).
9. Anderson, L. L., Bowerman, A. N., Melampy, R. M.: In: Advances in Neuroendocrinology, pp. **345**. Ed. A. V. Nalbandov. Urbana: University of Illinois Press 1963.
10. — Bland, K. P., Melampy, R. M.: Recent Progress in Hormone Research. New York: Academic Press 1969.
11. Melampy, R. M., Anderson, L. L.: 8th Biennial Symposium on Animal Reproduction. Urbana Illinois (In Press). J. Anim. Sci. **27**, Suppl. 1, 77 (1968).

12. Schomberg, D. W.: In: The Gonads. Ed. K. McKerns. New York: Appleton Century Crofts 1969.
13. Loeb, L.: Proc. Soc. Exp. Biol. (N. Y.) **20**, 441 (1923).
14. Andreoli, C.: Acta Endocr. (Kbh.) **50**, 65 (1965).
15. Neill, J. D., Johansson, E. D. B., Nobil, E.: Endocrinology **84**, 464 (1969).
16. Witbank, J. N., Casida, L. E.: J. Anim. Sci. **15**, 134 (1956).
17. Caldwell, B. V.: In: Advances in the Biosciences, Vol. **4**, p. 399—415. New York: Pergamon Press, Braunschweig: Vieweg 1970.
18. Moor, R. M., Rowson, L. E. A.: J. Reprod. Fertil. **11**, 307 (1966).
19. Inskeep, E. K., Butcher, R. L.: J. Anim. Sci. **25**, 1164 (1966).
20. Butcher, R. L., Barley, D. A., Inskeep, E. K.: Endocrinology **84**, 476 (1969).
21. Howe, G. R.: Endocrinology **77**, 412 (1965).
22. Fisher, T. V.: Amer. J. Anat. **121**, 425 (1967).
23. Barley, D. A., Butcher, R. L., Inskeep, E. K.: Endocrinology **79**, 119 (1966).
24. Clemens, J. A., Minaguhi, H., Meites, J.: Proc. Soc. expt. Biol. (N. Y.) **127**, 1248 (1968).
25. du Mesnil du Buisson, F.: Ann. Biol. Anim. Biochem. Biophys. **1**, 105 (1961).
26. Denamur, R.: J. Anim. Sci. **27**, Suppl. 1 (1968).
27. Caldwell, B. V., Mazer, R. S., Wright, P. A.: Endocrinology **89**, 3 477 (1966).
28. Rowson, L. E. A., Moor, R. M.: Proc. V. Int. Cong. Anim. Repro. Trento (1964).
29. Anderson, L. L., Butcher, R. L., Melampy, R. M.: Endocrinology **69**, 571 (1961).
30. — Melampy, R. M.: J. Anim. Sci. **21**, 1018 (1962).
31. Melampy, R. M., Anderson, L. L., Kragt, C. L.: Endocrinology **74**, 501 (1964).
32. Butcher, R. L., Chu, L. Y., Melampy, R. M.: Endocrinology **70**, 442 (1962)
33. Heckel, G. P.: Surg. Gynec. Obstet. **75**, 379 (1942).
34. Inskeep, E. K., Oloufa, M. M., Howland, B. E., Pope, A. L., Casida, L. E.: J. Anim. Sci. **22**, 159 (1963).
35. Neill, J. D., Day, B. N.: Endocrinology **74**, 355 (1964).
36. Caldwell, B. V., Moor, R. M., Wilmut, I., Polge, C., Rowson, L. E. A.: J. Reprod. Fertil. **18**, 107 (1969).
37. Cavazos, L. F., Anderson, L. L., Belt, W. D., Henricks, D. M., Kraeling, R. R., Melampy, R. M.: Biol. Reprod. **1**, 83 (1969).
38. Moor, R. M., Hay, M. F., Short, R. V., Rowson, L. E. A.: J. Reprod. Fertil. (in press) (1970).
39. Duby, R. T., McDaniel, J. W., Black, D. L.: Acta Endocr. (Kbh.) **60**, 611 (1969).
40. Bland, K. P., Donovan, B. T.: In: Advances in Reproductive Physiology, Vol. 1, p. 179. Ed. A. McLaren. London: Logus Press 1966.
41. Nalbandov, A. V., Cook, B.: Reproduction. Ann. Rev. Physiol. **30**, 245 (1968).

42. Wilson, R. F., Nalbandov, A. V., Krider, J. L.: J. Anim. Sci. **8**, 558 (1949).
43. Nalbandov, A. V.: Fertil. and Steril. **3**, 100 (1952).
44. Bland, K. P., Donovan, B. T.: J. Endocr. **41**, 95 (1968).
45. Ginther, O. J.: J. Amer. vet. med. Ass. **153**, 1656 (1969).
46. McCracken, J. A., Caldwell, B. V.: J. Reprod. Fertil. **20**, 139 (1969).
47. Brown, J. B., Matheu, C. D.: Recent Progr. Hormone Res. **18**, 337 (1962).
48. Anderson, L. L., Butcher, R. L., Melampy, R. M.: Nature (Lond.) **198**, 311 (1963).
49. du Mesnil du Buisson, F., Rombauts, P.: C. R. Acad. Sci. (Paris) **256**, 4984 (1963).
50. — Arch. Anat. Microscop. Morphol. Exptl. **56**, 127 (1969).
51. Anderson, L. L., Melampy, R. M., Chen, C. L.: Arch. Anat. Microscop. Morphol. Exptl. **56**, 373 (1967).
52. Loeb, L.: Amer. J. Physiol. **83**, 202 (1927).
53. Caldwell, B. V., Moor, R. M., Lawson, R. A. S.: J. Reprod. Fertil. **17**, 567 (1968).
54. Duby, R. T., McDaniel, J. W., Spillman, C. H., Black, D. L.: Acta Endocr. (Kbh.) **60**, 611 (1969).
55. Godding, J. R., Harrison, F. A., Heap, R. B., Linzell, J. L.: J. Physiol. (Lond.) **191**, 129 (1967).
56. Harrison, F. A., Heap, R. B., Linzell, J. L.: J. Endocr. **40**, 43 (Abstract) (1968).
57. McCracken, J. A., Caldwell, B. V., Tillson, S. A., Thorneycroft, I. H., Scaramuzzi, R. J.: Biology of Reproduction. Proc. 2nd Annual Meeting. (1969).
58. Niswender, G. D.: J. Anim. Sci. **27**, Suppl. 1, 135 (1968).
59. Denamur, R., Martinet, J., Short, R. V.: Acta Endocr. (Kbh.) **52**, 72 (1966).
60. Kaltenbach, C. C., Graber, J. W., Niswender, G. D., Nalbandov, A. V.: Endocrinology **82**, 753 (1968).
61. du Mesnil du Buisson, F., Leglise, P. C.: C. R. Acad. Sci. (Paris) **257**, 261 (1963).
62. Andreson, L. L., Dyck, G. W., Mori, H., Henricks, D. M., Melampy, R. M.: Amer. J. Physiol. **212**, 1188 (1967).
63. Heap, R. B., Perry, J. S., Rowlands, I. W.: J. Reprod. Fertil. **13**, 537 (1967).
64. Denamur, R., Martinet, J., Short, R. V.: Collogue sur le Physiologie de la Reproduction des Mammiferis, C.N.R.S. Paris (1967).
65. Bogdanove, E. M.: Endocrinology **79**, 1011 (1966).
66. Robson, J. M.: J. Physiol. (Lond.) **90**, 435 (1937).
67. Keyes, D. L., Nalbandov, A. V.: Endocrinology **80**, 938 (1967).
68. Denamur, R., Mauleon, P.: C. R. Acad. Sci. (Paris) **257**, 527 (1963).
69. Gardner, M. L., First, N. L., Casida, L. E.: J. Anim. Sci. **22**, 132 (1963).
70. Kialtenbach, C. C., Niswender, G. D., Zimmermann, D. R., Wiltbank, J. N.: J. Anim. Sci. **23**, 999 (1964).
71. Choudary, J. B., Greenwald, G. S.: J. Reprod. Fertil. **16**, 333 (1968).

72. Greenwald, G. S.: Endocrinology **76**, 1213 (1965).
73. Woody, C. O., First, N. L., Pope, A. L.: J. Anim. Sci. **26**, 139 (1967).
74. — Ginther, O. J., Pope, A. L.: J. Anim. Sci. **27**, 1383 (1968).
75. — Ginter, O. J.: J. Anim. Sci. **27**, 1387 (1968).
76. Spies, H. G., Zimmerman, D. R., Self, H. L., Casida, L. E.: J. Anim. Sci. **19**, 101 (1964).
77. de Jongh, S. E., Wolthuis, O. L.: Acta Endocr. (Kbh.) Suppl. **90**, 125 (1964).
78. Armstrong, D. T., Hansel, W.: J. Dairy Sci. **42**, 533 (1959).
79. Donaldson, L. E., Hansel, W., VanFleck, L. D.: J. Dairy Sci. **48**, 331 (1965).
80. Simmons, K. R., Hansel, W.: J. Anim. Sci. **23**, 136 (1964).
81. Ginther, O. J., Woody, C. O., Mahajao, S., Janakiraman, K., Casida, L. E.: J. Anim. Sci. **25**, 923 (1966).
82. Mazer, R. S., Wright, P. A.: Endocrinology **83**, 1065 (1968).
83. Lukaszewska, J. H., Hansel, W.: Endocrinology **86**, 261 (1970).
84. Bradbury, J. T., Brown, W. E., Gray, L. A.: Recent Progr. Hormone Res. **5**, 151 (1956).
85. Kiracofe, G. H., Spies, H. G.: J. Reprod. Fertil. **12**, 217 (1966).
86. Malven, P. V., Hansel, W.: J. Reprod. Fertil. **9**, 207 (1965).
87. Williams, W. F., Johnston, J. O., Lauterbach, M., Jagan, B.: J. Dairy Sci. **50**, 555 (1967).
88. Kiracofe, G. H., Spies, H. G.: J. Anim. Sci. **23**, 908 (1964).
89. Schomberg, D. W.: Proc. 3rd Intern. Congr. Endocrinol., Mexico City, Excerpta Med. Intern. Congr. Ser. (In Press) (1968).
90. McCracken, J. A., Glew, M., Scaramuzzi, R. J.: J. Clin. Endocr. In Press. (1970).
91. Pharriss, B. B., Wyngarden, L. J., Gutknecht, G. D.: Proc. Soc. Exp. Biol. and Med. **130**, 92 (1969).
92. Blatchley, J. R., Donovan, B. T.: Nature (Lond.) **221**, 1065 (1969).
93. Kirton, K. T., Pharriss, B. B., Ferbes, A. D.: Proc. Soc. Exp. Med. (1970). In Press.
94. Gutknecht, G. D., Cornette, J. C., Pharriss, B. B.: Biol. J. Repro. (1970). In Press.
95. Duncan, G. W., Bowerman, A. M., Anderson, L. L., Hearn, W. R., Melampy, R. M.: Endocrinology **68**, 199 (1961).
96. Stormshak, F., Kelly, H. E.: J. Anim. Sci. **26**, 952 (1967).
97. Channing, Cornelia P.: Nature Vol. **210**, 1266 (1966).
98. Schomberg, D. W.: J. Endocr. **38**, 359 (1967).
99. — In: The Gonads. Ed. K. W. McKearns (1970).
100. Moor, R. M., Rowson, L. E. A.: J. Reprod. Fertil. **12**, 539 (1966).
101. — — Nature, Lond. **201**, 522 (1964).
102. — J. Anim. Sci. **27**, Suppl. 1, 77 (1968).
103. — Rowson, L. E. A.: J. Endocr. **34**, 497 (1966).
104. — — Hay, M. F., Caldwell, B. V.: J. Endocr. **43**, 301 (1969).
105. Denamur, R., Martinet, J.: C. R. Seanc. Soc. Biol. **149**, 2105 (1955).
106. Asdel, S. A., Hammond, J.: Amer. J. Physiol. **103**, 600 (1933).
107. Bartke, A.: J. Reprod. Fertil. In Press (1970).

Animal Models of Inborn Errors of Steroidgenesis and Steroid Action

ALLEN S. GOLDMAN[1]

Division of Experimental Pathology, Children's Hospital of Philadelphia; and the Department of Pediatrics, University of Pennsylvania School of Medicine, Philadelphia, Pennsylvania

With 24 Figures

Mr. Chairman, Ladies and Gentlemen:

I would like to express my pleasure in having the opportunity to address this conference, and would like to thank the organizing Committee of this Colloquium for making it possible.

Congenital adrenal cortical hyperplasia is a disease of man, and is a classical example of an inborn error of metabolism due to an inherited autosomal recessive defect of one or more steroid-synthesizing enzymes [5].

Since its earliest description in 1865 [17], the study of this disease has yielded fundamental information to investigators within several disciplines. The physician has learned of the clinical consequences of the steroidal abnormalities, and has been occupied with the correction of these derangements by replacement of the appropriate missing steroid. The surgeon has developed the repair and reconstruction of genital organs. Steroid biochemists have uncovered the enzymatic pathways of steroid biogenesis by studying the variety of interesting compounds produced in this disease. The geneticist has utilized the findings of the biochemist to clarify the genetic nature of the hereditary defect. The psychologist has studied psychosexual development in man, and has learned of the possible influence of hormones on man's sexual behavior. Recently,

[1] Recipient of Career Development Award (HD-13,628) from the U.S.P.H.S. — Supported by a research grant (P-355) from the American Cancer Society, and (AM-10,521 and HD-4863) from the U.S.P.H.S.

we have become interested in the biochemical embryology resulting from the study of this disease, because we have observed an association of genital malformations in the female fetus (i.e. virilization) with defects in glucocorticoidogenic enzymes, and of genital malformations in the male fetus (i.e. failure of masculinization) with defects in androgenic enzymes [4].

Two other inherited diseases are associated with failure of masculinization of genetic males, human testicular feminization, and rat male pseudohermaphroditism. Testicular feminization has recently been suggested to be due to an autosomal recessive defect in a target organ steroid-activating enzyme, testosterone 5α-reductase [65, 72]. The pseudohermaphrodite rat has a sex-linked defect, the enzymatic nature of which has hitherto not been defined [2]. The present discussion will concern itself primarily with our use of selective inhibitors of steroidogenic enzymes to produce these genital and enzymatic defects in precise animal models of congenital adrenal hyperplasia in order to study the role of these enzymes in the control of fetal sexual differentiation. I also will just mention some of our recent preliminary studies of testicular feminization, and the pseudohermaphrodite rat which point to the possible involvement of two other enzymes in masculine differentiation of the male fetus.

I. Congenital Adrenal Hyperplasia

There are five steroidogenic enzymes in congenital adrenal hyperplasia which are associated with genital malformations [4] (Table 1). Those enzymes associated with virilization of females are: 21-hydroxylase, 11β-hydroxylase, and 3β-hydroxysteroid dehydrogenase, and those observed with failure of masculinization of males are: cholesterol desmolase and Δ^5-3β-hydroxysteroid dehydrogenase. It is theoretically possible that a defect in 17α-hydroxylase is associated with failure of masculinization of males, but this defect has only been observed in women.

Girls born with a defect in 21- or 11β-hydroxylase have enlargement of the clitoris, and varying degrees of labial fold fusion (Fig. 1) [4]. The degree of virilization of girls with a defect of 3β-hydroxysteroid dehydrogenase is less, and consists only of clitoral hypertrophy, while those with a defect in either cholesterol desmolase, or 17α-hydroxylase have normal female genitalia.

Table 1. *Defects of steroidogenic and steroid activating enzymes*

Enzyme	Species	Location	Virilization females	Incomplete masculinization males
Steroidogenic				
21-OHase	Human	A	++	—
11β-OHase		A	++	—
3β-ol, Δ^{5-4}		A, T	+	+
chol. des.		A, T	—	+
17-OHase		A, T	—	+[a]
17β-ol	Rat pseudo	T	—	+
Steroid activating				
5α-red.	Human	TON	—	+

[a] Suspected but not observed. A = Adrenal, T = Testis, TON = Target organ nuclei.

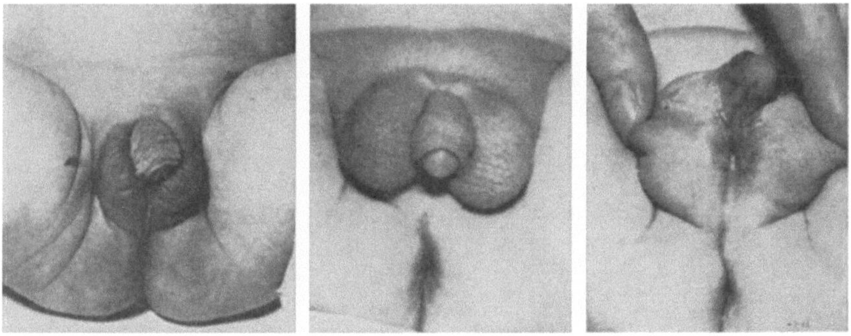

a b c

Fig. 1. Genital changes in congenital adrenal hyperplasia. Girl (left) with a genetic deficiency in activity of 21-steroid hydroxylase has clitoral hypertrophy and fusion of labia majora. Boy (center and right) with a deficiency of 3β-hydroxysteroid dehydrogenase has severe hypospadias

The genitalia of boys with congenital adrenal hyperplasia due to a defect in 21- or 11-hydroxylase are normal. Boys with deficient 3β-hydroxysteroid dehydrogenase and cholesterol desmolase fail to masculinize normally and have hypospadies or feminine external

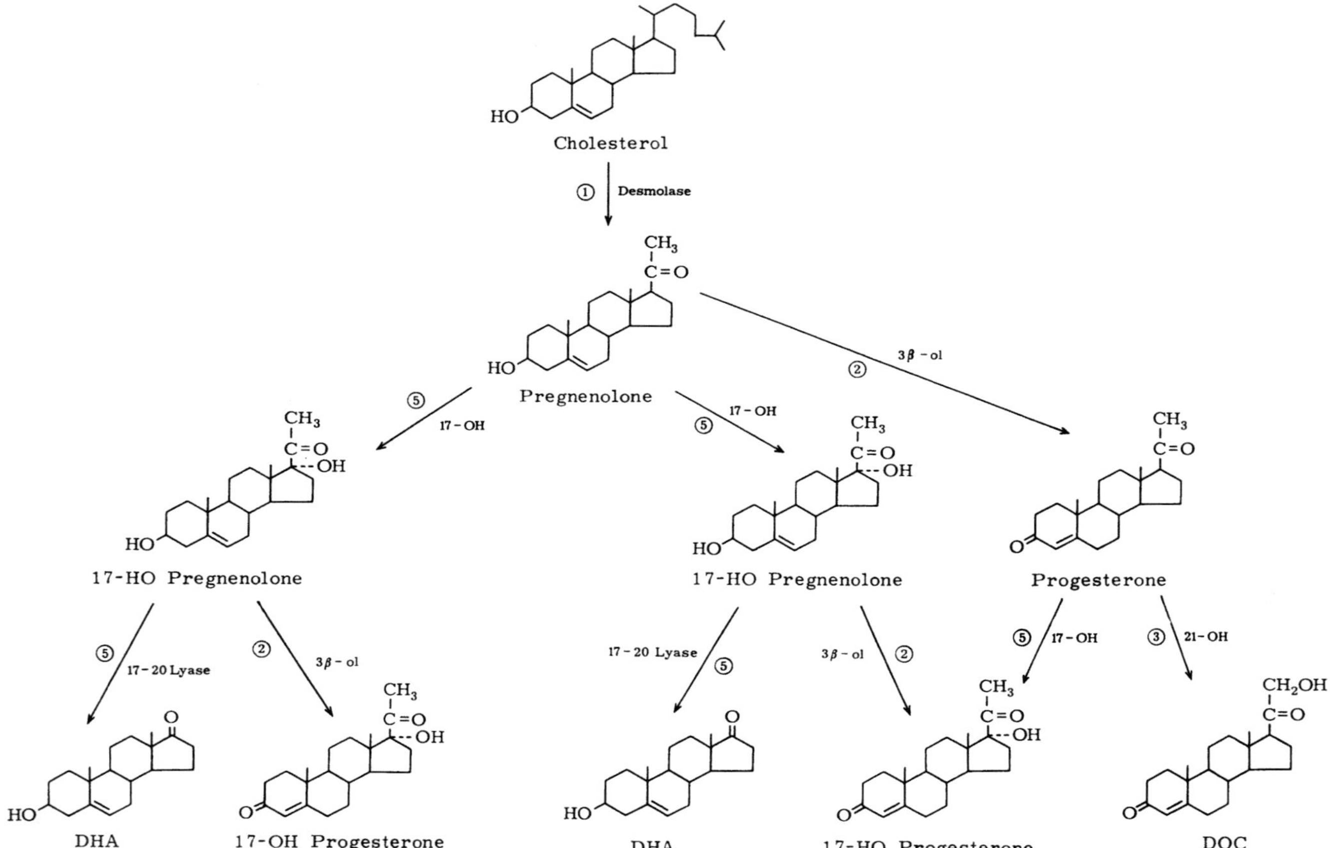

Cholesterol
① Desmolase
Pregnenolone
⑤ 17-OH
⑤ 17-OH
② 3β-ol
17-HO Pregnenolone
17-HO Pregnenolone
Progesterone
⑤ 17-20 Lyase
② 3β-ol
17-20 Lyase ⑤
3β-ol ②
⑤ 17-OH
③ 21-OH
DHA
17-OH Progesterone
DHA
17-HO Progesterone
DOC

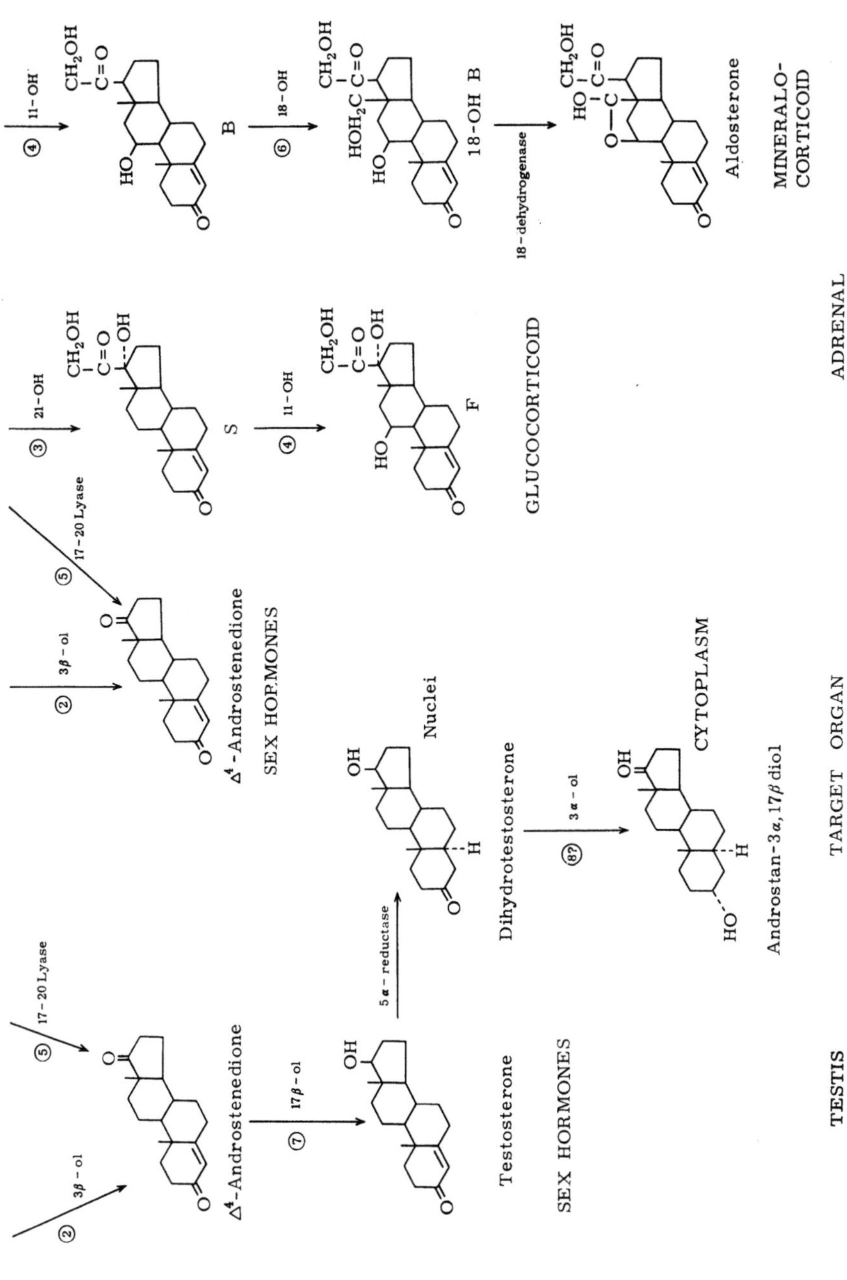

Fig. 2

(Legend on p. 394)

genitalia (Fig. 1). Hypospadias is a defect causing the male penis to resemble female genitalia. This is caused by the absence of the ventral portion of the shaft of the penis and displacement of the opening of the urethra from the tip down onto the shaft.

An examination of the enzymatic steps in steroidogenesis has suggested hypotheses for the understanding of the etiology of the clinical manifestations and sexual malformations (Fig. 2). The fetal adrenal gland synthesizes the three classes of steroids shown, while the fetal testis utilizes only the pathway leading to testosterone. The synthesis of cortisol (corticosterone in the rat), the most potent natural glucocorticoid, is regulated by a pituitary feedback mechanism involving ACTH and cortisol. Cortisol inhibits ACTH secretion and any enzymatic block which decreases cortisol secretion results in increased release of ACTH. All forms of congenital adrenal hyperplasia have been attributed to decreased cortisol synthesis and as a consequence ACTH secretion is increased. High plasma ACTH levels have been directly demonstrated in congenital adrenal hyperplasia due to deficient 21-hydroxylase [18]. The net result of this increased release of ACTH is to drive the adrenal to make increased amounts of all steroids just prior to the chemical step performed by the missing enzyme. On the basis of such abnormal urinary steroidal patterns, deficient activities of steroidogenic enzymes are usually inferred. A direct demonstration of impaired enzymatic activity in adrenal cortical disease has only been possible in two cases of 21-hydroxylase deficiency and in our case of a boy with 3β-hydroxysteroid dehydrogenase deficiency, because of the

---

Fig. 2. A simplified scheme of steroid biosynthesis occurring in adrenal and testis and androgen metabolism in target organ nuclei. In congenital adrenal hyperplasia excess steroidal precursors accumulate above the indicated defect: (*1*) cholesterol desmolase (*2*) Δ^5, 3β-hydroxysteroid dehydrogenase (*3*) 21-steroid hydroxylase (*4*) 11β-steroid hydroxylase (*5*) 17α-steroid hydroxylase. (*6*) A defect in 18-hydroxylase or 18-hydroxysteroid dehydrogenase has been described but is not associated with adrenal enlargement. The rat does not utilize 17α-hydroxylation in corticoidogenesis and produces compound B as its major glucocorticoid. We have recently shown a defect in 17β-ol dehydrogenase (*7*) in the pseudohermaphrodite rat. In testicular feminization defective target organ 5α-reductase (*8*) has been shown and we have observed inhibition of 3α-dehydrogenase by cyproterone acetate obvious difficulties in obtaining suitable tissue from human subjects [49].

Experimental embryologic evidence of Jost and others in the rabbit, mouse, and rat has indicated that removal of the fetal gonad leads to feminization of the external genitalia in either sex [59]. Testosterone induces masculine wolffian duct and penile development in castrated male fetuses. Fetal decapitation as a form of hypophysectomy impairs fetal testicular function, yielding hypospadias. Thus, normal wolffian duct and penile differentiation are brought about by fetal testicular production of androgen (testosterone ?), which organizes male development. Absence of fetal testicular function allows female development, and impaired testicular function leads to partial masculine development or hypospadias. Defects in cholesterol desmolase, the dehydrogenase system, or the 17α-hydroxylase system, as in congenital adrenal hyperplasia, all lead to deficient fetal testicular production of androgen. This would be expected to result in hypospadias or feminine external genitalia, depending on the degree of the block in testosteronogenesis.

In the 21- and 11β-hydroxylase defects, the prounounced virilization of the female fetus has been attributed to excess ACTH-driven adrenal production of androstenedione with peripheral conversion to testosterone [11, 27], while the milder virilization in the 3β-hydroxysteroid dehydrogenase defect has been attributed to excess ACTH-driven adrenal production of dehydroepiandrosterone (DHA). Girls with defects of the desmolase and 17α-hydroxylase have normal genitalia, presumably because their adrenals can produce no androgens (or estrogens).

A. 3β-Hydroxysteroid Dehydrogenase

1. Development of Dehydrogenase and Genital Differentiation

In order to answer the question whether the failure of normal masculine differentiation is due to a deficiency of one of the testosteronogenic enzymes in the fetal testes, we have studied the normal development of the dehydrogenase in steroidogenic endocrine tissues of the human fetus [56]. The steroid secretory cells (histologically, the Leydig cells) of the testis have activity of the dehydrogenase at the time of the formation of the normal penis,

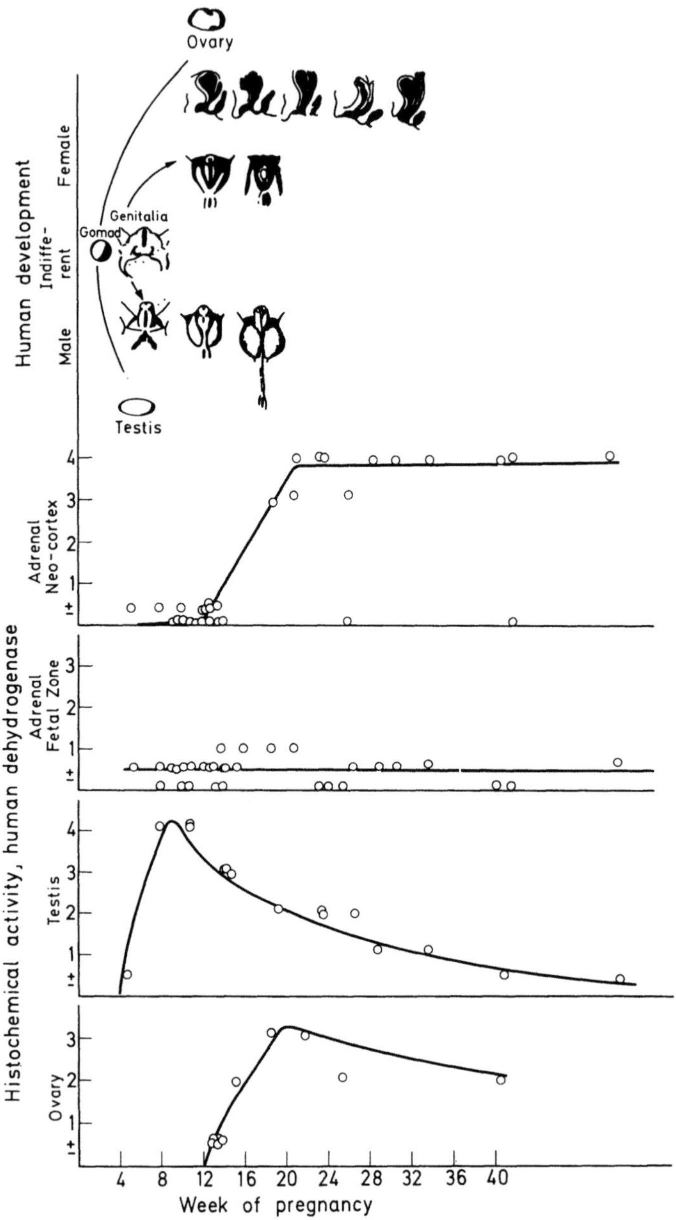

Fig. 3. Development of histochemical activity of 3 β-hydroxysteroid dehydrogenase in the human fetus compared with that of normal gonadal and genital differentiation

before any other tissue within the fetus of either sex (Fig. 3). Activity of the dehydrogenase next appears in the adrenal cells of either sex when the clitoris and vagina differentiate. A similar early onset in activity of the dehydrogenase in the testes of rat fetuses has also been observed [55] (Fig. 4). Thus, in both man and the rat, differentiation of the testes and penis occurs earlier than that of the

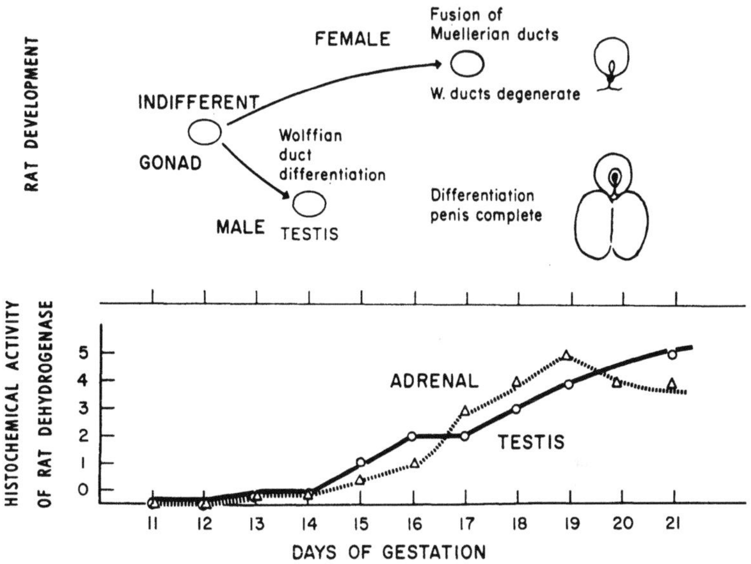

Fig. 4. Development of histochemical activity of 3 β-hydroxysteroid dehydrogenase in the rat fetus compared with that of gonadal and genital differentiation

clitoris and vagina, and coincides with the onset of strong activity of the dehydrogenase in the fetal testis. The time of the differentiation of the female genitalia coincides with the onset of strong activity of this enzyme in the adrenals of both species. Adrenal enzyme activity normally does not influence genital development, but when defective, may give rise to virilization of females because of excess adrenal production of androgens. Male genital differentiation, however, is complete before this excess androgen is produced. These findings, considered with JOST's observation that impaired testicular function leads to hypospadias, suggest that normal human penile differentiation is brought about through the activity of the dehydrogenase system in the fetal testis, at a critical period in fetal development.

2. Inhibition of Dehydrogenase

The creation of an experimental model of the genetic deficiency in the dehydrogenase system in order to test our hypothesis concerning the role of steroidogenic enzymes in sexual differentiation

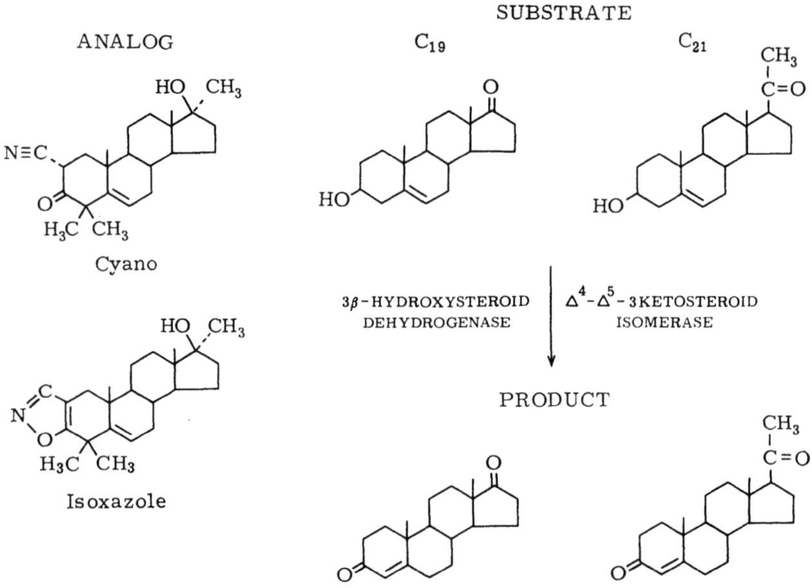

Fig. 5. Cyano- and isoxazole analogs of a C-19 substrate of 3β-hydroxysteroid dehydrogenase Δ^{5-4}, 3-ketosteroid isomerase system

became possible after the report that certain analogs of a C-19 product of the dehydrogenase and isomerase, (2α-cyano-4,4,17α-trimethylandrost-5-en-17β-ol-3-one (cyano-ketone)) and 17β-hydroxy-4,4,17α-trimethylandrost-5-ene-(2,3d)-isoxazole (iso azole) inhibited the dehydrogenase system from Pseudomonas testosteroni at very low concentrations (Fig. 5) [25, 26]. These compounds produce adrenal cortical hyperplasia in rats [8, 58, 75] by blocking the normal biosynthesis of adrenal steroid hormones [54, 66]. We have shown that the cyano-analog also produces luteal and Leydig cell hyperplasia. Our studies in the rat, rabbit, dog, and cow have shown that these analogs inhibit the dehydrogenase system in all steroid-producing tissues in an unusual fashion. The analog does not affect the activity of any other steroid-synthesizing enzyme we have tested, i.e., either cholesterol desmolase, 11- and 21-steroid hydroxylases, or of the 17β- or 3α-hydroxysteroid dehydrogenase in the liver, or of the testosterone 5α-reductase or 3α-hydroxysteroid dehydrogenase in prostatic tissue.

a) *Mechanism of Inhibition.* Before describing the unusual type of inhibition produced by the analogs, a word about inhibition in general may be appropriate. Competitive inhibition is usually reversible, and may be produced by specific binding of a catalytic site by substances structurally related to the substrate. In such reversible inhibition, there is an equilibrium between free inhibitor and the enzyme, so enzymatic activity would be expected to return to normal after the inhibitor has been lost from the tissue. However, if the affinity of a competitive inhibitor for an enzyme site is sufficiently great, the inhibition may appear to be irreversible because the inhibitor is not lost from the tissue. Webb has suggested that enzyme inhibitor systems of this type be called "mutual depletion systems", since a high enzyme concentration or a low dissociation constant would favor mutual depletion of free enzyme and free inhibitor [88]. The high affinity of these inhibitors for the enzyme permit titration of the activity of the enzyme, since the amount of enzyme inactivated is directly proportional to the amount of inhibitor added.

The isoxazole is a competitive inhibitor of the adrenal dehydrogenase and isomerase, and the cyano-analog is a competitive inhibitor of the isomerase (Fig. 6) [29, 30, 70, 71]. Dialysis for 24 hours fails to restore activity of either enzyme inhibited by the isoxazole or of the isomerase inhibited by the cyano-analog [36]. These observations indicate that the cyano and isoxazole analogs are inhibitors of the mutual depletion type. The cyano-analog is noncompetitively bound to the mammalian dehydrogenase, and its binding is so tight that its inhibition can be called stoichiometric. This unusual kind of inhibition is characterized kinetically by the

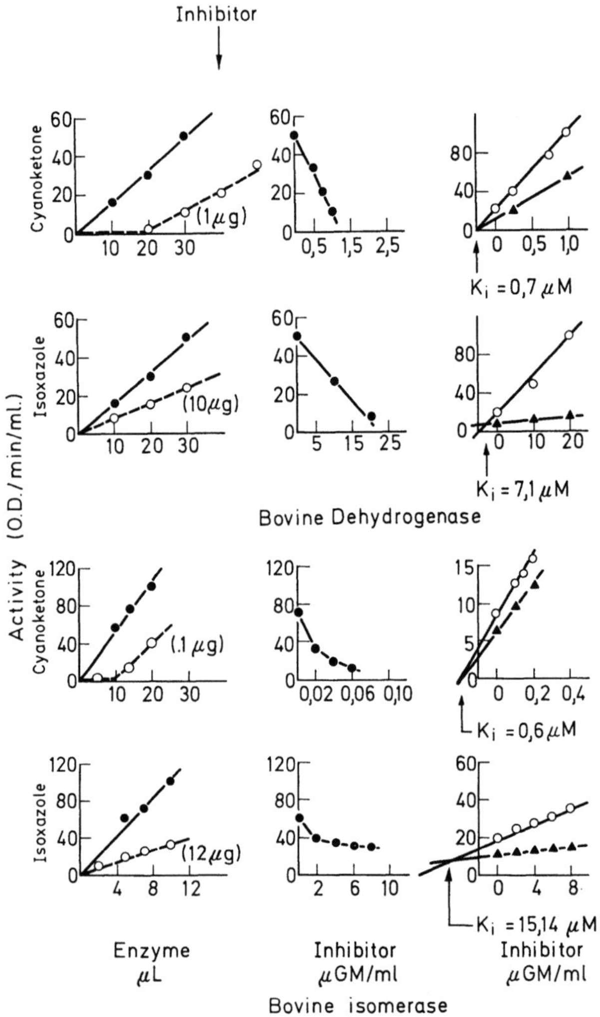

Fig. 6. Kinetic studies of bovine dehydrogenase and isomerase. Enzyme activity in OD units/min/ml $\times$ 10^{-3} is plotted against enzyme concentration in μL/ml on the left (without analog: solid line; with analog: broken line); enzyme activity in OD units/min/ml $\times$ 10^{-3} is plotted against concentration of analog in μGm/ml in the middle: and the reciprocal of enzyme activity (1/OD units/min/ml) is plotted against concentration of test steroid in μGm/ml on the right

fact that the effective dissociation constant of the enzyme-inhibitor complex is so small, that at concentrations of the analog inadequate to cause complete inhibition of the enzyme, almost all of the inhibitor is enzyme bound [89]. Fig. 6 demonstrates this stoichiometric inhibition by showing that enzymatic activity in the presence of inhibitor is prevented until enough enzyme is added to remove all free inhibitor, whereupon enzymatic activity increases at the same slope as the control. Enzymatic activity in the presence of usual inhibitors will pass through the origin at a lesser slope than the control. These analogs are among the most tightly bound enzyme inhibitors known.

b) *Specificity of Inhibition.* The structural similarity of these analogs to the substrate of the enzyme indicate the site of its action to be at or near the active site of the enzyme. The Δ^5-configuration of the cyano-analog is structurally similar to the substrate of the dehydrogenase, but the 3-ketone is more similar to the product of this enzyme. In a preliminary experiment, we have found that this analog prepared with the 3β-hydroxy configuration does not inhibit the mammalian or bacterial enzyme [33]. Thus, the cyano-analog is truly a product analog and cannot be said to be bound precisely to the active site (competitive), but must be bound overlapping the active site. This may account for the noncompetitive nature of the inhibition produced by the cyanoketone according to LINEWEAVER and BURK kinetics.

The specificity of tissue binding of the cyano-analog has been studied by determining the degree of inhibition of a bacterial 3β-ol-dehydrogenase assay system produced by ethyl acetate extracts of various tissue protein preparations to which a certain amount of inhibitor has been added [30]. Only protein preparations containing activity of the dehydrogenase yield extracts which do not inhibit the bacterial enzymatic assay system. Consequently, these preparations which include suitable control human adrenal tissue bind the inhibitor completely. However, the hyperplastic adrenal cortical tissue of a hypospadic male with congenital adrenogenital syndrome, in whom we have demonstrated directly a selectively deficient activity of 3β-hydroxysteroid dehydrogenase, yields extracts which completely inhibit the bacterial enzymatic assay system, and thus does not bind the inhibitor. Therefore, one might expect that binding of the analog to the enzyme in vivo, is highly specific and would occur only in tissues containing the dehydrogenase.

3. Animal Model

We have administered these analogs to pregnant rats to determine whether they would produce in their offspring both the biologic defects and deficient activity of the dehydrogenase characteristic of infants with this form of the adrenogenital syndrome [55, 40]. The experimental fetuses have severe adrenal-cortical hyperplasia and deficient activity of the dehydrogenase in the adrenals and testes (Fig. 7). The experimental males have severe hypospadias (Fig. 8), and the experimental females have marked clitoral hypertrophy (Fig. 9, Table 2).

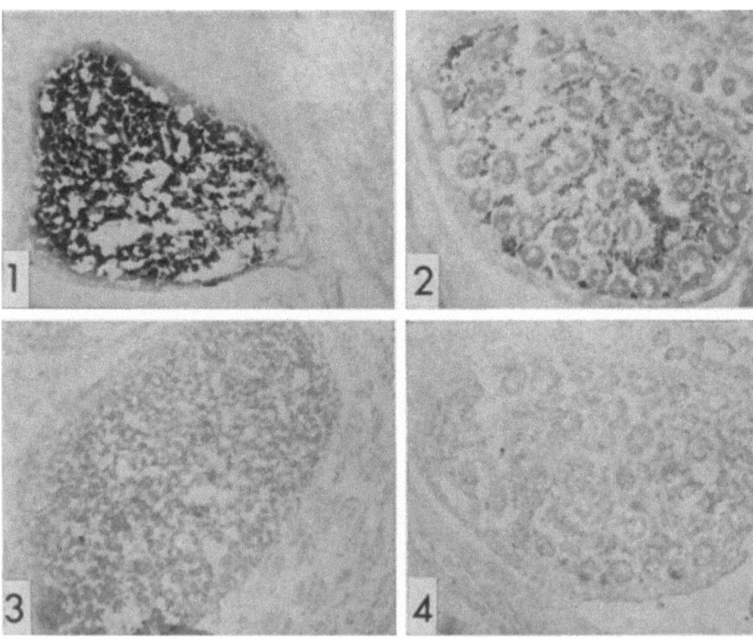

Fig. 7. Histochemical activity of 3β-hydroxysteroid-dehydrogenase (black staining) in adrenal (*1*) and testis (*3*) of fetus from mother given vehicle and in experimental adrenal (*2*) and experimental testis (*4*) of fetus of mother given cyano-analog

Fig. 8. Phallus and scrotum of normal male fetus (left). Urethra (*U*) is located in center of tip of penis and anus (*A*) at base of scrotal raphe. Anourethral distance is 3 mm. Phallus of experimental males of pregnant rats treated with 10603 and inhibition of 17α-steroid hydroxylase (left middle); with aminoglutethimide and inhibition of cholesterol desmolase (right middle); and with cyanoanalog and inhibition of Δ^5, 3β-hydroxysteroid dehydrogenase (right). Note shorter anourethral distances and smaller glans. $\times$ 6

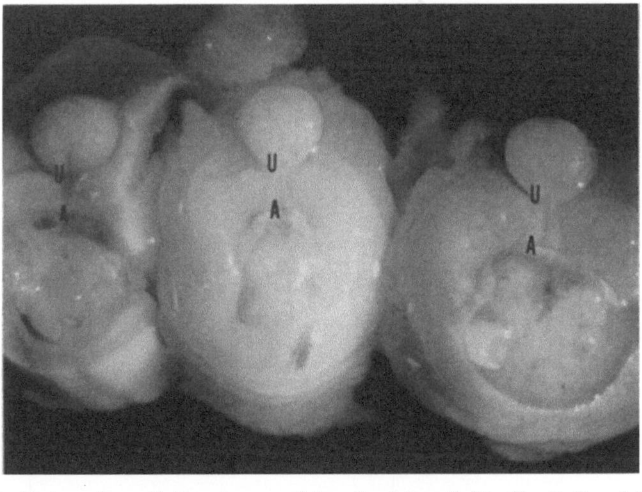

Fig. 9. External genitalia of normal female fetus (left). Genitalia of female fetus of pregnant rat treated with cyanoanalog and inhibition of Δ^5, 3β-hydroxysteroid dehydrogenase (middle); and with metyrapone and inhibition of 11β-steroid hydroxylase (right). Note increased anourethral distance and enlargement of clitoris in experimental fetuses $\times$ 8

Table 2. *Animal models*

Enzyme	Inhibitor	Virilization of females due to adrenal defect			Incomplete masculinization of males due to testis defect			Enzyme defect demonstrated	
		Expected	Observed	Corticosterone reversible	Expected	Observed	Testosterone reversible	Adrenal	Testis
Chol. des.	Aminoglutethimide	NO	+	+	+	+	+	+	...
3β-ol, Δ^{5-4}	Cyano analog	+	+	+	+	+	+	+	+
	Isoxazole analog	+	+	+	+	+	+	+	+
	Estradiol-17β	+	+	+	+	+	+	+	+
	Progestins	+	+	?	+	+	?	?	?
11β-OHase	Metyrapone	+	+	+	NO	NO	—	+	—
17α-OHase	SU 10603	NO	NO	—	+	+	?	+	+
C_{17-20} Lyase	SU 8000	NO	NO	—	+	+	?	+	+
17β-ol	X-chromosome defect pseudohermaphrodite	...	...	...	+	+	?	+	+
								Genitalia	
5α-red.	Cyproterone	—	—	—	+	+	NO	NO	
3α-ol								+	

a) *Critical Periods.* The critical developmental period for the production of hypospadias in the experimental male fetuses by a single dose of analog corresponds to the beginning of the activity of the dehydrogenase in the testis (Fig. 10) [56]. On the other hand,

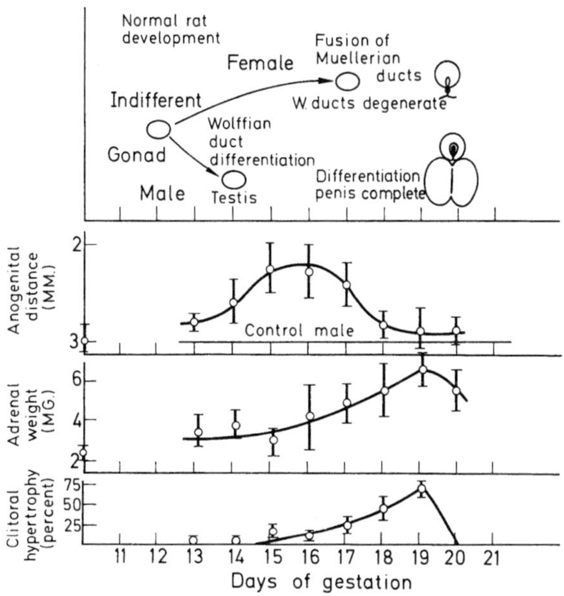

Fig. 10. Critical periods for the production of hypospadias, adrenal hyperplasia, and clitoral hypertrophy by the inhibitor. Vertical bars indicate one standard deviation

the critical period for the production of adrenalcortical hyperplasia and of clitoral hypertrophy by the inhibitor has a developmental pattern which is almost identical to that of the enzymatic activity in the normal fetal adrenal. The cogent implication of these data, that these effects resulted from direct inhibition of the fetal enzyme, has been experimentally verified by the production of this defect in the fetus into which the analog is directly injected through the uterine wall during the critical period of development [31].

b) *Prevention of Genital Defects.* Testosterone given to the pregnant animal completely prevents the production of hypospadias by the inhibitor without affecting the degree of adrenal hyperplasia or inhibition of the dehydrogenase (Fig. 11) [56]. Corticosterone, on the other hand, does not affect the production of

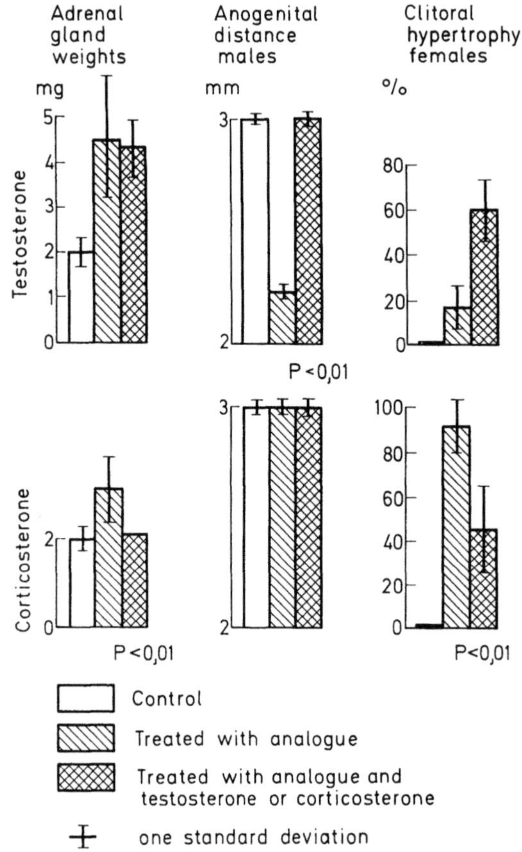

Fig. 11. Effect of testosterone and corticosterone on some actions of cyano-analog in the rat fetus

hypospadias, but completely prevents the production of adrenal hyperplasia and clitoral hypertrophy [57]. These findings support the conclusion that inhibition of fetal testicular dehydrogenase system results in hypospadias, while inhibition of the fetal adrenal dehydrogenase system leads to clitoral hypertrophy.

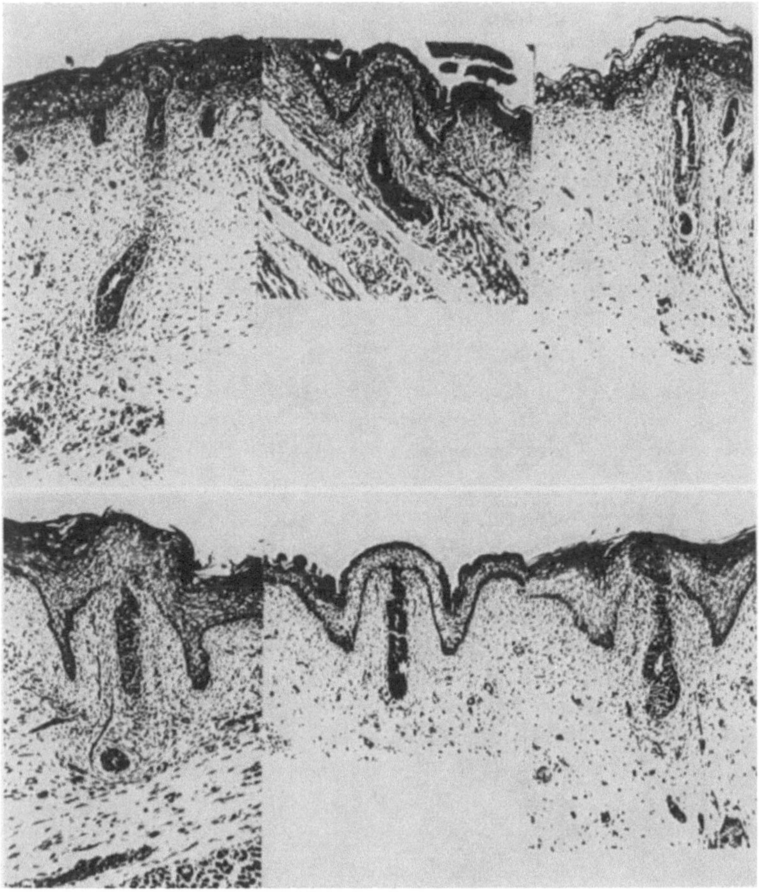

Fig. 12. Top: Mammary gland anlagen of male fetuses. The mother animals were treated with DMSO (left); cyano-analog (middle); and both cyano-analog and testosterone (right). Note nipple induced by cyano-analog is prevented by testosterone. × 100. Bottom: Mammary gland anlagen of female fetuses. The mother animals received DMSO (left); cyano-analog (middle); and both cyano-analog and corticosterone (right). Note that the inhibition of nipple development produced in females by the cyano-analog is evident in the markedly decreased epidermal thickening at the juncture of the primary glandular bud and the surface, and in the circular epidermal invagination. Corticosterone prevents these changes. × 100

c) *Mammary Gland Differentiation.* The development of mammary glands has been studied in fetuses of pregnant rats treated with the cyano-analog in conjunction with Dr. F. NEUMANN, Schering A/G., Berlin [52]. In male fetuses, the analog causes the development of nipples, which are normally suppressed by fetal testicular function (Fig. 12). In light of the previous demonstration of the feminizing effect of the analog on the male external genitalia, this observation suggests that inactivation of the testicular dehydrogenase system leads to insufficient Δ^4, 3-ketoandrogen (testosterone) production, thus allowing the development of the nipple in the male. In female fetuses, the analog inhibits nipple development, presumably because of adrenal overproduction of DHA in response to inhibition of the adrenal dehydrogenase system. Much support for this hypothesis is given by the findings that testosterone prevents the appearance of nipples in male fetuses of analog-treated females, while corticosterone prevents the inhibition of nipple development in experimental female fetuses [67]. The work of Dr. NEUMANN with cyproterone-acetate (1,2α-methylene-6-chlor-$\Delta^{4,6}$-pregnadien-17α-ol-3-20-dione-17α-acetate) also supports our conclusions concerning sexual differentiation in the male fetus [69]. The predominant action of cyproterone-acetate is thought to be a blockade of androgen action at the peripheral target organs, although this steroid is also a potent progestin and a weak androgen. Maternal administration of cyproterone-acetate produces hypospadias and nipples in male fetuses. Testosterone does not prevent the production of these anomalies by cyproterone-acetate.

d) *Effect of Dehydroepiandrosterone* (*DHA*) *in Genital Development.* The hypothesis that the virilization of female fetuses in experimental and human congenital adrenal hyperplasia due to deficient activity of the dehydrogenase system is due to the adrenal overproduction of 3β-hydroxyandrogens, primarily dehydroepiandrosterone, was tested by determining the effects of administration to pregnant rats of large doses of DHA and its 3 sulfoconjugated form, DHA-SO_4 [44]. DHA produced the same degree of clitoral hypertrophy and the same slight increase in anourethral distance as observed in the model of congenital adrenal hyperplasia due to inhibition of the dehydrogenase (Fig. 13). DHA-SO_4 did not affect genital development. In a preliminary experiment, 2 μgm,/100 ml of unesterified DHA only has been identified in the plasma of cyano-analog-treated rats, while none was found in that of control animals [37]. These observations not only provide evidence in support of the proposed mechanism of virilization in human congenital adrenal hyperplasia due to deficient activity of the dehydrogenase system, but also suggest that the reason normal newborn girls are not virilized by the high levels of DHA-SO_4 in umbilical plasma [22] is a lack of androgenicity of the sulfated form of DHA.

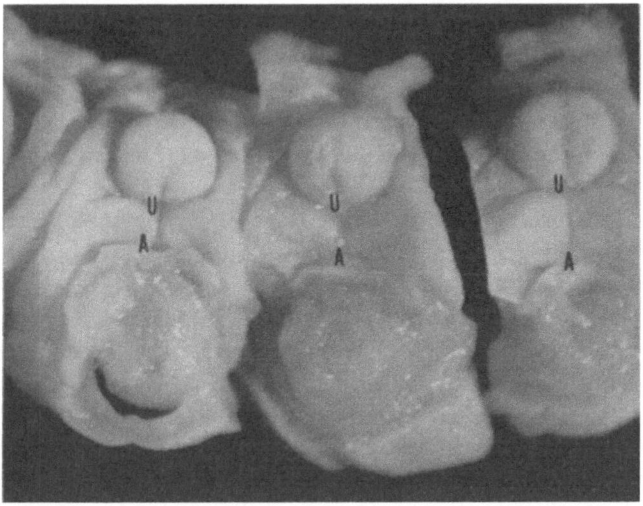

Fig. 13. Genitalia of female fetus from control mother (left); cyano-ketone-treated mother, 120 mg/kg on days 15—20 (middle); and DHA-treated mother, 100 mg/kg on days 13—20 (right). Clitoral enlargement and a slight increase of anourethral distance is apparent in both experimental fetuses. (A, anus; u, urethra) × 10

These studies, then, have demonstrated an experimental model of congenital adrenal hyperplasia due to a deficient activity of the dehydrogenase system produced by inhibition. In the male fetus, the inhibitor blocks production of testosterone in the fetal Leydig cells, and thereby prevents penis formation, and allows nipples to remain. In the female, the urogenital sinus, clitoris, and nipple anlagen are virilized in response to the blockade of fetal adrenal corticosterone biosynthesis by the inhibitor. This blockade presumably leads to an abnormal adrenal overproduction of 3β-hydroxysteroids (unesterified dehydroepiandrosterone), which in turn, stimulate growth of the clitoris and inhibit nipple development in the female fetus.

e) *Early Effectiveness of Analog.* It is remarkable that these analogs are effective when injected in the mother considerably before the differentiation of the fetal enzyme. Teratogenic susceptibility in the rat usually occurs abruptly after the separation of the three germ layers about eight days after copulation, and rises to a maximum at 9 to 10 days [91]. Most teratogens are without effect when injected before day 8, and have only slight effectiveness on day 8. Fertilization occurs in the rat within 8 to 20 hours after copulation. The cleaving egg makes its way down the tube from days 2 to 4, and the early blastula implants on days 5 and 6.

All malformations characteristic of the cyano-analog are produced when it is injected in pregnant rats on days 3 to 8 of gestation (pre-implantation) or to females 6 or 29 days prior to copulation (pre-ovulation) (Fig. 14) [39]. Thus, the analog is an effective teratogen when injected considerably before the development of demonstrable activity of the dehydrogenase. Enzyme inhibitory levels of analog are detected in analog-treated animals by a bacterial enzyme assay procedure in the injected leg 20 minutes and one hour after injection, and in the blood one hour after injection. Two days after injection, analog is not detected in the injected leg, liver, spleen, blood, fat, or intestines. Inhibitory levels of analog added directly to adrenals or ovaries are not extractable, but these levels added directly to liver, fat, and spleen are completely recovered. Thus, levels of analog sufficient to saturate enzyme-active sites in vitro may be tightly and selectively bound to maternal enzyme-containing tissues, while a depot of free analog in other maternal tissues, if any, would be much less than this. These findings suggest that either newly formed fetal enzyme may be directly inactivated by inhibitory levels of analog made available from maternal enzyme-bound stores during enzyme catabolism, or the fetal enzyme-producing machinery may be damaged by sub-inhibitory levels of analog either released from maternal non-endocrine tissues, or immediately taken up and retained by the rat egg or blastula.

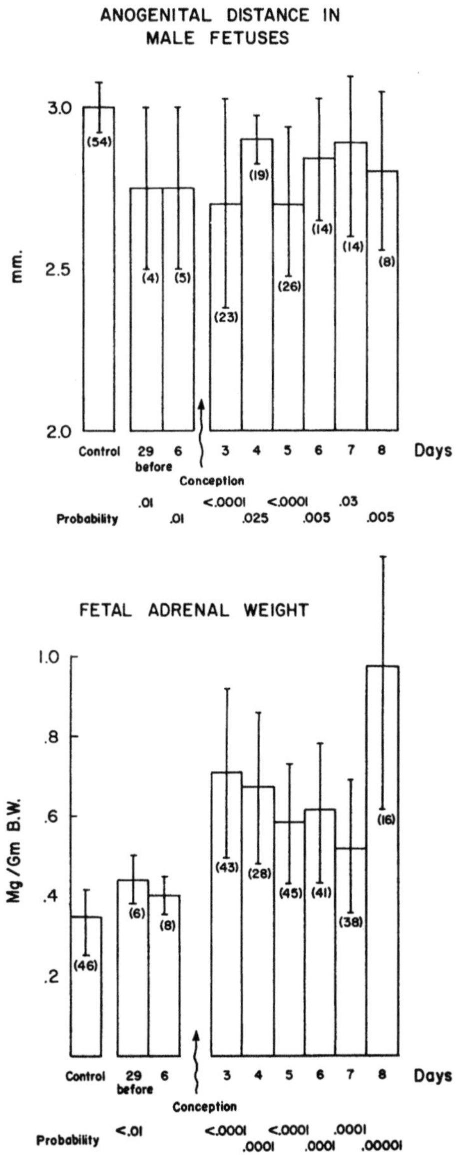

Fig. 14. Effect of cyano-analog on male fetal anogenital distance and on fetal adrenal weight when a single doses of 60 mg/kg was injected in the mother before differentiation of the dehydrogenase in utero. Numbers of fetuses are in parentheses

f) *Persistent Enzymatic Defect.* Virtually all offspring of pregnant rats receiving several doses of the cyano-analog die shortly after birth, as do most human infants with this form of the disease [31]. However, we have been able to study the development of maturing offspring from pregnant rats given a single dose of the analog. All experimental offspring have a significant degree of enlargement of their adrenals (Fig. 15), and of the steroid-secreting

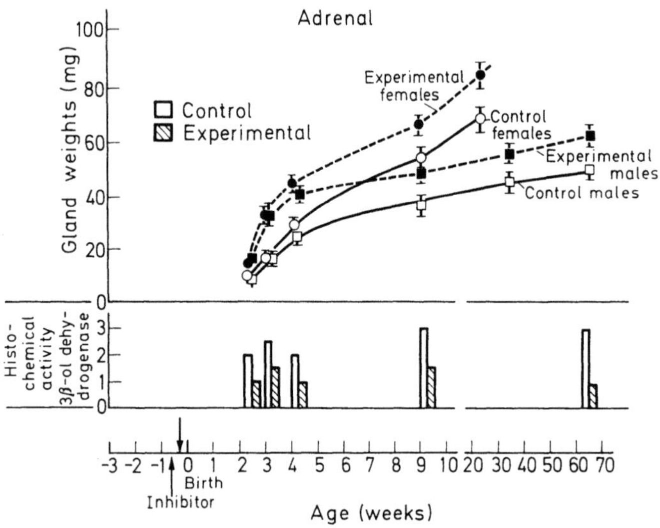

Fig. 15. Persistence of adrenal enlargement and inhibition of histochemical activity of the dehydrogenase after injection of cyano-analog during gestation

cells of the testes (Fig. 16) throughout their postnatal development into adulthood for at least one year after the single maternal injection during gestation. Associated with this hyperplasia, there is a markedly diminished activity of the dehydrogenase system. These data indicate to us that the analog affects activity of the dehydrogenase in two ways. The first effect is the inactivation of the maternal or fetal enzyme at the time of the single injection of the analog. The second effect is a persistently reduced level of the dehydrogenase after this inhibition.

g) *Specific Retention of Isoxazole.* The distribution of the isoxazole analog having a ^{14}C-label in the C-2α position of the isoxazole ring has been studied in the rat [42] to test whether the persistent enzymatic defect is correlated with persistence of label.

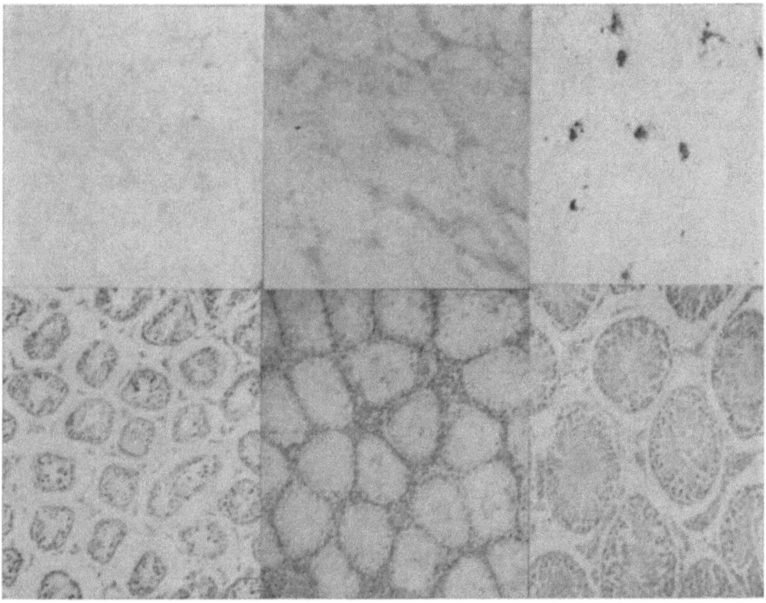

Fig. 16. Sections of testes of mature male offspring stained with hematoxylin and eosin (top), and for activity of 3β-hydroxysteroid dehydrogenase (bottom). Note tubules with normal spermatogenesis and strong enzymatic activity in Leydig cells of mature control male (left); tubules with atrophic spermatogonia and hypercellular stromal tissue with diminished enzymatic activity of male treated on 19th day of gestation with dose of 60 mg/kg (middle), and tubules with no spermatogonial maturation, small amount of stroma with slight enzymatic activity of mature male treated on 16th day of gestation with doses of 60 mg/kg (right). $\times$ 100

Twenty-four hours after a single dose, label is present in the adrenals, ovaries, and liver. Beyond 48 hours, label is retained only by the adrenals and ovaries, where it remains for at least 13 weeks. Inhibition of enzymatic activity in vivo is directly correlated with uptake of label, demonstrating the titration phenomenon in vivo

(Fig. 17). Label extracted from adrenal homogenates has similar inhibitory capacity in a bacterial 3β-hydroxysteroid dehydrogenase assay system, and identical mobility in three thin-layer

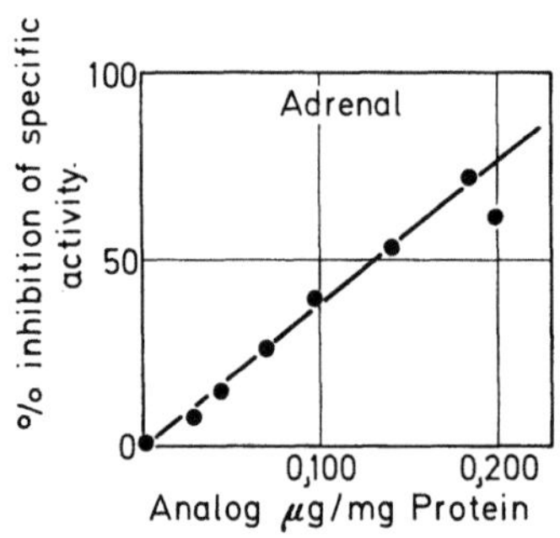

Fig. 17. Plots of percent inhibition of activity of the dehydrogenase (μg product/min/mg protein of experimental adrenals divided by that of control adrenals) against ^{14}C content/mg protein after a single dose of ^{14}C-isoxazole

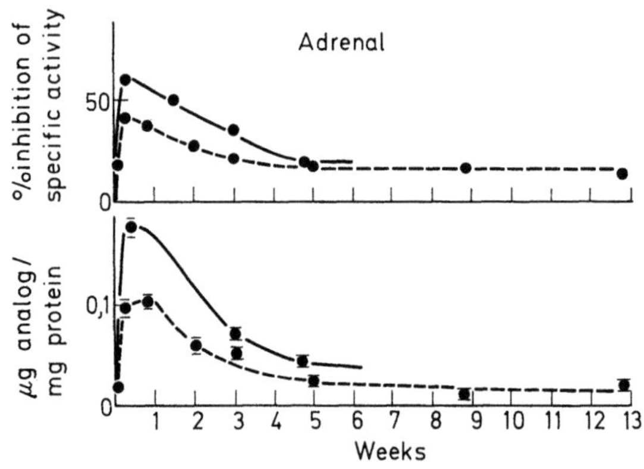

Fig. 18. Plot percent inhibition of activity of the dehydrogenase and of ^{14}C in the adrenal against time in weeks after injection of the isoxazole-^{14}C at a dose of 2 mg/kg (broken line) and of 12 mg/kg (unbroken line)

chromatographic systems to that of standard ^{14}C-isoxazole. Return of enzymatic activity correlates with loss of label, and the degree of enzymatic inhibition and amount of label in both glands attain constant low values after about one month (Fig. 18).

Similar findings of ^{14}C-uptake and persistence of label and inhibition of the dehydrogenase system were obtained selectively in enzyme-containing tissues of offspring of ^{14}C-isoxazole-treated pregnant rats (Fig. 19) [50]. The offspring have significant adrenal enlargement, inhibition of the dehydrogenase, and ^{14}C content at

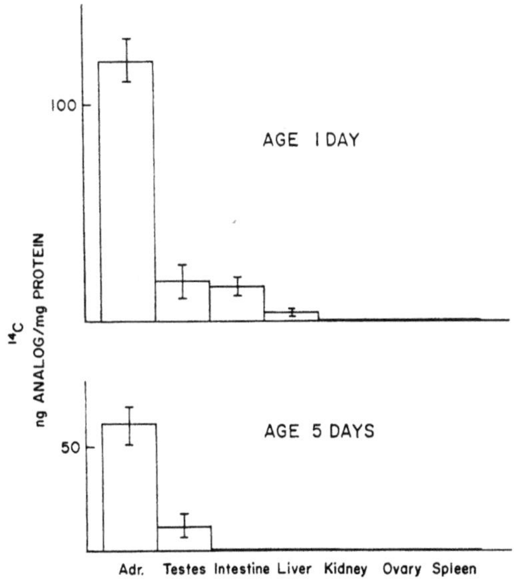

Fig. 19. ^{14}C content of fetal tissues at one and five days of age. Ordinate: ^{14}C in ng ^{14}C-isoxazole/mg protein. Vertical bars: $\pm$ one standard deviation

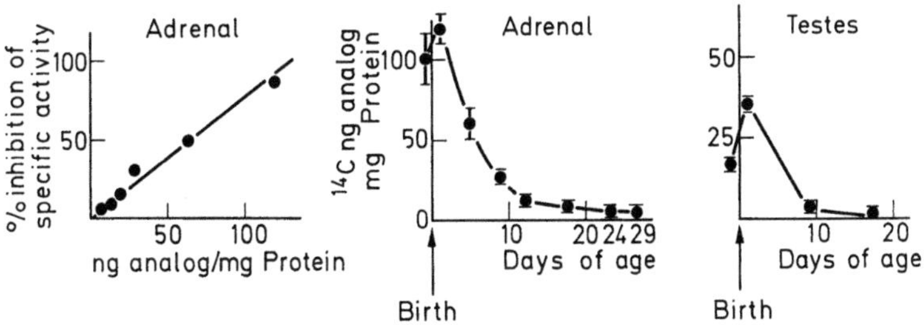

Fig. 20. Plots of percent inhibition of specific activity of fetal adrenal dehydrogenase against ^{14}C content of fetal adrenal (left) of adrenal (middle) and testicular (right) ^{14}C content/mg protein against days of age after a single maternal dose of ^{14}C-isoxazole

least until 12 days of age (Fig. 20). These observations demonstrate the specificity of action of this analog, and indicate that its long-term effects are a function of its selective and tight binding to the active sites of the dehydrogenase and isomerase.

The exceptional delay in complete loss of label and return of the level of activity of the dehydrogenase to normal raises interesting questions. The delay in loss of label probably depends on the interplay of the rates of release of analog by enzyme degradation, and of recapture of the analog by enzyme synthesis. The relatively long lag in restoration of enzymatic activity can be explained by a very slow enzyme turn-over, or by the fact that the presence of analog-bound enzyme retards synthesis of new enzyme, or both. Since most enzymes have rapid turnover [80], it is difficult to postulate a long turnover in the dehydrogenase-isomerase system. However, consistent with the hypothesis of a long enzyme turnover is the observation that the loss of label and restoration of activity level off at about one month, the same time required for total activity to be restored in one remaining ovary after hemiovariectomy [78]. Recent theories of enzyme inhibition also suggest that an enzyme inhibitor may also depress synthesis of that enzyme [16].

h) *Effect of ACTH on Defect of Dehydrogenase.* To study the role of the pituitary-adrenal axis in the persistent defect in activity of 3β-hydroxysteroid dehydrogenase and $\Delta^{5,4}$, 3-ketosteroid isomerase produced by 17β-hydroxy-4,4,17α-trimethyl-androst-5-ene-(2,3d) isoxazole, the effects of large doses of ACTH and multiple doses of inhibitor on the level of this enzyme system and uptake of ^{14}C has been studied in control and inhibitor-treated rats [41]. ACTH produces in normal animals a significant rise in total activity of the dehydrogenase, and in inhibited animals a rise in uptake of ^{14}C by the adrenal, and in the specific and total activity of the dehydrogenase system. Multiple doses of inhibitor increase adrenal size and uptake of ^{14}C by the adrenal (and ovary) but do not affect the level of specific or total activity of the noninhibited dehydrogenase (Fig. 21). Since uptake of label by endocrine tissues after administration of ^{14}C-labelled isoxazole has been attributed to specific binding of the isoxazole to the active site of the dehydrogenase, increased uptake of label can be attributed to synthesis of new enzyme. The adrenal of normal and ^{14}C-isoxazole-treated animals is capable of further enlargement, and of synthesizing more dehydrogenase in response to exogenous ACTH, and to multiple doses of inhibitor.

Since adrenal enlargement does not occur in inhibitor-treated hypophysectomized animals [75, 8, 58], the increase in adrenal size and uptake of ^{14}C in response to multiple doses of inhibitor is

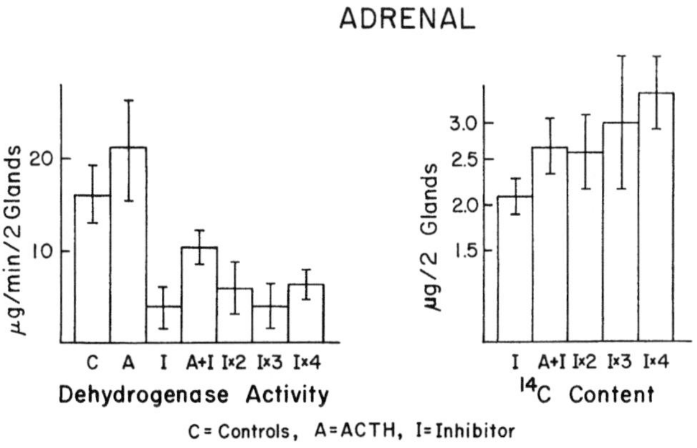

Fig. 21. Effect of ACTH and multiple daily doses of ^{14}C-isoxazole analog on activity of adult female rat adrenal dehydrogenase (left ordinate) and content of ^{14}C (right ordinate). ACTH produces a significant rise in activity of control and inhibited animals but multiple doses of ^{14}C-isoxazole analog do not affect level of non-inhibited dehydrogenase (left). However, multiple doses of ^{14}C-isoxazole analog do produce a significant rise in level of bound ^{14}C-analog--dehydrogenase (right)

probably due, at least in part, to an increased secretion of ACTH produced by the inhibition of the dehydrogenase. A curious point in the present observations is that although multiple doses of the inhibitor nearly doubled the size of the adrenal, and produced a 50 % increase in the uptake of ^{14}C-label (presumably isoxazole-bound by the dehydrogenase), the amount of active (unbound) dehydrogenase remains about the same (i.e., about 30—50 % of the control level). Since adrenal enlargement is only observed at or above a dose of inhibitor which produces 50 % inhibition of the dehydrogenase [48, 42], it is possible that this acute control of adrenal dehydrogenase level is only brought into play with a loss of 50 % or more of the amount of enzyme. We have shown that the levels of adrenal 3β-hydroxysteroid dehydrogenase and circulating corticosterone level in rats either 5 hours or 6 weeks after a single dose of the cyanoanalog are both about 45—50 % of control values [53]. Two other groups of investigators have observed similar degree of depression of corticosterone levels and enzymatic activity after multiple doses of the cyano-ketone in the rat [66] and guinea pig [12]. These findings imply that the pituitary is sensitive only to a drop in circulating corticosterone level of 50 % or more. Although there is no apparent explanation for this hypothesis, it is consistent with previous failures to find an increased excretion of steroidal precursors in heterozygous parents of children with congenital adrenal hyperplasia [15]. If heterozygotes had less than a 50 % reduction in enzymatic activity, and if more than 50 % reduction were needed for an ACTH response, no derangement of steroidal metabolism would be observed. Thus, these results indicate that although the pituitary-adrenal axis may raise the level of the adrenal dehydrogenase acutely when there is greater than 50—70 % inhibition of this enzymatic activity, it does not play a role in the persistence of the defect at this reduced level.

i) *Reversal of Enzymatic Defect.* The data obtained with the analogs suggest that the blockade of the enzyme's active site also interferes with the mechanisms controlling the level of the enzyme, either through persistent depression of enzyme biosynthesis, or increased enzyme degradation, or both. This is a striking observation because of the fact that other procedures, which reduce the production of steroid hormones, markedly increase the activity of this enzyme system.

We have investigated whether phenol-detergent (SDS) extracts of nucleic acids may correct in vivo the persistent defect in activity of the dehydrogenase system produced by the analog [35]. A partial, but highly significant, correction of the depressed level of the activity of this enzyme in the adrenals was produced by extracts of rat or bovine adrenal RNA, DNA, or a mixture of both. Boiling of these extracts or treating them with the appropriate nuclease reduced this correction. Bovine thymus DNA was ineffective. These results have suggested that adrenal nucleic acids may be capable of producing a partial correction of the defect of adrenal 3β-hydroxysteroid dehydrogenase created by this analog.

j) *Normal Development of Dehydrogenase in Tissue Culture.* Our studies of the dehydrogenase have also led to another interesting biochemical embryological model. Although animal cell cultures rarely perform differentiated functions of the tissue of origin, an understanding of the factors controlling the level or appearance of the dehydrogenase would be greatly served by a system which allows this enzyme to appear normally in tissue culture. Cultures of adult steroidogenic tissues or organs, and of fetal adrenals and testes have been reported to have steroidogenic enzymes, steroidogenesis, and endocrine function [61, 73, 85]. PICON has shown that fetal testes explanted on day $14^1/_2$ when activity of the dehydrogenase is not demonstrable, or at $15^1/_2$ or $16^1/_2$ days when activity is first demonstrable, all have activity in organ culture at a time corresponding to day 21 [74]. She concluded that fetal testes develop and maintain activity of the dehydrogenase in organ culture at a time when they normally do in the fetus.

Our early observation that the rat fetal ovary has no demonstrable histochemical activity of the dehydrogenase [56] has been recently confirmed by SCHLEGEL and co-workers, who have shown that activity of this enzyme in the rat ovary is first demonstrable in the interstitial cells at nine days of age, and in the theca interna at ten days of age [81]. These observations are consistent with previous studies showing that the rat fetal ovary is not endocrinologically active, does not affect the prenatal growth or differentiation of female reproductive tracts, and is uninfluenced morphologically by fetal decapitation (hypophysectomy) or gonadotrophin [77, 23].

Activity of the dehydrogenase is demonstrable after nine days in culture of cells derived from term rat fetal ovaries (Fig. 22) [51]. Since this is the time at which the dehydrogenase appears in the ovaries of the intact postnatal animal, this enzyme apparently differentiates normally in tissue culture according to a timetable which is inherent in ovarian cells, and independent of extra-ovarian influence. Thus, a model has been provided for future study of the differentiation of a specific enzyme in mammalian tissue culture, and further suggests that the differentiation of other mammalian enzymes may be sought in tissue culture.

k) *Estrogens, Progestins, and Inhibition of Dehydrogenase.* In a review of various reported effects of certain progestins and estrogens on human and rat genital development, we have presented evidence

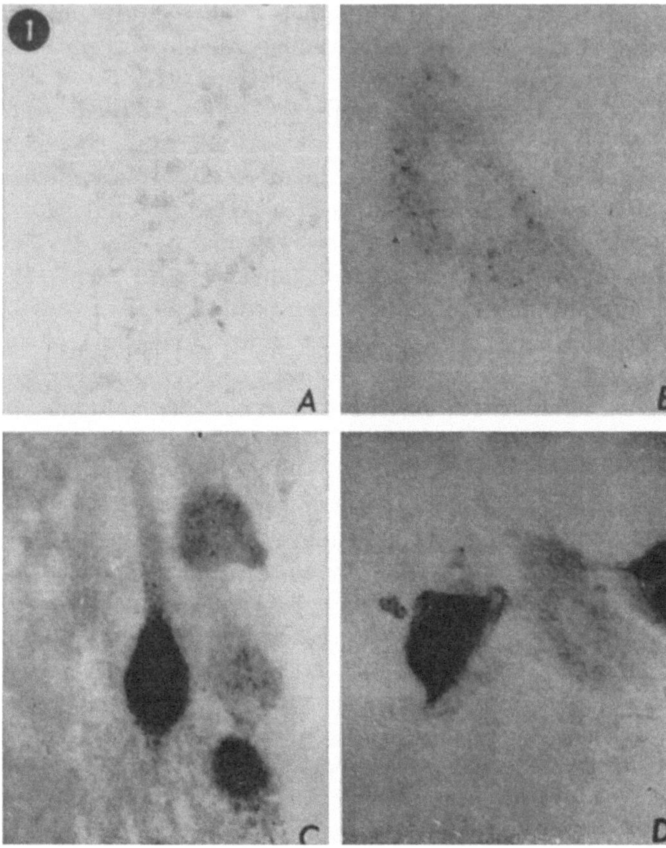

Fig. 22. Histochemical activity of 3β-hydroxysteroid dehydrogenase in cells of fetal ovary grown in culture for a time corresponding to 6 days of age, *A* (X200); to 10 days of age, *B* (X900); to 13 days of age, *C* (X900); and to 16 days of age, *D* (X900). Note granules representing activity present in *B*, *C*, and *D*, and not in *A*

that suggests that when these agents are given at the time of urethral fold fusion, they may block penile differentiation in males, or produce labial fusion in females [48]. When given after this time, they may produce only clitoral hypertrophy in females. Thus, these agents may produce the same effects on genital differentiation at the same critical periods as the analog in our experimental studies. Therefore, we investigated whether these effects of estrogens and progestins may be produced through inhibition of the dehydrogenase and the isomerase.

The natural products of this enzyme system (androstene-dione, 17-hydroxyprogesterone) are simple competitive inhibitors with an affinity for the active sites of both enzymes about equal to that of the substrate [29, 36, 63]. The progestins (medroxyprogesterone and norethindrone, and 17-hydroxyprogesterone caproate), which are synthetic derivatives of progesterone, are also competitive inhibitors of the bacterial enzyme with an affinity for the active site about ten times that of the substrate. Estradiol-17β, however, produces the same stoichiometric pattern of inhibition of the bacterial and bovine adrenal enzyme as does the analog which produces experimental congenital adrenal hyperplasia and deficient activity of 3β-hydroxysteroid dehydrogenase in rats. Thus, it appears that the active site of the bacterial 3β-hydroxysteroid dehydrogenase is susceptible to end-product inhibition, which can be pseudo-irreversible if there is a nucleophilic cyano or isoxazole group at the C-2α position of the steroidal inhibitor. The applicability of this conclusion to mammalian 3β-hydroxysteroid dehydrogenase is as yet undemonstrated, but it is suggested by the fact that the concentrations of these agents required to inhibit the bacterial 3β-hydroxysteroid dehydrogenase are in the same relative proportion as are the minimum effective doses required to produce the genital malformations in fetal rats [48]. Recently, a case of a boy having reversible congenital adrenal hyperplasia with "salt-losing" crises has been attributed to maternal administration of medroxyprogesterone [64].

l) *Experimental Model of Dehydrogenase Defect Produced by Estradiol-17β.* We have administered large doses of estradiol-17β to pregnant rats to determine whether this steroid will mimic the action of the analog in mammals as it does in the bacterial 3β-hydroxysteroid dehydrogenase system [34]. At term, the pregnant animals have severe adrenal and ovarian hyperplasia and reduced activity of the dehydrogenase in these glands in proportion to dose. The fetuses have adrenal hyperplasia, clitoral hypertrophy or hypospadias, and deficient histochemical activity of the enzyme in the adrenals and testes (Table 2). This proposed mechanism of action of estradiol-17β on fetal genital differentiation is supported by the fact that the production of hypospadias by estradiol-17β is prevented by testosterone, while that of adrenal and clitoral enlargement is prevented by corticosterone [38].

B. 11β-Steroid Hydroxylase

At this point we turn to 11β-steroid hydroxylase. Metyrapone has been shown to block 11β-hydroxylation in the biosynthesis of adrenal corticoids and to produce patterns of excess urinary 11-desoxysteroids both experimentally [14] and in humans [84]. This compound most probably achieves its effects through competitive inhibition of the activity of 11β-hydroxylase [20]. This mitochondrial enzyme is essential in mammalian steroid-producing tissues for the placement of a hydroxyl group in the 11β-position, and therefore is essential in the adrenal for the synthesis of glucocorticoids and aldosterone.

We have produced congenital adrenal cortical hyperplasia in the offspring of pregnant rats treated with metyrapone (Table 2) [32]. Female offspring also have clitoral hypertrophy (Fig. 9). These effects are reduced by corticosterone. Basal corticosterone production and the increased corticosterone production due to the addition of substrates of 11β-steroid hydroxylase in quartered adrenal glands are markedly reduced in mature females treated with metyrapone. These experiments have shown that an inhibitor of 11β-steroid hydroxylase produces the same type of congenital adrenal cortical hyperplasia and clitoral hypertrophy which occur spontaneously in the hypertensive form of congenital adrenal hyperplasia due to a genetic deficiency of this enzyme.

C. Cholesterol Desmolase

A genetic defect in cholesterol desmolase has been postulated to be the cause of congenital lipoid adrenal hyperplasia, characterized by enlarged adrenal cortices containing excess cholesterol and many multinucleated, lipid-filled giant cells without excessive 17-ketosteroids, hypospadias, and Leydig cell hyperplasia in genetic males, and normal genitalia in genetic females [76]. Aminoglutethimide is a drug whose primary action in the adult rat is thought to produce a blockade of steroidogenesis by inhibition of the cholesterol desmolase system [9, 10, 13, 19, 62, 60, 28].

1. Experimental Congenital Lipoid Adrenal Hyperplasia

Daily intramuscular injections of 100 mg/kg of this drug have been given to pregnant rats [43]. Aminoglutethimide produces considerable maternal and fetal adrenal enlargement, with multinucleated, lipid-filled giant cells. There is marked accumulation of cholesterol in maternal adrenals and ovaries, and in fetal adrenals and testes. Experimental male fetuses have hypospadias (Fig. 8, Table 2) and an increased number of Leydig cells. There is no defect in activity of maternal or fetal Δ^5, 3β-hydroxysteroid dehydrogenase, but experimental maternal and fetal adrenals have a severe defect in the conversion of cholesterol 1,2 H^3 to pregnenolone 1,2 H^3. Aminoglutethimide, paradoxically, produces severe virilization of female fetuses (Fig. 9), indicating excess adrenal androgenesis despite the steroidogenic blockade of the fetal adrenals produced by this drug, thereby complicating the animal model of the human disease. Testosterone prevents the production of hypospadias by aminoglutethimide, while corticosterone prevents the adrenal enlargement, adrenal cholesterol accumulation, clitoral enlargement, and increase in female anourethral distance [47]. These findings demonstrate the production of the adrenal and testicular defects of human congenital lipoid adrenal hyperplasia in fetal rats by a drug whose predominant action is inhibition of the cholesterol desmolase system. Thereby, experimental evidence has been provided in support of the postulated enzymatic defect in the human disease.

D. 17α-Hydroxylase

SU 10603 and SU 8000 (Ciba) have been shown to be selective inhibitors of 17α-hydroxylase and C_{17-20} lyase in mature animals of several species [28]. We have administered these inhibitors to pregnant rats and have found that they produce hypospadias (Fig. 8, Table 2) [45]. This drug does not affect adrenal size, since the rat adrenal does not utilize these enzymes in the biosynthesis of corticosterone, its primary glucocorticoid. We have found a marked reduction in the activity of the hydroxylase and lyase in the testes of fetuses of pregnant rats treated with SU 10603. These findings indicate that these drugs probably block the formation of the fetal penis by inhibiting the testicular 17α-hydroxylase, and C_{17-20} lyase thereby providing experimental evidence for this predicted defect.

II. Testicular Feminization

A. 5α-Reductase

I should like now to turn to testicular feminization, a disease proposed to be due to a defect of an enzyme involved in the utilization of a steroid hormone in the target organ [65, 72]. This disease of genetic and gonadal males is characterized by fully feminine development of external genitalia, breasts, a vagina without uterus, and an unresponsiveness to testosterone. These changes suggest that there has been a complete lack of androgen-dependent development in the fetus.

Recent studies have indicated that testosterone is converted in target organs to dihydrotestosterone, a metabolite which in certain bioassay systems, appears to be about twice as active as testosterone itself [21, 79]. Thus, these studies have suggested a possible role in sexual differentiation of testosterone 5α-reductase in peripheral secondary sex organs of the fetus. Although 5α-reductase, which performs this reduction, is located in the liver, it has been shown that after intravenous administration of testosterone 1,2-^{3}H, appreciable amounts of dihydrotestosterone are recovered only in organs of accessory reproduction in the male rat, and the only detectable testosterone metabolite recovered in prostatic nuclei, a presumed site of action of this hormone, is dihydrotestosterone [6, 7]. Prostatic nuclei contain the 5α-reductase and dihydrotestosterone, rather than testosterone, is the predominant androgen bound to nuclear chromatin within 15 minutes after testosterone administration. [1] Dihydrotestosterone is bound to nuclear histones, but is converted by prostatic cytoplasm to 5α-androstan-3α-17β-diol [90]. Thus, it has been suggested that dihydrotestosterone is the active androgen in target organs. Moreover, it may be that the formation of the dihydrotestosterone-nucleohistone complex is responsible for the known increases in DNA, RNA, DNA and RNA polymerases, and protein synthesis produced in target organs by testosterone, and as such, may be an example of derepression.

1. Cyproterone-acetate

Cyproterone-acetate has a dose-dependent inhibitory effect on the masculinization of female fetuses caused by the administration of testosterone propionate [68]. However, the feminization of male fetuses provoked by cyproterone-acetate is not, or is only slightly, antagonized by the additional administration of testosterone propionate. The mechanism of action of cyproterone-acetate is unknown, but this agent has been shown to block the binding of testosterone to a cytoplasmic protein, and of dihydrotestosterone to a nuclear histone in prostatic tissue [24, 86]. We have recently found that cyproterone is a highly effective competitive inhibitor of the conversion of testosterone to 5α-androstan-3α, 17β-diol by homogenates of prostatic tissue (Table 2). We have also observed that dihydrotestosterone is somewhat more effective than testosterone in virilizing female rat fetuses, while the 3α, 17β-diol is about equally effective (Fig. 23). Both these steroids also effect a significant correction of the hypospadias and virilization of females produced by cyproterone to a similar degree as testosterone. These results support the suggestion that dihydrotestosterone is the active androgen, and further suggest that cyproterone-acetate may produce a model of testicular feminization by inhibiting the formation of the diol as well as the formation of dihydrotestosterone-nucleohistone complex. They furthermore suggest that the recently observed failure of dihydrotestosterone to correct certain androgen-dependent defects in patients with testicular feminization [87] may be due to the fact that the human defect may consist of a defect in the testosterone and dihydrotestosterone-binding proteins, as well as the reductase and 3α-dehydrogenase.

III. Pseudohermaphrodite Rat

Lastly, I should like to turn to the pseudohermaphrodite rat. This inherited form of male pseudohermaphroditism in the rat has been described by Stanley and Gumbreck [82]. The affected animals have absence of wolffian and mullerian derivatives, a 1—2 mm vagina, well-developed Leydig cells, nipples, adrenal and pituitary enlargement, excess secretion of DHA of both adrenal and testicular origin, defective secretion of testosterone, and the adults are unresponsive to testosterone [2]. The defect is carried by the female, who passes it to half her male progeny, and it is generally assumed that the defect is sex-linked, i.e. carried on the X-chromosome. Stanley and co-workers have postulated that this is a model of testicular feminization, and as such, might be expected to have a target organ defect in testosterone 5α-reductase, since

ANDROGEN METABOLISM

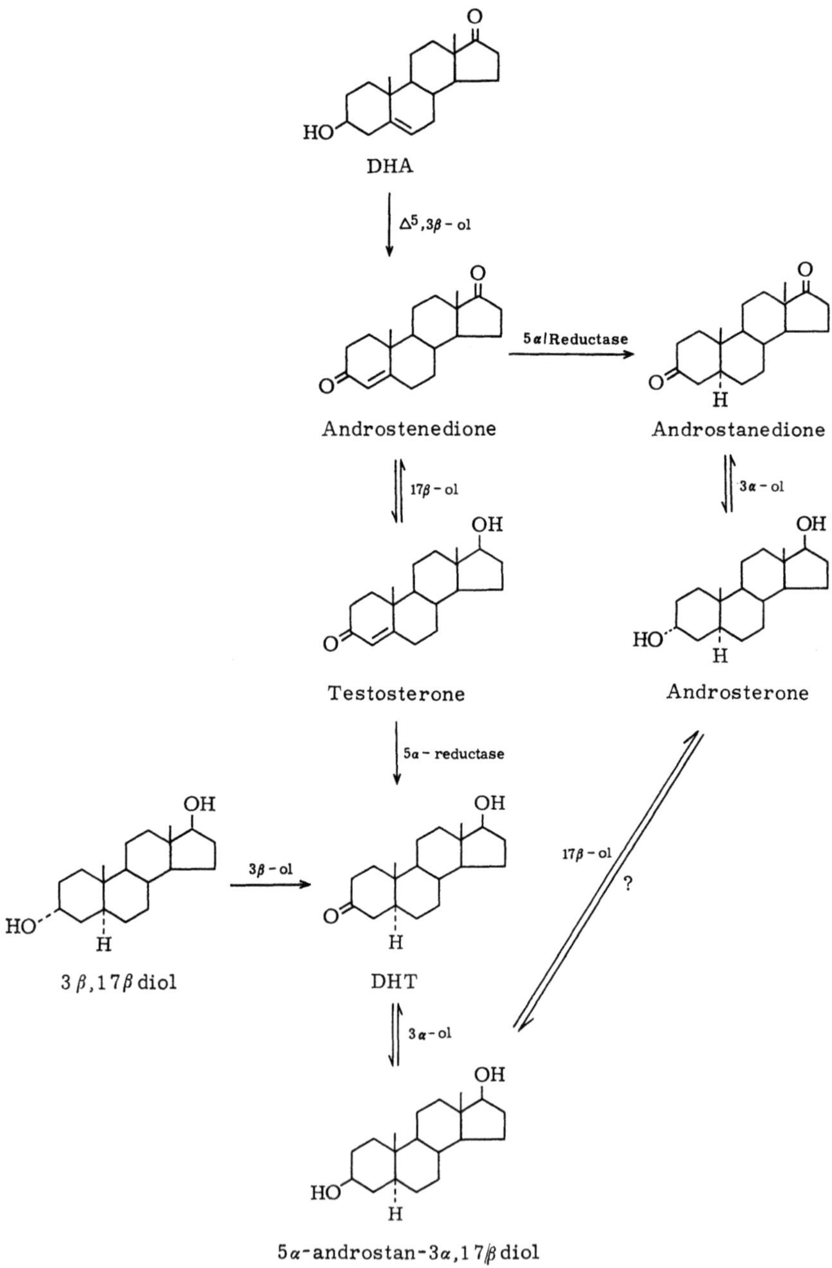

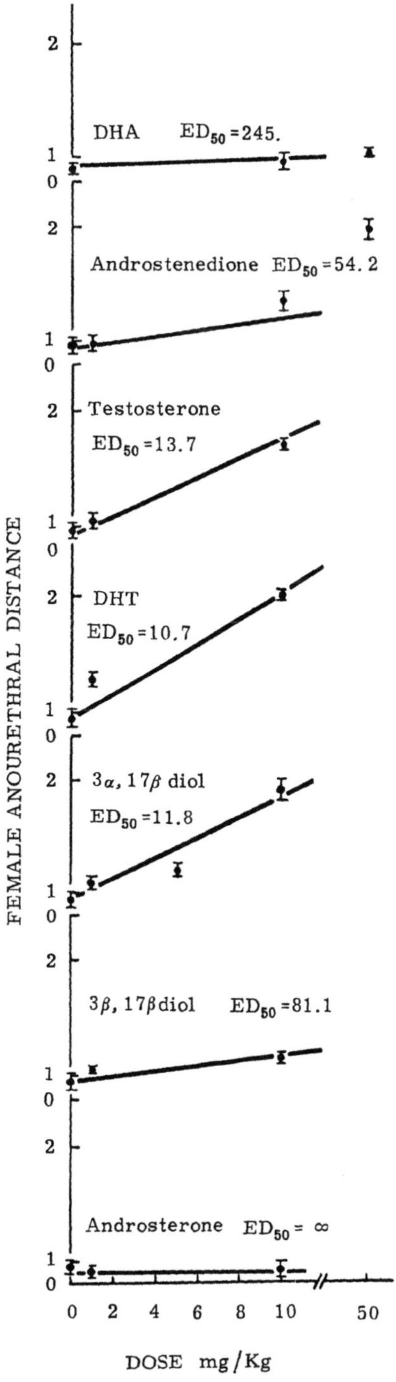

Fig. 23. Pathways of androgen formation and metabolism (left) and effect of various androgens administered from days 13 to 20 on anourethral distance of female rat fetus (mm) (right). ED_{50}: maternal dose calculated to raise anourethral distance in female fetus to half that of normal male. Note that dihydrotestosterone and the 3α-diol are slightly more androgenic than testosterone

this defect has recently been proposed to be the etiology of the human disease. Bardin, et al. have reported that the pseudohermaphrodite rat does not have a reduced capacity to synthesize DHT in the skin and preputial glands [3, 33]. We have confirmed this observation and have shown that the pattern of formation of DHT and 5α-androstan-3α, 17β-diol by the pseudohermaphrodite rat is unlike that of its normal male, but resembles that of the normal female [33]. Thus, testosterone unresponsiveness of the pseudohermaphrodite rat is associated with a defect in peripheral 5α-reductase as in human testicular feminization.

The occurrence of adrenal and pituitary enlargement [81] and defective testosterone production [2] had suggested to us that the pseudohermaphrodite rat may not be a model of testicular feminization, but may have a defect in one of the steroidogenic enzymes. Our first approach to this problem was an investigation of the serum level of corticosterone in these rats. The pseudohermaphrodite rat has an elevated level of corticosterone compared to normal littermate males, and the degree of elevation is about the same as that of the enlargement of the adrenals [46]. In other words, there is no defect in the glucocorticoid series of steroidogenic enzymes, and thus, the failure of normal embryonic masculinization must not involve either cholesterol desmolase, or the dehydrogenase system, but another testosteronogenic enzyme. We have confirmed this by observing an increase in the level of both adrenal and testicular 3β-hydroxysteroid dehydrogenase in the pseudohermaphrodite rat.

A. 17β-Hydroxysteroid Dehydrogenase

When we incubated either tritiated pregnenolone, dehydroepiandrosterone, progesterone, or 17-hydroxyprogesterone with appropriate cofactors and control testicular tissue, we obtained testosterone and testosterone metabolites as the primary end products as has been reported many times in the literature (Fig. 24) [46]. However, when we incubated these steroids with testicular tissue of the pseudohermaphrodite rat, we found that the end product was primarily an androstenedione metabolite, androsterone. Incubation of androstenedione and appropriate cofactors with control testicular homogenates gave testosterone and testosterone metabolites, but with pseudohermaphrodite testicular tissue, only androstenedione and androsterone. In other

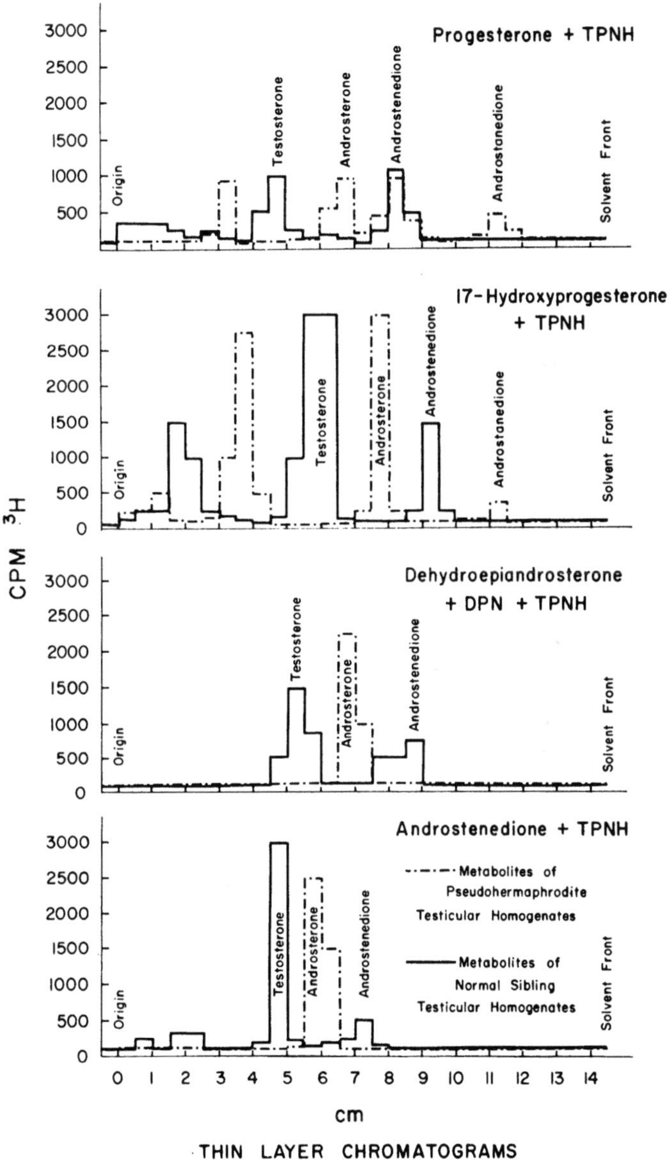

Fig. 24. Androgen biosynthesis by testicular homogenates of Pseudohermaphrodite rats (broken line) and of normal male siblings (solid line). Principal products of normal rats are androstenedione, testosterone and testosterone metabolites, whereas pseudohermaphrodite makes primarily a metabolite of androstenedione, i.e., androsterone

words, the pseudohermaphrodite rat has a defect in 17β-hydroxysteroid dehydrogenase. A defect in this enzyme would give rise to insufficient testosterone production during fetal life, and thereby the pseudohermaphrodite rat would fail to undergo any testosterone dependent development. The failure of wolffian duct development in this rat associated with a deficiency of fetal testosterone production suggests that wolffian duct development in the rat is like other species, testosterone-dependent, and that the reason duct development is not altered by steroidogenic enzyme inhibition or anti-androgenic treatment is that the blockade of testosteronogenesis by these agents is incomplete. In this case nature's experimentation has again far outdone that of man.

An interesting suggestion made by this 17β-hydroxysteroid dehydrogenase defect is that postnatal testosterone responsiveness may be set by testosterone in fetal life. The lack of testosterone responsiveness in the pseudohermaphrodite rat cannot be due simply to the absence of testosterone and the presence of some unusual androgen, such as androsterone. Support for the hypothesis that testosterone responsiveness is set in fetal life is given by the observation that mature offspring of rat mothers treated with cyproterone during gestation are not testosterone-responsive (F. Neumann, personal communication).

One last point about this interesting animal is the fact that the available genetic evidence suggests that this defect is sex-linked recessive. Thus, the verification of this defect as one in 17β-hydroxysteroid dehydrogenase, which controls testosteronogenesis and thereby fetal masculine development, will identify the first gene for an enzyme on the X-chromosome in the rat.

In summary, by selective inhibition of certain fetal steroidogenic enzymes (cholesterol desmolase, 3β-hydroxysteroid dehydrogenase, 3-ketosteroid isomerase, 11β-hydroxylase, and 17α-hydroxylase), animal models have been produced with the precise sexual malformations and enzymatic deficiencies of the human genetic defects in congenital adrenal hyperplasia. Inhibition of rat fetal testicular desmolase, dehydrogenase, isomerase, or 17α-hydroxylase has led to insufficient androgen production, which, in turn, produces hypospadias and, in the case of the dehydrogenase and isomerase, abnormal appearance of nipples in male fetuses. The effect of inhibition of the desmolase and 17α-hydroxylase on the development of nipples has not yet been studied. Inhibition of the adrenal dehydrogenase, isomerase, and 11β-hydroxylase leads to ACTH-mediated enlargement and excess production of dehydroepiandrosterone or androstendione, respectively, which in turn produces varying degrees of clitoral hypertrophy and labial fold fusion. The pseudohermaphrodite rat has a sex-linked defect which also results in failure of masculinization due to deficient fetal testosteronogenesis. We have shown that this rat has a selective deficiency in a testosteronogenic enzyme, 17β-hydroxysteroid dehydrogenase. It is hoped that further studies utilizing these unique systems for the study of the role of enzymes in the control of fetal sexual differentiation will provide further fundamental information about normal fetal sexual differentiation and further clarify the sexual anomalies occurring in human adrenogenital syndrome.

References

1. Anderson, K. M., Liao, S.: Selective retention of dihydrotestosterone by prostatic nuclei. Nature (Lond.) **219**, 277 (1968).
2. Bardin, C. W., Allison, J. E., Stanley, A. J., Gumbreck, L. G.: Secretion of Testosterone by the Pseudohermaphrodite Rat. Endocrinology **84**, 435 (1969).
3. — Bullock, L., Gram, T. E., Gillette, J. R.: End organ insensitivity to testosterone in the pseudohermaphrodite rat. 51st meeting, Endocrine Society, New York 1969.
4. Bongiovanni, A. M., Eberlein, W. R., Goldman, A. S., New, M.: Disorders of Adrenal Steroid Biogenesis. Recent Prog. Hormone Res. **23**, 375—449 (1967).
5. — Root, A. W.: The adrenogenital syndrome. New Engl. J. Med. **268**, 1283—1289, 1342—1351, 1391—1399 (1963).
6. Bruchovsky, N., Wilson, J. D.: The conversion of testosterone to 5α-androstan-17β-ol-3-one by rat prostate in vivo and vitro. J. biol. Chem. **243**, 2012 (1968).
7. — — The Intranuclear Binding of Testosterone and 5α-Androstan-17β-ol-3-one by Rat Prostate. J. biol. Chem. **243**, 5953 (1968).
8. Burnham, D. F., Beyler, A. L., Potts, G. O.: Adrenal blocking action of a new steroid. Fed. Proc. **22**, 270, Abstr. 653 (1963).
9. Camacho, A. M., Brough, A. J., Cash, R., Wilroy, R. S.: Adrenal toxicity associated with the administration of an anticonvulsant drug. J. Pediat. **68**, 852 (1966).
10. — Cash, R., Brough, A. J., Wilroy, R. S.: Inhibition of adrenal steroidogenesis by aminoglutethimide (Elipten) and the mechanism of action. J. Amer. med. Ass. **202**, 20 (1967).

11. CAMACHO, K. M., MIGEON, C. J.: Testosterone excretion and production rate in normal adults and in patients with congenital adrenal hyperplasia. J. clin. Endocr. **26**, 893—896 (1966).
12. CASTELLS, S., BRANSOME, E. D., JR.: Comparison of the effects of ACTH and an inhibitor of 3β-hydroxysteroid dehydrogenase on the synthesis of adrenal microsomal proteins. J. clin. Endocr. **29**, 536 (1969).
13. CASH, R., BROUGH, A. J., COHEN, M. N. P., SATOH, P. S.: Aminoglutethimide (Elipten-Ciba) as an inhibitor of adrenal steroidogenesis: Mechanism of action and therapeutic trial. J. clin. Endocr. **27**, 1239 (1967).
14. CHART, J. J., SHEPPARD, H.: Studies on adrenocortical inhibitors. Proc. Inter. Cong. Horm. Ster. **1**, 399—411 (1964).
15. CHILDS, B., GRUMBACH, M. M., VAN WYK, J. J.: Virilizing adrenal hyperplasia: A genetic and hormonal study. J. clin. Invest. **35**, 213 (1956).
16. CLINE, A. L., BOCK, R. M.: Translational control of gene expression. Cold Spr. Harb. Symp. quant. Biol. **31**, 321 (1966).
17. DE CRECCHIO, L.: Morgagni **7**, 151 (1865).
18. DEMURA, J., WEST, C. D., NUGENT, C. A., NAKAGAWA, K., TYLER, F. H.: A sensitive radioimmunoassay for plasma ACTH levels. J. clin. Endocr. **26**, 1297 (1966).
19. DEXTER, R. N., FISHMAN, L. M., NEY, R. L. LIDDLE, G. W.: Inhibition of adrenal corticosteroid synthesis by aminoglutethimide: Studies of the mechanism of action. J. clin. Endocr. **24**, 473 (1967).
20. DOMINGUEZ, O. V., SAMUELS, L. T.: Mechanism of inhibition of adrenal steroid 11β-hydroxylase by Methopyrapone (Metopirone). Endocrinology **73**, 304—309 (1963).
21. DORFMAN, R. I., SHIPLEY, R. A.: Relative activities of androgens. In: Androgens, ch. 8, p. 116. New York: John Wiley and Sons, Inc. 1956.
22. EBERLEIN, W. R.: Steroids and Sterols in Umbilical Cord Blood. J. clin. Endocr. **25**, 1101 (1965).
23. EGUCHI, Y., MORIKAWA, Y.: Changes in pituitary-gonadal interrelations during perinatal days in the rat. Anat. Rec. **161**, 163 (1968).
24. FANG, S., LIAO, S.: Antagonistic action of anti-androgens on the formation of a specific dihydrotestosterone-receptor protein complex in rat ventral prostate. Mol. Pharmacol. **5**, 420 (1969).
25. FERRARI, R. A., ARNOLD, A.: Inhibition of β-hydroxysteroid dehydrogenase. I. Structural characteristics of some steroidal inhibitors. Biochim. biophys. Acta (Amst.) **77**, 349—356 (1963).
26. — — Inhibition of β-hydroxysteroid dehydrogenase: II. Kinetics and pH effect. Biochim. Biophys. Acta (Amst.) **77**, 357—364 (1963).
27. FRASIER, S. D., HORTON, R., ULSTROM, R. A.: Androgen metabolism in congenital adrenal hyperplasia due to 11β-hydroxylase deficiency. Pediatrics **44**, 201 (1969).
28. GAUNT, R., STEINETZ, B. G., CHART, J. J.: Pharmacologic alteration of steroid hormone functions. Clin. Pharmacol. Therapeut **9**, 657 (1968).
29. GOLDMAN, A. S.: Inhibition of 3β-hydroxysteroid dehydrogenase from Pseudomonas testosteroni by various estrogenic and progestinic steroids. J. clin. Endocr. **27**, 320—324 (1967).

30. GOLDMAN, A. S.: Stoichiometric inhibition of various 3β-hydroxysteroid dehydrogenases by a substrate analogue. J. clin. Endocr. **27**, 325—332 (1967).
31. — Experimental congenital adrenocortical hyperplasia: Persistent postnatal deficiency in activity of 3β-hydroxysteroid dehydrogenase produced in utero. J. clin. Endocr. **27**, 1041—1049 (1967).
32. — Experimental model of congenital adrenal cortical hyperplasia produced in utero with an inhibitor of 11β-steroid hydroxylase. J. clin. Endocr. **27**, 1390 (1967).
33. — Unpublished observations.
34. — Production of congenital adrenocortical hyperplasia in rats by estradiol-17β and inhibition of 3β-hydroxysteroid dehydrogenase. J. clin. Endocr. **28**, 231 (1968).
35. — Elevation of level of 3β-hydroxysteroid dehydrogenase in rats by extracts of DNA or RNA. J. clin. Endocr. **28**, 1127 (1968).
36. — Further studies of steroidal inhibitors of Δ^5, 3β-hydroxysteroid dehydrogenase and Δ^5-Δ^4, 3-ketosteroid dehydrogenase in Pseudomonas testosteroni and in bovine adrenals. J. clin. Endocr. **28**, 1539 (1968).
37. — Inhibitors of adrenal steroid synthesis. Excerpta Med., Third Int. Congr. Endocrinol., 184, 817 Progress in Endocrinology. Mexico (1969).
38. — Prevention of anatomic defects in congenital adrenal hyperplasia produced by estradiol-17β in rats. J. clin. Endocr. **29**, 30 (1969).
39. — Congenital effectiveness of an inhibitor of 3β-hydroxysteroid dehydrogenase administered before implantation of the rat blastula. Endocrinology **84**, 1206 (1969).
40. — Maternal and fetal effects of two inhibitors of 3β-hydroxysteroid dehydrogenase and Δ^5-3-ketosteroid isomerase. Endocrinology **85**, 325 (1969).
41. — Effect of ACTH on enzymatic defect produced by a competitive inhibitor of 3β-hydroxysteroid dehydrogenase. Endocrinology **86**, 583 (1970).
42. — Specific retention of an inhibitor of 3β-hydroxysteroid dehydrogenase in enzyme-containing tissues of the rat. Endocrinology **86**, 678 (1970).
43. — The production of congenital lipoid adrenal hyperplasia in rats by an inhibitor of cholesterol side-chain cleavage enzymes. Endocrinology **86**, 1245 (1970).
44. — Virilization of the external genitalia of the female rat fetus by dehydroepiandrosterone. Endocrinology **87** (1970) (in press).
45. — Production of hypospadias by inhibition of fetal testicular 17α-steroid hydroxylase in utero. Endocrinology (in press).
46. — The pseudohermaphrodite rat: Failure to masculinize in utero due to a defect in 17β-hydroxysteroid dehydrogenase. 52nd Meeting Endocrine Socity. St. Louis (1970).
47. — Experimental congenital lipoid adrenal hyperplasia: Prevention of anatomic defects produced by aminoglutethimide. Endocrinology **87** (1970) (in press).
48. — BONGIOVANNI, A. M.: Induced genital anomalies. Ann. N. Y. Acad. Sci. **142**, 755—767 (1967).

49. Goldman, A. S., Bongiovanni, A. M., Yakovac, W. C., Prader, A.: Study of Δ^5, 3β-hydroxysteroid dehydrogenase in normal, hyperplastic and neoplastic adrenal cortical tissue. J. clin. Endocr. **24**, 894 (1964).
50. — Kenneck, C. Z.: Persistence of label in rat offspring after maternal administration of a ^{14}C-labelled inhibitor of 3β-hydroxysteroid dehydrogenase. Endocrinology **86**, 711 (1970).
51. — Kohn, G.: Normal development in tissue culture of rat ovarian 3β-hydroxysteroid and glucose-6-phosphate dehydrogenases. Proc. Soc. exp. Biol. (N. Y.) **133**, 478 (1970).
52. — Neumann, F.: Differentiation of the mammary gland in experimental congenital adrenal hyperplasia due to inhibition of Δ^5, 3β-hydroxysteroid dehydrogenase in rats. Proc. Soc. exp. Biol. **132**, 237 (1969).
53. — Winter, J. S. D.: The activities of various steroidogenic enzymes in two forms of experimental adrenal hyperplasia. J. clin. Endocr. **27**, 1717 (1967).
54. — Yakovac, W. C., Bongiovanni, A. M.: Persistent effects of a synthetic androstene derivative on activities of 3β-hydroxysteroid dehydrogenase and glucose-6-phosphate dehydrogenase in rats. Endocrinology **77**, 1105 (1965).
55. — — — Production of congenital adrenal hyperplasia, hypospadias, and clitoral hypertrophy (Adrenogenital Syndrome) in rats by inactivation of 3β-hydroxysteroid dehydrogenase. Proc. Soc. exp. Biol. (N. Y.) **121**, 757 (1966).
56. — — — Development of activity of 3β-hydroxysteroid dehydrogenase in human fetal tissues and in two anencephalic newborns. J. clin. Endocr. **26**, 14 (1966).
57. — — Experimental congenital adrenal cortical hyperplasia: Prevention of adrenal hyperplasia and clitoral hypertrophy by corticosterone. Proc. Soc. exp. Biol. (N. Y.) **122**, 1214 (1966).
58. Harding, H. R., Potts, G. O.: The blocking effect of cyanotrimethylandrostanolone on the induction of liver tryptophan pyrrolase (TPO) activity by ACTH. Fed. Proc. **23**, 356 (1964).
59. Jost, A.: Embryonic sexual differentiation (morphology, physiology, abnormalities). In: Hermaphroditism, Genital Anomalies, and Related Disorders: Jones, H. W., Scott, W. W., eds.), p. 15. Baltimore: Williams & Wilkins Comp. 1958.
60. Kahnt, F. W., Neher, R.: Über die adrenale Steroidbiosynthese in vitro. II. Bedeutung von Steroiden als Hemmstoffe. Helv. Chim. Acta **49**, 123—133 (1966).
61. Kahri, A.: Histochemical and electron microscopic studies on the cells of the rat adrenal cortex in tissue culture. Acta Endocr. **52**, 108 (Suppl.) (1966).
62. Kowal, J.: A possible system for studying ACTH during complete steroidogenic blockade. Clin. Res. **15**, 4 (1967).
63. — Forchielli, E., Dorfman, R. I.: The Δ^5-3β-hydroxysteroid dehydrogenase of bovine corpus luteum and adrenal. I. Properties, substrate specificity and cofactor requirements. Steroids **3**, 531—549 (1964).

64. LIMBECK, G. A., RUVALCABA, R. H. A., KELLEY, V. C.: Simulated congenital adrenal hyperplasia in a male neonate associated with medroxyprogesterone therapy during pregnancy. Amer. J. Obstet. Gynec. **103**, 1169 (1969).
65. MAUVAIS-JARVIS, P., BERCOVICI, J. P., GAUTHIER, F.: In vivo studies on testosterone metabolism by skin of normal males and patients with the syndrome of testicular feminization. J. clin. Endocr. **29**, 417 (1969).
66. MCCARTHY, J. L., RIETZ, C. W., WESSON, L. K.: Inhibition of adrenal corticosteroidogenesis in the rat by cyanotrimethylandrostenolone, a synthetic androstane. Endocrinology **79**, 1123—1129 (1966).
67. NEUMANN, F., GOLDMAN, A. S.: Prevention of mammary gland defects in experimental congenital adrenal hyperplasia due to inhibition of 3β-hydroxysteroid dehydrogenase. Endocrinology **86**, 1169 (1970).
68. — KRAMER, M.: Antagonism of androgenic and anti-androgenic agents in their action on the rat fetus. Endocrinology **75**, 423—433 (1964).
69. — VON BERSWORDT-WALLRABE, W., ELGER, W., STEINBECK, H., HALM, J. D., KRAMER, M.: Aspects of androgen-dependent events as studied by anti-androgens. Recent Progr. Hormone Res. **25**, (1969) (in press).
70. NEVILLE, A. M., ENGEL, L. L.: Inhibition of 3α- and 3β-hydroxysteroid dehydrogenases and of steroid Δ-isomerase by anabolic steroids. J. clin. Endocr. **28**, 49 (1968).
71. — — Inhibition of steroid Δ-isomerase of the bovine adrenal gland by substrate analogues. Endocrinology **83**, 873 (1968).
72. NORTHCUTT, R. C., ISLAND, D. P., LIDDLE, G. W.: An explanation for the target organ unresponsiveness to testosterone in the testicular feminization syndrome. J. clin. Endocr. **29**, 422 (1969).
73. ORTIZ, E., ZAAIJER, J. J. P., PRICE, D.: Organ culture studies of hormone secretion in endocrine glands of fetal guinea pigs. IV. Androgens from fetal adrenals and ovaries and their influence on sex differentiation. K. Ned. Akad. Wetenschap. Ser. C **70**, 4 (1967).
74. PICON, R.: Activite Δ^5-3β-hydroxysteroide deshydrogenasique du testicule foetal de Rat in vitro. Arch. Anat. micr. Morph. exp. **56**, 3—4, 281 (1967).
75. POTTS, G. O., BURNHAM, D. F., BEYLER, A. L.: Inhibitory action of two new steroids on the adrenal cortex of rats. Fed. Proc. **22**, 166, Abstr. 35 (1963).
76. PRADER, A., GURTNER, H. P.: Das Syndrom des Pseudohermaphroditismus masculinus bei kingenitaler Nebennierenrinden-Hyperplasie ohne Androgenüberproduktion (adrenaler Pseudohermaphroditismus masculinus). Helv. paediat. Acta **10**, Fasc. **4**, 397 (1955).
77. PRICE, D., ORTIZ, E., ZAAIJER, J. J. P.: Organ culture studies of hormone secretion in endocrine glands of fetal guinea pigs. III. The relation of testicular hormone to sex differentiation of the reproductive ducts. Anat. Rec. **151**, 27—42 (1967).
78. RUBIN, B. L., HAMILTON, J. A., KARLSON, T. J., TUFARO, R. I.: Effect of the age of the rat at the time of hemicastration on the weight, estimated secretory activity and Δ^5-3β-hydroxysteroid dehydrogenase activity of the remaining ovary. Endocrinology **77**, 909 (1965).

79. SAUNDERS, F. J.: Some aspects of relation of structure of steroids to their prostate-stimulating effects. Biology of the Prostate and Related Tissues. Nat. Cancer Inst. Monogr. **12**, 139 (1963).
80. SCHIMKE, R. T.: Studies on the roles of synthesis and degradation in the control of enzyme levels in animal tissues. Bull. Soc. Chim. biol. (Paris) **48**, 1009—1030 (1966).
81. SCHLEGEL, R. J., FARIAS, E., RUSSO, N. C., MOORE, J. R., GARDNER, L. I.: Structural changes in the fetal gonads and gonaducts during maturation of an enzyme, steroid 3β-ol-dehydrogenase, in the gonads, adrenal cortex and placenta of fetal rats. Endocrinology **81**, 565 (1967).
82. STANLEY, A. J., GUMBRECK, L. G.: Male pseudohermaphroditism with feminizing testis in the male rat — a sex-linked recessive character. 46th Meeting, Endocrine Soc., San.Fran. (1964).
83. — — EASLEY, R. B.: F.S.H. content of male Pseudohermaphrodite rat pituitary glands together with the effects of androgen administration and castration in gland and organ weights. 48th Meeting, Endocr. Soc., Chicago (1966).
84. STEIKER, D. D., BONGIOVANNI, A. M., EBERLEIN, W. R., LEBOEUF, G.: Adrenocortical and adrenocorticotropic function in children. J. Pediat. **59**, 884—889 (1961).
85. STEINBERGER, A., STEINBERGER, E.: In vitro culture of rat testicular cells. Exp. Cell Res. **44**, 443—452 (1966).
86. STERN, J. M., EISENFELD, A. J.: Androgen accumulation and finding to macromolecules in seminal vesicles: Inhibition by cyproterone. Science **166**, 233 (1969).
87. STRICKLAND, A. L., FRENCH, F. S.: Absence of response to dihydrotestosterone in the syndrome of testicular feminization. J. clin. Endocr. **29**, 1284 (1969).
88. WEBB, J. L.: The kinetics of enzyme inhibition. Enzyme and Metabolic Inhibitors, Vol. I, Ch. 3, p. 49. New York: Academic Press 1963.
89. WERKHEISER, W. C.: Specific binding of 4-amino folic acid analogues by folic acid reductase. J. biol. Chem. **236**, 888 (1961).
90. WILSON, J. D., WALKER, J. D.: The conversion of testosterone to 5α-androstan-17β-ol-3-one (dehydrotestosterone) be skin slices of man. J. clin. Invest. **48**, 371 (1969).
91. WILSON, J. G.: Embryological considerations in teratology. In: Teratology, Principles and Techniques, (J. G. WILSON and J. WARKANY, eds.), Ch. 10, p. 25. Chicago: Univ. of Chicago Press 1965.

Hormonal Effects on Socio-Sexual Behavior in Dogs

FRANK A. BEACH

University of California, Dept. of Psychology, Berkeley, CA (USA)

With 15 Figures

Introduction

Because mine is the only paper in this very stimulating symposium to deal with the topic of behavior I wish it were possible to present a comprehensive review of the current state of knowledge with respect to hormonal control of behavioral phenomena in general. Since the very extensive literature on this subject precludes so ambitious an undertaking I have chosen a much more modest objective which is to describe in some detail a few experiments which my associates and I have conducted to investigate hormonal factors affecting socio-sexual behavior in dogs.

Several different investigations will be used as a basis for illustrating general theoretical and methodological problems encountered in all studies of hormone-behavior relationships. Furthermore, results we have obtained with dogs will be compared with data provided by other experiments on different species of mammals to provide a framework for a comparative approach to the central issues involved.

Effects of Testosterone on Copulatory Behavior in Male Dogs

The general design of the first experiment was deceptively simple. It involved comparing sexual responses shown by male dogs before and after castration, and then determining whether any postoperative changes which were observed could be reversed by administering exogenous androgen to the castrated males (BEACH, 1970a). Although the procedure is easily stated its execution was complicated by methodological problems.

Methodological Problems

An essential preliminary step was to become thoroughly familiar with the normal copulatory pattern of the domestic dog. We were quite surprised to discover that this had not been quantitatively described in the scientific literature. Before any experimental manipulation of endocrine factors could be undertaken we were compelled to devise an objective and quantifiable method of describing and scoring the behavior which we hoped to modify by experimental treatment. It was also essential that the standardised system of scoring and recording behavior be thoroughly learned and understood by all participants in the experiment so that inter-observer consistency would be high.

Once the details of classifying and measuring component responses making up the mating pattern had been determined, it was necessary that our experimental subjects be observed in a series of preoperative mating tests. No two male dogs have identical mating patterns and preliminary standardization tests are essential for determining the extent and nature of individual differences in a normal population. Furthermore the same male's sexual performance shows some variability from one test to the next and it is necessary to ascertain the magnitude of such temporal variation before carrying out any experimental manipulations such as changing the hormonal status of the subjects.

In addition to the foregoing requirements it was of course necessary to include a group of control males which were subjected to the same behavioral tests as the experimental dogs but were not castrated or treated with hormones.

Procedure

Observations of a large number of completed matings revealed that the male dog's pattern can be adequately described and measured in terms of 8 discrete responses to the female. Records made during every preoperative and postoperative test included the frequency and time of occurrence of each of these 8 basic responses. They will be described in the presentation of experimental results.

All males were given several series of preoperative observations, each series consisting of 5—10 mating tests distributed over a period of several weeks. After their normal or preoperative responsiveness and potency had been measured, members of the experimental group were castrated, and testing continued at intervals of several months for from 3 to $5^1/_2$ years after operation. Two of the operates were adrenalectomised 3 years and 8 months after castration, and then retested for mating performance. Others were treated with testosterone propionate at different intervals after castration.

Results

The results of this experiment came as a surprise to me. One of the quantitative measures of sexual responsiveness was the speed with which a male began to copulate when a receptive female was made available. This measure is called the "mount latency". It consists of the number of seconds elapsing between the female's entrance into the test arena and the male's first copulatory mount.

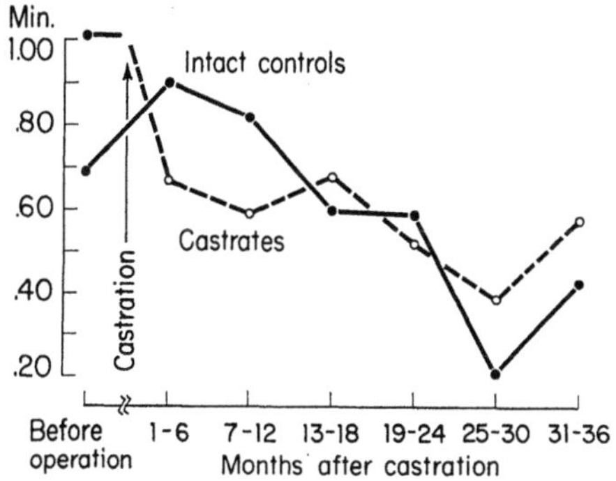

Fig. 1. Average mount latency

Figure 1 shows that the average mount latency declined in the course of this experiment. I do not know why mount latencies became shorter, but we can see in this graph the value of having included a control group of unoperated males. The most important result reflected in these curves is that for at least 3 years after operation castrated males showed average latencies which were as short as those of the normal control dogs.

When they copulate, male dogs mount the female from the rear, clasp the forelegs around her body just anterior to her pelvis, and then execute thrusting movements of their hindquarters which serve to direct the penis toward the vulva. The mount-and-thrust response often is followed by dismounting, and usually it is repeated several times before the male succeeds in inserting his penis into the vagina. The rate at which successive mounting responses occur is one measure of the intensity of the male's sexual excitement. Figure 2 shows that this measure did not deteriorate over a period of 3 years after operation. Results presented in the first 2 figures seem to show that loss of testicular hormone has no effect on the

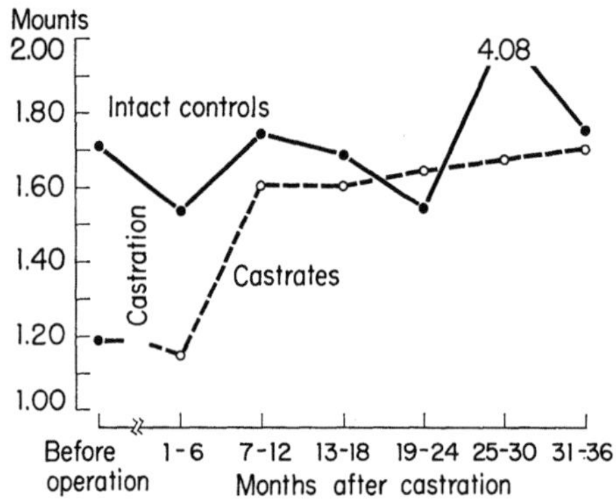

Fig. 2. Average number of mounts/minute

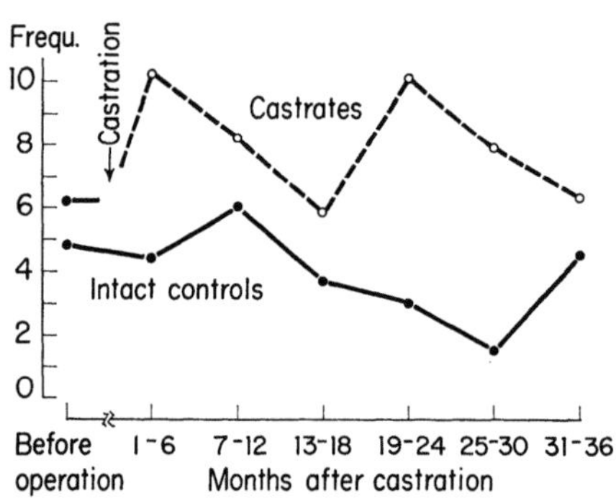

Fig. 3. Average number of mounts preceding insertion

male dog's copulatory performance, and it can be added at this point that the behavior of castrates was not changed by subsequent adrenalectomy. However, evidence that castration does affect sexual ability is provided by other behavioral measures.

The male dog usually does not achieve insertion the first time he mounts the female. Ordinarily several mounts with thrusting precede the occurrence of intromission. Figure 3 shows that castration was followed by an increase in the average number of mount-and-thrust responses that were executed before the achievement of insertion. This increase might reflect interference with coordination of the male's behavior, but I suspect it indicates difficulty in achieving the partial state of erection necessary for insertion. In any event castration clearly has a deleterious effect.

Fig. 4. Male and female in coital lock

As soon as intromission occurs the male begins to thrust much more rapidly and continues to do so for approximately 15—25 seconds. During the period of rapid thrusting the penis becomes fully erect. When the copulatory organ is completely erect there is marked enlargement of the bulbus glandis, a specialized region of the penis which swells inside of the vagina and prevents the male from withdrawing until detumescence begins. As soon as the erection is complete males usually dismount, lifting one foreleg over the female's back and turning around so that the penis is reflected backward between the male's rear legs and is retained within the vagina. At this point the pair is said to be "locked." The position of a locked pair is illustrated in Fig. 4.

Average duration of locking in our normal males was about 15 minutes and this measure was clearly affected by castration. Castrated males rarely lost the ability to lock with females, but their locks tended to be relatively short. This is illustrated in Fig. 5. It is clear that erection can be achieved for at least 3 years after castration, but it is not maintained for normal periods of time. This effect of castration can be reversed by administration of testosterone propionate. Replacement therapy was effective even when it was delayed until several years after operation. This effect is demonstrat-

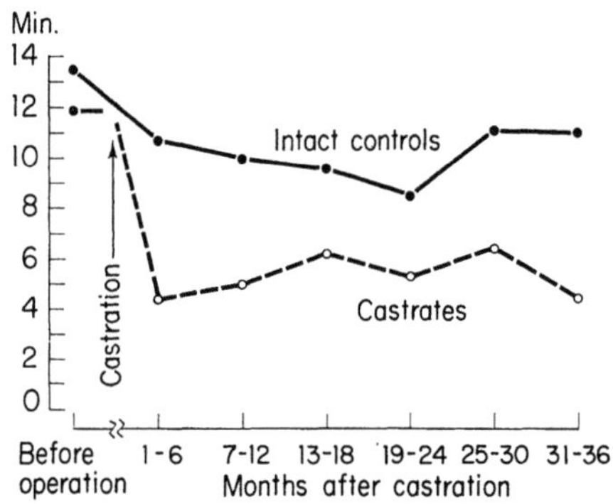

Fig. 5. Average duration of lock

ed in figure 6 which shows the duration of every lock achieved by one male over a period of 5 years and 9 months after castration and for purposes of comparison includes the record of one unoperated male.

Gonadectomy had still other effects upon the male dogs copulatory performance but our results can be summarized by the statement that castration did not appear to reduce the male's excitability or responsiveness to the receptive female, but did produce a definite, though partial decline in potency.

Comparison with Other Species

The discovery that male dogs continue to copulate for years after castration was quite unexpected. Many experiments on rats, guinea pigs, mice, hamsters and rabbits had shown that castration is followed within a few weeks or at the longest within a few months by virtually complete loss of mating reactions. When castrated males of these species are injected with testosterone propionate this

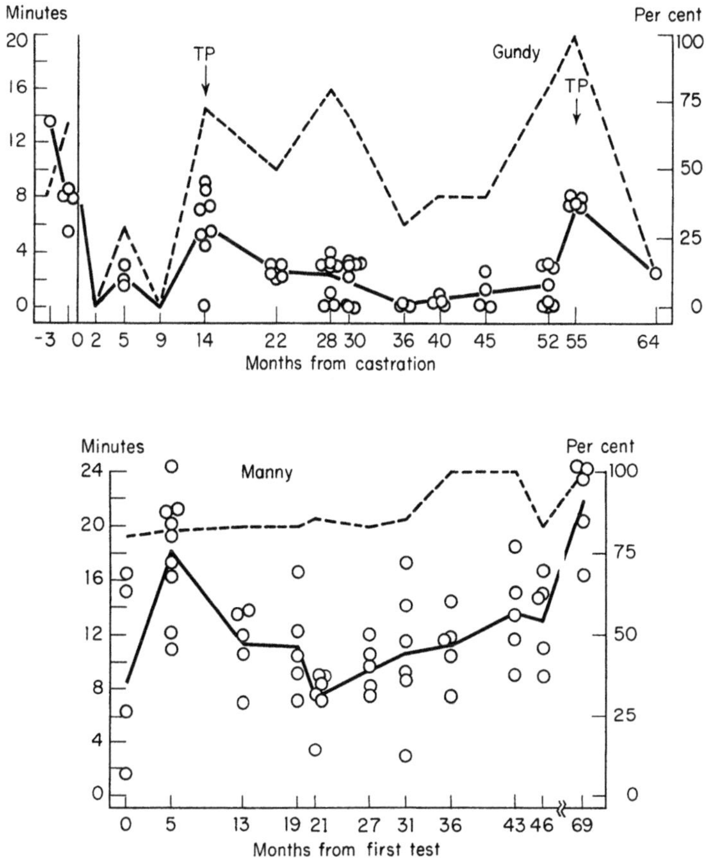

Fig. 6. Duration of each lock by a castrate (Gundy) and an intact male (Manny) over a period of 5+ years. Each circle represents one lock. Duration indicated on left abscissa and average for each series indicated by solid line. Dash line shows percent of tests in which at least one lock occurred

treatment restores the full copulatory pattern which continues to appear as long as therapy is maintained and then disappears when treatment is discontinued. This effect is reflected in Fig. 7 which reveals the ejaculatory performance of 3 groups of castrated male rats under the influence of different amounts of exogenous androgen (Beach and Holz-Tucker, 1949).

Results of one experiment dealing with copulation in male cats before and after castration indicated that sexual performance in experienced males of this species lasts much longer after removal of the gonads than is the case in rats, mice and guinea pigs (ROSENBLATT and ARONSON, 1958). Taken together with our results on

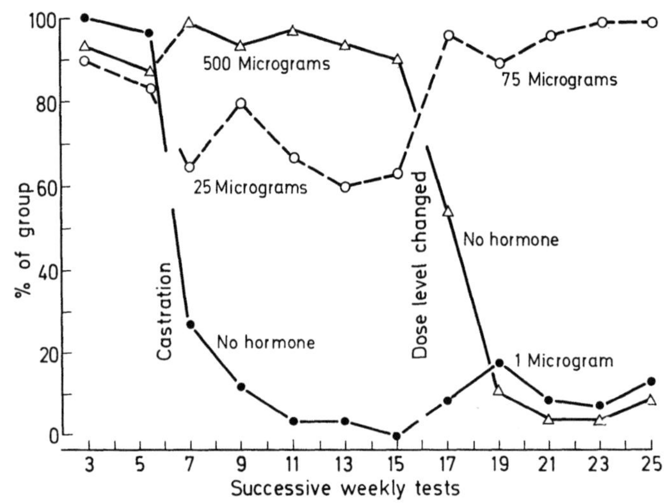

Fig. 7. Ejaculatory performance of castrated male rats receiving different amounts of testosterone propionate (BEACH and HOLZ-TUCKER, 1949)

dogs this suggests that male carnivores may be less dependent upon testosterone than are male rodents.

The next obvious question is whether male primates are even more emancipated from hormonal control over sexual performance. Unfortunately the literature contains no satisfactory account of the effects of castration in adulthood upon mating behavior in male monkeys or apes. There is one very important study of castration in infancy. This is being conducted at the Regional Primate Center in Portland, Oregon. A group of male monkeys castrated at birth has been carefully observed through the infantile and juvenile periods. These animals are not yet mature, but up to this time their pre-adult sexual behavior has been entirely normal and indistinguishable from that shown by intact males of the same age (GOY, 1968).

We shall have to wait until experiments are carried out to measure copulatory activity before and after castration in adulthood, but it may be that male monkeys will be found to be even less dependent upon testicular secretions than are male dogs. If this proves to be the case, it would be reasonable to speculate that men whose testes are removed after adulthood has been reached would not suffer any pronounced reduction in sexual responsiveness or capacity.

Effects of Ovarian Hormones on Behavior of Female Dogs

The effects of ovarian hormones on socio-sexual behavior in female dogs can be summarized as follows. When they are in estrus females tend to seek out the company of males and to remain near them (LeBoeuf, 1967). Most females in heat solicit the attention of males and behave in such a way as to promote the occurrence of social interaction. Under the influence of estrogen and progesterone females display a readiness to engage in coitus with males — i.e. they become sexually receptive (Beach and Merari, 1968). However, even when the effect of ovarian hormones upon behavior is at its height, most females are not indiscriminately receptive, but tend to accept some males much more readily and enthusiastically than others (Beach and LeBoeuf, 1967).

In all of our experiments on the behavior of females we have attempted to fulfill five criteria some of which were mentioned in describing the experiment on males. They are sufficiently important to deserve restatement at this point. (1) Behavior should be described objectively and in quantitative terms. (2) The population tested should be large enough to provide an adequate measure of individual differences. (3) The number and duration of tests should be sufficient to permit statistical estimates of the reliability or consistency of the behavioral measures. (4) The stimulus situation should be controlled so far as possible by holding environmental conditions constant and by providing the experimental subject with a stimulus animal displaying the desired behavioral and physiological characteristics. (5) It is essential that no experimental manipulations be introduced until the normal behavior pattern is well known and can be described quantitatively. None of our experiments has ever completely fulfilled all of these ideal requirements and none is likely to do so in the future. Nevertheless they should always be considered and realised in so far as practical limitations allow.

Social Attraction to Other Dogs

One simple but effective way to measure a female dog's tendency to approach males and to engage in social interaction with them is to arrange the test situation so that the female is free to control the frequency and duration of social contacts. The male should be able to react to the female if and when she seeks him out, but he should be prevented from pursuing her if she desires to avoid him. We have achieved these conditions by chaining the male to a stake in the center of an open field. The male can move about, but only within the circumference of a circle 2 meters in diamter. One female at a

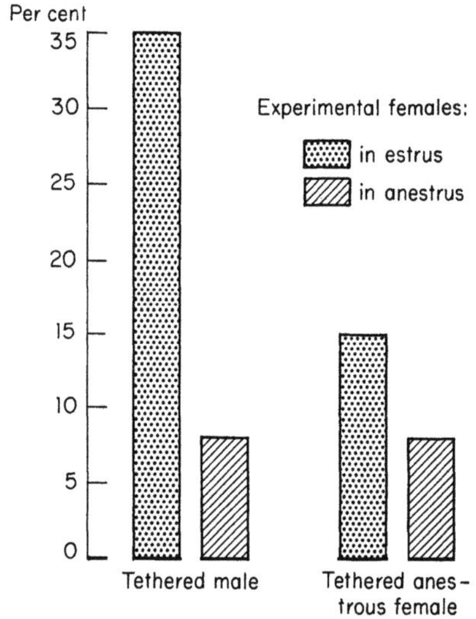

Fig. 8. Percent of test time females spent visiting males and anestrous females (based on LeBoeuf, 1967)

time is released in the open area for 5 minutes and records are made showing the frequency with which she visits the tethered male. Each entrance into the male's circle is scored as a "visit," and the time spent inside the circle is recorded as well as the frequency of each type of response shown by each animal toward the other.

In one experiment using this method the same females were tested when they were in estrus and again when they were in anestrus (LEBOEUF, 1967). Some of the results are illustrated in Fig. 8. There was approximately 4 times as much visiting when females were in estrus. It was of course necessary to discover whether ovarian hormones had any effect upon the tendency to visit females as well as males. Figure 8 demonstrates two relevant

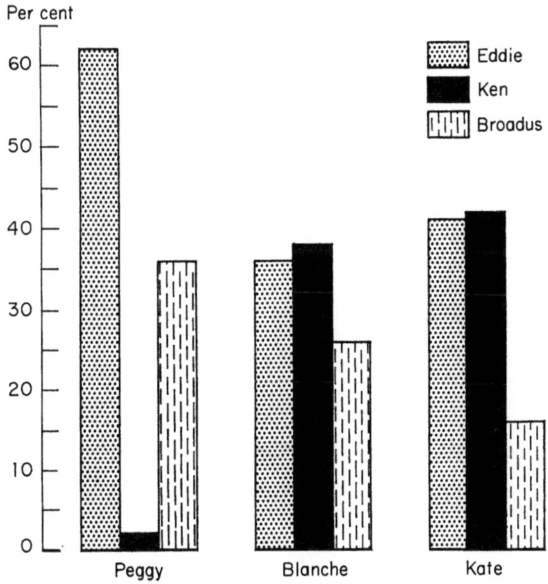

Fig. 9. Percent of total visiting time devoted by each female to three different males

findings. (1) During estrus, females showed some increase in the tendency to visit tethered females. (2) The increase in visits to males was much more pronounced than the increase in visits to other females. In other words, when they are in heat females are more sociable but they prefer to visit males rather than females.

The visiting time of estrous females to males was not divided equally among all possible hosts. Most females showed distinct preferences, visiting some males for fairly long periods of time and visiting others very briefly or not at all. This is illustrated in Fig. 9.

Three males were tethered at different locations in a field and females could choose between them. Some males were clearly more attractive than others, and the pattern of preference was different for each female.

In another experiment we examined the visiting behavior shown by females during successive stages of induced estrus (BEACH and MERARI, 1970). Our subjects received 3 injections of estrogen

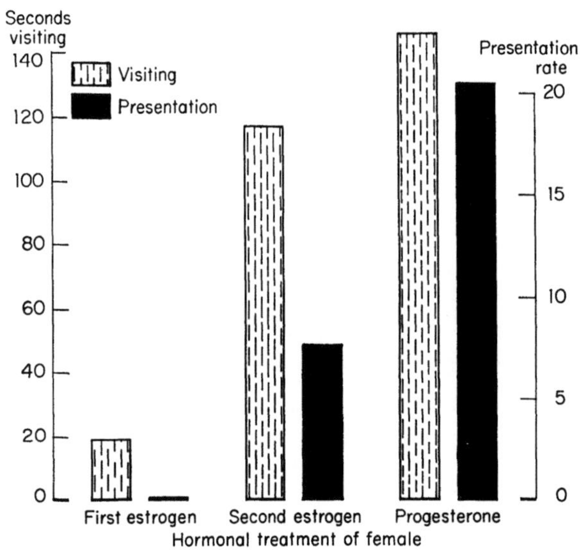

Fig. 10. Visiting and sexual presentation shown to tethered males by females during successive stages of induced estrus (BEACH and MERARI, 1970)

followed by one injection of progesterone. The results are illustrated in Fig. 10 which demonstrates a progressive increase in visiting time.

This same figure reveals another effect of the hormone treatment. It not only intensified the female's tendency to visit tethered males, but also produced a new response which constitutes one form of sexual soliciation. This is the sexual presentation. When she presents, the bitch places herself in front of the male and deviates her tail laterally, exposing her vulva. Frequency of presentation increased under the effects of estrogen and increased even further when progesterone was added.

Behavior in Mating Tests

In the standardised mating tests used at our field station one male and one female are placed together in an enclosure measuring 25 × 25 feet and surrounded by a solid board fence. The male is left in the arena for 5—10 minutes and then the female is released, timers are started and recording begins.

Each type of response shown by the male toward the female is given a separate symbol and is recorded every time that it occurs (Beach and LeBoeuf, 1967). In reaction to each of the male's acts the female can make any one of several distinct responses. If she is strongly negative to the male she may attack him. Each time this occurs she receives a score of minus 3. A less intense form of rejection termed "threat" consists of snarling, snapping or baring her teeth at the male. Each time she threatens the male the female is given a score of minus 2. When the female simply avoids the male or evades his mating attempts she is given a score of minus 1. A female that is willing to copulate can exhibit a variety of positive responses. For each of these she receives a score of plus 1 to plus 3 depending upon various criteria.

The practice of scoring every act shown by the male toward the female and her responses to them is especially useful for the quantification of sexual preferences. The system is applied to a series of tests conducted throughout the female's estrous period. Then scores on all tests with each male are combined in the form of an Interaction Matrix. Two such matrices are illustrated in Fig. 11. Such a matrix allows us to calculate a ratio between the male's activity and the female's responses, and this ratio or coefficient constitutes a quantitative expression of the degree to which she rejects or accepts the male.

Individual differences in patterns of female preference for the same males as sexual partners are reflected in Table 1 which presents average Rejection Coefficients for 5 females tested with 5 males in 2 estrous periods (Beach and LeBoeuf, 1967). It is clear that the sexual popularity of males varied with different females, and that although some males were more acceptable than others to all females, no individual male was strongly rejected by every female.

Male *Broadus* Dates *Jan. 5-16, 1966* Locks *4*
Female *Peggy* Tests *4* Intros. *9*

♂ \ ♀	-3	-2	-1	0	1	2	3	4	Σ
1			/						1
2					/				1
3			//		卌 卌 // 卌				19
4			////	卌	////	卌 //			20
5			/	//	//	///	卌 ///		16
6			/			///	卌 卌 卌		19
7								卌 ////	9
Σ			9	7	24	13	23	9	85

Per Cent Lock *100*

Rejection Coefficient *0*

Male *Ken* Dates *Jan. 6-16, 1966* Locks *2*
Female *Peggy* Tests *9* Intros. *2*

♂ \ ♀	-3	-2	-1	0	1	2	3	4	Σ
1	///	////							7
2	/	卌 ///							9
3	卌 卌	卌 //	/	//	//				22
4	///	卌 ////	//	/		/	/		17
5		//	/			/	/		5
6						卌	卌 卌		15
7								//	2
Σ	17	30	4	3	2	7	12	2	77

Per Cent Lock *22*

Rejection Coefficient *82*

Fig. 11. Representative interaction matrices for one female with two males. Peggy accepts Broadus and rejects Ken (Beach and Merari, 1967)

I have been strongly impressed with the consistency of such preferences from one heat period to the next. In our colony there are some pairs which have been tested repeatedly since 1964, and females which rejected certain males 6 years ago still reject them in 1970. Males that were accepted then remain acceptable today.

Table 1. *Average rejection coefficients shown by 5 females toward 5 males in 2 successive estrous periods* (BEACH *and* LEBOEUF, *1967*)

	Female					
Male	Peggy	Spot	Blanche	Kate	Dewey	Mean
Broadus	0	4	6	0	0	2
Eddie	0	12	58	12	35	23
John	10	10	15	2	34	14
Clark	33	61	55	5	46	40
Ken	81	33	59	30	0	41
Mean	25	24	39	10	23	

Comparison with Other Species

The full significance of the evidence concerning sexual behavior and ovarian hormones in female dogs is not apparent until it is compared with knowledge of similar phenomena in other mammalian species. Many investigations have been conducted to explore this relationship in small mammals such as the rat, guinea pig, mouse, hamster, chinchilla and others. The results are practically unanimous in showing that estrogen and progesterone are responsible for the display of sexually receptive behavior, and that when these hormones are present in appropriate combination they are effective in producing receptivity in almost all females.

Females of various infrahuman primate species appear to be less dominated by hormones secreted by their ovaries. There is no doubt that for most female monkeys and apes estrogen increases the tendency to solicit and accept the sexual attentions of a male. Furthermore, when she is in such a condition the female is a more exciting sexual partner as far as males are concerned. Nevertheless individual differences are marked. Some heterosexual pairs of rhesus monkeys show no interest in each other and do not engage in copulatory behavior no matter what the female's hormonal condition. When such a male and female are given different partners both of them may promptly begin to show mating responses (MICHAEL, 1968). The general picture for infrahuman primates seems to be one in which individual differences are very great, and although an influence of ovarian hormones undoubtedly exist, it often is blurred or attenuated by nonhormonal factors.

Comparing the female dog with female rodents and monkeys we see that she belongs in an intermediate position. A bitch is never sexually receptive except when in physiological estrus. In this respect she resembles female rodents. On the other hand, many female dogs in full estrus are nonreceptive to some males and entirely receptive to others. Seen in this light the female dog is somewhat like the female primate because nonhormonal factors apparently can override or at least strongly modify the primary behavioral effects of ovarian hormones.

Clinical findings with respect to human females clearly suggest that ovarian secretions are even less influential in *Homo sapiens* than in lower primates. It cannot be concluded that they are without effect, but it is fairly certain that in comparison with other species a woman's gonadal hormones exert relatively little control over sexual desire and capacity for response to sexual stimulation.

Developmental Effects of Testosterone

Gonadal hormones have more than one type of effect upon behavior. Experiments described thus far dealt with temporary, reversible effects, — effects present when the hormone is present and absent when it is removed. Such effects have been termed "activational" (YOUNG, 1961) but I prefer the more neutral, descriptive designation "concurrent". A quite different class of hormonal effects may be identified as "developmental." Developmental effects are exerted during restricted periods in ontogeny and ordinarily cannot be produced at any earlier or later time. In contrast to concurrent effects, developmental effects are permanent in the sense that once established they do not disappear even if the hormone responsible for their occurrence is eliminated and is never replaced.

A number of experiments on rodents have shown that the kinds of sexually dimorphic behavior which an animal will show in adulthood are influenced by the presence or absence of testosterone before and/or immediately after birth. Female guinea pigs whose mothers are injected with TP during the second trimester of pregnancy are born with male type genitals. When they become adult these masculinized females do not display normal female mating behavior after injections of estrogen and progesterone. At the same time they do exhibit masculine sexual tendencies (PHOENIX, GOY, GERALL, and YOUNG, 1959). Similar results can be produced in female rats (GRADY, PHOENIX, and YOUNG, 1965; GERALL and WARD, 1966).

Under normal circumstances male mammals are stimulated by testosterone secreted by the fetal testis and in some species such as the rat this stimulation continues for a few days postpartum before the testis becomes relatively inactive. Male rats deprived of the effects of endogenous testosterone prenatally by treatment of the mother an antiandrogen (cyproterone acetate) show feminine behavior in adulthood when injected with estrogen and progesterone (NEUMANN and ELGER, 1965). Males castrated on the day of birth exhibit similar behavior (WHALEN, 1968).

It has been theorized (YOUNG, 1961) that testosterone exerts an "organizing" action upon the developing nervous system, tending to suppress the development of neural mechanisms capable of mediating feminine mating responses and to stimulate development of mechanisms for male behavior. Without questioning the validity of the behavioral observations, I have elsewhere presented reasons for questioning the appropriateness of the organizational concept (BEACH, 1970b). At the present state of knowledge it seems advisable to adopt a conservative point of view and employ less interpretative terms such as "developmental effects".

We have been investigating the behavior of female dogs exposed to testosterone stimulation both before birth and immediately thereafter. Although the investigation is far from completion results obtained to date are most illuminating.

Method

We have injected testosterone propionate into pregnant females at different stages of gestation and then studied the behavior of their female offspring. In other cases we have given androgen to the mother during pregnancy and continued treatment of the female offspring by implanting pellets of crystalline testosterne beneath the skin on the day of birth.

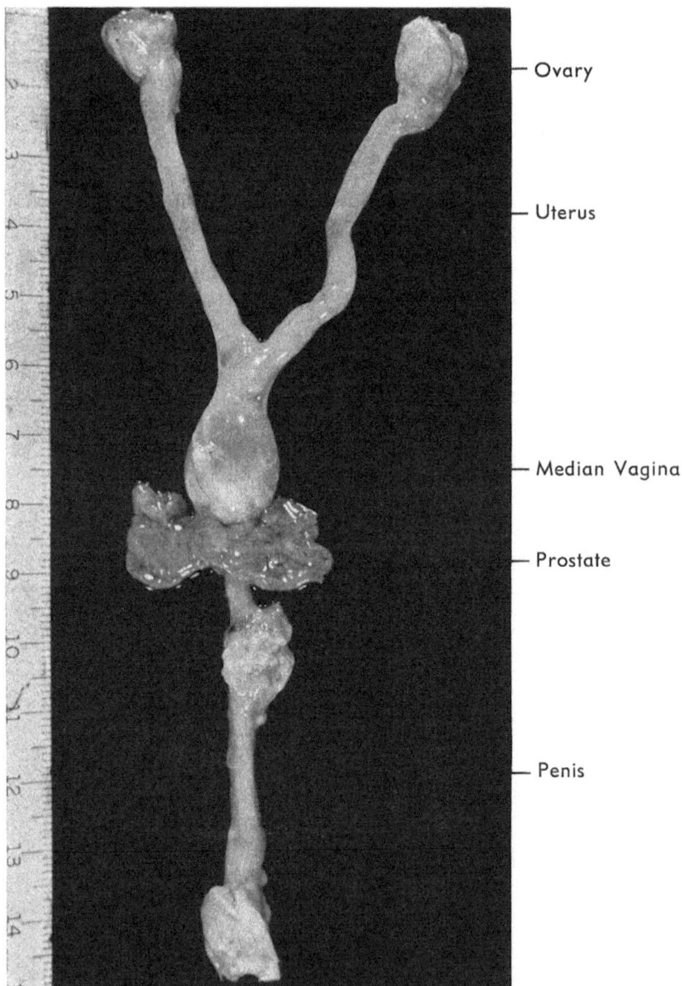

Fig. 12. Entire reproductive tract of infant female whose mother was injected with TP on Days 28—37 of pregnancy

Reproductive Anatomy

Evidence of the effectiveness of the experimental treatment is provided by modification of the anatomy of the reproductive tract. Fig. 12 is a photograph of the entire tract of a 10-day old female whose mother was given injections of TP from the 28th through the 37th day pregnancy. Fig. 13 shows the penis of a normal adult male, the vulva of a normal female during anestrus and the penis of a genetic female exposed to androgenic stimulation in fetal life and given a testosterone implant at the time of birth. Fig. 14 shows

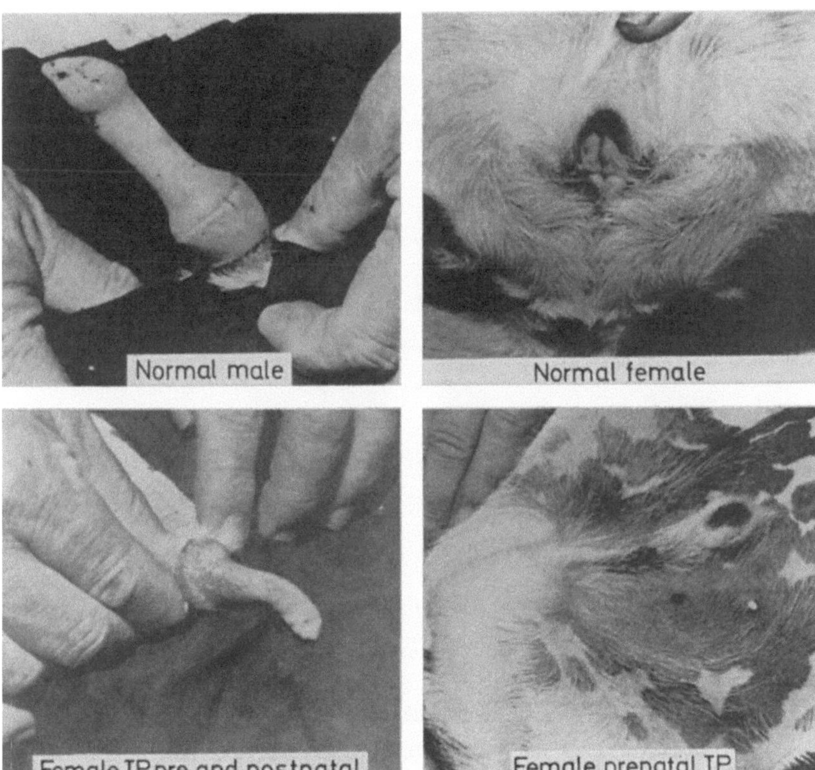

Fig. 13. External genitalia of a normal male, a normal female, a female treated with testosterone in utero and a female exposed to androgenic stimulation pre- and postnatally

the penis of the same female in a fully erect condition induced by massage. This animal showed ejaculatory reflexes under the influence of masturbation.

We have found marked differences in the genital anatomy of females which received the same hormonal treatment during development, but it is obvious that we have successfully masculinized female dogs at least as far as the external genitals are concerned.

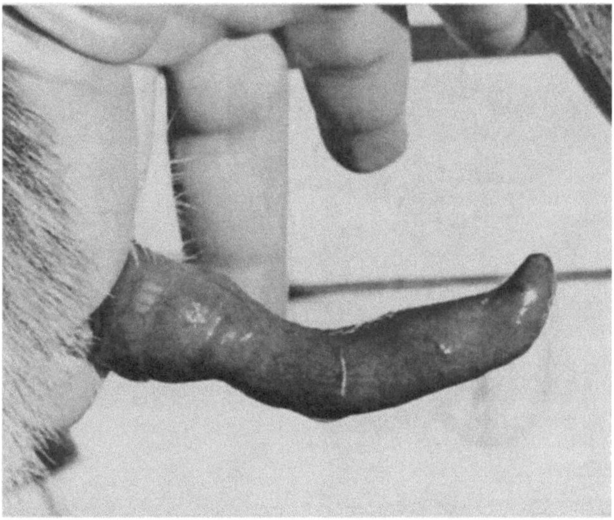

Fig. 14. Tumescent penis of female treated with androgen pre- and postnatally

Female Sexual Behavior

To examine the capacity of masculinized females for the display of feminine mating behavior we conducted standard mating tests similar to those already described. Each animal was injected with the same combination of estrogen and progesterone that we use to induce sexual receptivity in normal females. Members of 2 control groups were injected with the same hormones in the same amounts. One control group consisted of spayed females which had previously served in mating tests and had shown good receptivity. The other control group consisted of male dogs which had been castrated as adults approximately 8 months before the tests began. All control and experimental animals were tested with highly experienced, unoperated males previously used as breeders.

One series of tests was conducted before the subjects were given any ovarian hormone. Then each animal was injected with estrogen and progesterone and a second test series was conducted.

Sexual Presentation. Table 2 illustrates the proportion of each group which showed the sexual presentation response. This behavior never appeared in members of any group prior to administration of ovarian hormones but after hormone treatment it was shown by nearly all of the control females and by 5 of 11 females given androgen before birth. Only 1 of the 11 females receiving androgen both before and after birth exhibited presentation, and none of the castrated males did so. Table 3 shows that the same group differences persisted when nonresponding individuals were eliminated.

Table 2. *Proportion of each group showing sexual presentation*

Group	*N*	Per cent of group showing response	
		No hormone	Estrogen + progesterone
Spayed females	12	0	92
Females given androgen in utero	11	0	45
Females given androgen in utero + neonatally	11	0	9
Castrated males	6	0	0

Table 3. *Proportion of tests in which sexual presentation was shown by individuals responding at least once*

Group	Animals responding	Per cent tests positive	
		No hormone	Estrogen + progesterone
Spayed females	11	0	52
Females given androgen in utero	5	0	37
Females given androgen in utero + neonatally	1	0	12
Castrated males	0	0	0

Tail Deviation. Table 4 shows that tail deviation, another indication of sexual receptivity, never occurred in tests conducted before administration of estrogen and progesterone. After treatment the response was shown by all of the control females and was next most common in prenatally androgenized females. Ovarian hormones produced very little effect on tail deviation in males or in females given androgen pre- and postnatally. Table 5 reveals that those individuals which did show tail deviation nevertheless were influenced by early androgen treatment.

Table 4. *Proportion of each group showing tail deviation*

Group	*N*	Per cent of group showing response	
		No hormone	Estrogen + progesterone
Spayed females	12	0	100
Females given androgen in utero	11	0	82
Females given androgen in utero + neonatally	11	0	9
Castrated males	6	0	17

Table 5. *Proportion of tests in which tail deviation was shown by individuals responding at least once*

Group	Animals responding	Per cent tests positive	
		No hormone	Estrogen + progesterone
Spayed females	12	0	74
Females given androgen in utero	9	0	55
Females given androgen in utero + neonatally	1	0	12
Castrated males	1	0	12

Positive Social Responses. There are a number of acts in addition to presentation and tail deviation by which a female dog can exhibit her sexual interest in a male and solicit his attention. One of these is "prancing" in which the female faces the male with her forelegs flexed and her sternum touching the ground. At the same time she may vocalize, uttering a series of short, sharp, high-pitched barks. Another response is "striking", in which the female touches or hits the male with one forepaw. Alternatively she may place both forefeet on his back, nudge his body with her head, investigate his genital region or other body areas, etc.

During every test each of these activities is recorded and scored separately, but in the final analysis they can be grouped together in one category which we call "Positive Social Responses" (PSR). A PSR is defined as any act by one dog which is clearly oriented to another individual and which has the effect of intiating or maintaining nonagonistic interaction between the members of the pair.

Table 6 reveals the effects of ovarian hormones on the frequency of positive social responses in the experimental and control groups. It is clear that estrogen and progesterone produced marked increases in the social responsiveness of control females and of those experimental females that had received prenatal androgen treatment. In contrast there was no such effect upon behavior shown by control males nor by experimental females given androgen both before and after birth.

Table 6. "*Positive social responses*" *shown by experimental animals to stimulus males in mating tests*[a]

Group	*N*	Mean responses per test	
		No hormone	Estrogen + progesterone
Spayed females	12	4.25	11.58
Females given androgen in utero	11	7.57	13.03
Females given androgen in utero + neonatally	11	5.73	2.96
Castrated males	6	5.04	5.64

[a] PSR defined in text.

Response of Normal to Experimental Animals

There is another way of measuring the effects of the experimental treatment. This is to study the responses of normal males to the masculinized females. Administration of estrogen and progesterone to normal females increases their stimulus value and causes males to mount them frequently and vigorously. When masculinized females and their controls were given the hormonal treatment which characteristically produces receptivity, the behavior they elicited from normal males was as shown in Table 7. Before the administration of ovarian hormones there was practically no mounting by the normal males. One control male was mounted once but occasional homosexual mounting is not at all uncommon. Following hormone-treatment 12 of 12 control females and 10 of 11 prenatally androgenized females were mounted. In contrast, the stimulus males mounted only one of the 11 females which had received androgen both pre- and postnatally. None of the control males were mounted during these tests. Table 8 indicates an even higher degree of selectivity shown by the stimulus males.

Table 7. *Proportion of each group mounted by stimulus males*

Group	*N*	Per cent of group mounted by stimulus males	
		No hormone	Estrogen + progesterone
Spayed females	12	0	100
Females given androgen in utero	11	0	91
Females given androgen in utero + neonatally	11	0	9
Castrated males	6	17	0

Table 8. *Proportion of tests in which stimulus males mounted those individuals that were mounted at least once*

Group	Animals mounted	Per cent of tests in which mounting occurred	
		No hormone	Estrogen + progesterone
Spayed females	12	0	60
Females given androgen in utero	10	0	46
Females given androgen in utero + neonatally	1	0	25
Castrated males	1	25	0

Many of the positive social responses displayed by receptive females to males are also shown by males when the female is in heat. Table 9 reveals the average frequency per test of PSRs shown by normal males toward members of each experimental and control group before and after injections of estrogen and progesterone. The

Table 9. "*Positive social responses*" *shown by stimulus males to experimental animals in mating tests*

Group	N	Mean responses per test	
		No hormone	Estrogen + progesterone
Spayed females	12	5.94	17.59
Females given androgen in utero	11	6.11	16.69
Females given androgen in utero + neonatally	11	7.16	7.53
Castrated males	6	4.96	8.14

PSR defined in text.

pattern of results is comparable to the one we have already seen in connection with other behavioral measures. Administration of ovarian hormones markedly increased the attractiveness or stimulus value of control females and of females given androgen prior to birth but had relatively little or no effect upon the response of normal males to castrated males or to pre- plus postnatally androgenized females.

The results we have obtained with dogs and the findings of other investigaors working with rodents can be summarized by stating that in addition to masculinizing the genitalia, androgenic stimulation of genetic females during early development results in a marked reduction in sensitivity to ovarian hormones. This insensitivity is reflected in failure to exhibit normal feminine sexual responses in adulthood when given amounts of estrogen and progesterone that are sufficient to evoke such behavior in normal females. There are, in addition, other modifications of behavior which are manifested without the administration of exogenous hormones.

Development Changes Observable in the Absence of Concurrent Hormonal Stimulation

This type of effect has been described in the case of female monkeys whose mothers were given testosterone propionate during gestation. Clear-cut sex differences exist in the social behavior of juvenile rhesus monkeys. For example, before puberty males engage in much more rough-and-tumble play, show more threat responses and more masculine sexual responses than females (HARLOW, 1965). Females exposed to androgenic stimulation *in utero* have been carefully observed throughout infancy and prepubescence. It is very clear that the social play of such females tends to be more masculine than that of normal female monkeys but less masculine than that of normal males (PHOENIX, GOY, and RESKO, 1968).

We have recently discovered a permanent effect of early androgen treatment upon a sexually dimorphic behavior pattern in female dogs. It appears that the experimental treatment tends to modify the frequency and style of urination behavior.

Urination Behavior

The urination behavior of normal adult male and female dogs differs in two ways. First, males urinate much more frequently than females. Second, when they urinate females adopt a squatting posture often lifting one foot off the ground. Adult males usually elevate one rear leg, abducting it at the hip and flexing the knee joint. These two postures are illustrated in Fig. 15.

Table 10 summarises our observations on urination behavior in the experimental females and their controls. As far as frequency of urination is concerned there was a clear gradient from feminine to masculine starting with the control females who urinated infrequently, continuing to females given androgen *in utero* who urinated more often, then to females receiving androgen before and after birth, and reaching the highest frequency in males castrated in adulthood.

The remainder of the table indicates the relative frequency with which the full feminine and masculine postures were adopted during the act of urinating. Intermediate postures have been omitted to clarify the presentation. With respect to posture, as with respect to frequency, the masculinizing effects of androgen treatment during development are clearly revealed. Females given androgen *in utero* were somewhat masculinized, and those treated before and after birth were even more like males.

Fig. 15. Characteristic feminine (top) and masculine (bottom) postures during urination

Table 10. *Urination, frequency and posture, in dogs*

Measure	Spayed females	Females given androgen in utero	Females given androgen in utero + neonatally	Castrated males
Number of animals	12	11	11	6
Average frequency urination per test	0.53	0.99	1.90	2.09
Per cent of total urinations in each posture				
Squat	84	70	36	0
Full leg elevation	0	21	48	70

Conclusion

In conclusion I wish to reiterate the three major facts I have endeavored to stress in the foregoing presentation.

The first point is that gonadal hormones can and do affect socio-sexual behavior in two different ways. The most commonly recognized effects are what I have called "concurrent effects". Behavior appears when the responsible hormones are present and disappears when they are withdrawn. The second category involves what I have termed the "developmental effects" of gonadal hormones. These are of two types. One type of modification produced by early manipulation of the hormone condition is revealed in adulthood in the form of altered sensitivity to the concurrent effects of ovarian hormones. A second type of premanent modification involves sexually-dimorphic behavior which appears in adulthood or earlier without the necessity of concurrent hormonal stimulation.

The second point is that the concurrent effects of gonadal hormones on socio-sexual behavior are subject to very great differences between species. What is true for the rat is not necessarily true for the cat or dog. And what is true for the carnivore is not necessarily true for the primate. Although the evidence is fragmentary and at some points really inadequate, there are some reasons to believe that the extent to which socio-sexual behavior depends upon current effects of gonadal hormones tends to be less extreme in carnivores than in rodents and perhaps even less marked in primates. Much more work must be done before the validity of this speculation can be determined, but if true it carries some very interesting implications for the interpretation of physiological control of human sexual activities.

My third point, which is secondary to the first two, is that study of behavior in various species is important and valuable, and that the proper conduct of such study demands the application of objective and quantitative measurement to the same degree as does experimentation in any other branch of biological science.

References

Beach, F. A.: Coital behavior in dogs. VI. Long-term effects of castration upon mating in the male. J. comp. physiol. Psychol. **70** (3) Part 2, 1 – 32 (1970a).

— Hormonal factors controlling the differentiation, development and display of copulatory behavior in the ramstergig and related species. In: E. Tobach (Ed.): Proceedings of the Conference on the Biopsychology of Development. New York: Academic Press 1970b (in press).

— Holz-Tucker, Marie, A.: Effects of different concentrations of androgen upon sexual behavior in castrated male rats. J. comp. physiol. Psychol. **42**, 433—453 (1949).

— LeBoeuf, B. J.: Coital behavior in dogs. I. Preferential mating in the bitch. Anim. Behav. **15**, 546—598 (1967).

— Merari, A.: Coital behavior in dogs. IV. Effects of progesterone in the bitch. Proc. nat. Acad. Sci. (Wash.) **61**, 442—446 (1968).

— — Coital behavior in dogs. V. Effects of estrogen and progesterone on mating and other forms of social behavior in the bitch. J. comp. physiol. Psychol. **70** (1) Part 2, 1—22 (1970).

Gerall, A. A., Ward, I. L.: Effects of prenatal exogenous androgen on the sexual behavior of the female albino rat. J. comp. physiol. Psychol. **62**, 370—375 (1966).

Goy, R. W.: Organizing effects of androgen on the behaviour of rhesius monkeys. In: R. P. Michael (Ed.): Endocrinology and human behaviour, pp. 12—31. London: Oxford University Press 1968.

Grady, K. L., Phoenix, C. H., Young, W. C.: Role of the developing rat testis in differentiation of the neural tissues mediating mating behavior. J. comp. physiol. Psychol. **59**, 176—182 (1965).

Harlow, H. F.: Sexual behavior in the rhesus monkey. In: F. A. Beach (Ed.): Sex and behavior, pp. 234—265. New York: John Wiley & Sons 1965.

LeBoeuf, B. J.: Interindividual associations in dogs. Behaviour **29**, 268—295 (1967).

MICHAEL, R. P.: Gonadal hormones and the control of primate behaviour. In: R. P. MICHAEL (Ed.): Endocrinology and human behaviour, pp. 69—93. London: Oxford University Press 1968.

NEUMANN, F., ELGER, W.: Physiological and psychical intersexuality of male rats by early treatment with an antiandrogenic agent (1, 2 a-methylene-6-chloroΔ^6-hydroxyprogesterone acetate). Acta endocr. (Kbh.) Suppl. **100**, 174 (1965).

PHOENIX, C. H., GOY, R. W., GERALL, A. A., YOUNG, W. C.: Organizing action of prenatally administered testosterone propionate on the tissues mediating mating behavior in the female guinea pig. Endocrinology **65**, 369—382 (1959).

— — RESKO, J. A.: Psychosexual differentiation as a function of androgenic stimulation. In: M. DIAMOND (Ed.): Perspectives in reproduction and sexual behavior, pp. 33—50. Bloomington: Indiana University Press 1968.

ROSENBLATT, J. S., ARONSON, L. R.: The decline of sexual behavior in male cats after castration with special reference to the role of prior sexual experience. Behaviour **12**, 285—338 (1958).

WHALEN, E. R.: Differentiation of the neural mechanisms which control gonadotrophin secretion and sexual behavior. In: M. DIAMOND (Ed.): Perspectives in reproduction and sexual behavior, pp. 303—340. Bloomington, Indiana: Indiana University Press 1968.

YOUNG, W. C.: The hormones and mating behavior. In: W. C. YOUNG (Ed.): Sex and internal secretions, Vol. 2, 3rd Ed., pp. 1173—1239. Baltimore, Md.: Williams & Wilkins 1961.

Summary

Heinz Gibian

Leiter der Experimentellen Forschung Pharma Schering AG., Berlin

Ladies and Gentlemen!

Originally I intended to give my concluding remarks in German. But in the meantime we have become aware of how well English was accepted so that I shall continue in English. Only my style will now suffer somewhat, but it will be revised for the printed paper.

I want to close with a short, and of course personally influenced, survey of the presentations.

Dr. Jost gave us an outstanding lecture in reproductive mamalian physiology and its developmental background, thus reminding us of, or providing us with, the necessary general information.

Dr. Elger added an example of a species specificity, thus warning us against uncritical generalizations: with female rat fetuses the Wolffian ducts are maintained by androgens as in other species; not so in the male fetus, where anti-androgens do *not* induce retrogression of the Wolffian ducts.

Dr. Schally refered to his work on releasing factors based on extracting more than 3/4 of a million hypothalami. Porcine LH-RF seems to have some inherent FSH-RF activity and is apparently active without much species specificity.

As Dr. White confirmed, both of these activities may possibly be attributed to the same chemical individual, a rather astonishing result, since two different hormones are being controlled thereby. This refers not only to their liberation but presumably also to their synthesis.

Dr. Schally stressed the essential rôle the hypothalamus may play within the mechanism of action of modern steroidal contraceptives.

Dr. Karg again gave us examples of species-specific deviations: prolactin, defined as luteotropic hormone (LTH) in rats and mice, bears no relation at all to the reproductive cycle of the cow.

Dr. Steinberger and Dr. Lipsett informed us about the biosynthetic pathways of male and female peripheral hormones; especially for the female organism a lot of data is already being accumulated by which actions of gonadotropins and steroids may be interpreted in terms of, or correlated to, biochemical parameters such as enzyme activities, secretion rates, biosynthetic patterns in different tissues or cell populations, or nucleic acid synthesis for transcription.

Dr. Bedford really fascinated us with his saga of the spermatozoa, telling us about their capacitation within the female tract and showing us wonderful pictures of the acrosomal reaction, the penetration of the sperm through the barriers of the granulosa cells, the zona pellucida, the vitelline membrane, ending with the process of syngamy within the ovum. Much biochemical work must be done to really complete all the morphological and physiological evidence!

Dr. Ruhenstroh-Bauer commented on differences between epididymal and ejaculated spermatozoa, apparently based on adsorption on a special protein fraction of seminal plasma.

Dr. Koester very clearly described those tubal processes which, being so nicely correlated and hormonally controlled, govern the transport of the ovum, i.e. the ciliary movements, peristaltics, and the flux of the secretions.

Dr. Beier's experiments into the composition of endometrial secretions are promising as to future insights into causal connections between morphological and biochemical processes in the uterus.

Dr. Brinster showed us biochemical results gained in experiments with his famous techniques for culturing fertilized ova in vitro. From the very first stages to the morula, the serial development of metabolic enzymes for energy and nutritional purposes can be observed.

Dr. Krieg discussed a topic which for decades has been of some interest regarding sterility and in recent times regarding control of fertility: the important question of immunological reactions, the appearance of sperm autoantibodies in the seminal plasma and in the serum of the male as well as of the female individual. Very stimulating is also the question why the fetus, being a foreign tissue *strictu sensu* is normally tolerated immunologically by the maternal organism; certain abortions may be based on lack of tolerance.

Dr. EDWARDS discussed the genetic factors which are operative during early mammalian development. Regarding the techniques, admirable experiments promise to produce more and more insight into inherent relevant facts from identification of single spermatozoa to the observation of the commencing genetic activity in late stages of embryonal cleavage. I was greatly impressed by the results of exchanging and implanting single cells into a blastocyst to give live chimaeric offsprings! The possible impact of the expression of fetal antigens for implantation was alluded to.

In his paper on the maintenance of pregnancy, Dr. JUNG also briefly discussed immunological problems, besides concentrating on the physiology of the muscular reactions of the pregnant uterus. The application of progestogens in problems of early pregnancy before mensis IV has been cited. From mensis V on sympathicomimetica were said to be the means of choice for sedation of the uterus at imminent abortion; β-adrenergic drugs are being developed in the direction of fewer side effects.

Dr. MOOR referred to the influence of the conceptus on the life span of corpora lutea in various species: in the human female a placental luteotrophin (either HCG or HPL) maintains the corpus luteum of pregnancy. On the other hand, in the ewe the conceptus acts locally as anti-luteolytic agent. Here the non-pregnant uterus induces luteal regression possibly via a luteolysin.

According to the paper of Dr. CALDWELL hysterectomy in several species prolongates the functional life span of corpora lutea. It could be shown that ectopic reimplantation of an excised uterine horn reverses this effects in pig, sheep, and several rodents; the mechanism of action is not yet quite clear. A luteolytic factor (luteolysin) could not yet definitely be isolated from the endometrium.

In recent times teratology has gained more and more importance, especially in the development of drugs, stimulated by regrettable events. Dr. GOLDMAN exemplified the correlation of selective inhibition of enzymes which are specific for steroid biosynthesis with morphological aberrations and fetal malformations, especially in connection with sexual differentiation.

The last paper of Dr. BEACH on the rôle of hormones regarding masculine and feminine behaviour will still be fresh in your mind so that I need not review it once more. Interesting were the phylogenetic aspects of the presentation.

Ladies and Gentlemen, may I amplify Dr. BEDFORD's title by saying that we have had a number of competent research workers telling us the saga of complete mammalian reproduction. Obviously we heard more biochemistry than was to be supposed at the early planning stages of this Symposium. Nevertheless, the lack of relevant knowledge in various respects and degrees has become apparent to us throughout.

I should like to repeat my introductory wish and hope that some of the present audience or some of the future readers of the proceedings, which I hope will be printed soon, (the earlier they are printed, the earlier the manuscripts of the distinguished lecturers will be delivered!), i.e. that one or other of them might feel stimulated to embark on biochemistry or the physiology of reproduction as an exciting, promising, and highly important scientific field. In any case I believe that we can all go home very much enriched, and therefore I want to thank all speakers very heartily on behalf of the scientific organizers. I should also like to thank all those who have so actively and effectively participated in the preparation of this Colloquium: my secretary, Frau VON LÜBTOW, my colleague, Dr. ELGER, the treasurer of this society, Dr. AUHAGEN, and his secretary, Frau HÜLSBECK.

MIX
Papier aus verantwortungsvollen Quellen
Paper from responsible sources
FSC® C105338

If you have any concerns about our products,
you can contact us on
ProductSafety@springernature.com

In case Publisher is established outside the EU,
the EU authorized representative is:
Springer Nature Customer Service Center GmbH
Europaplatz 3, 69115 Heidelberg, Germany

Printed by Libri Plureos GmbH
in Hamburg, Germany